Lecture Notes in Networks and Systems

Volume 149

The series "Lecture Notes in Networks and Systems" publishes the latest developments in Networks and Systems—quickly, informally and with high quality. Original research reported in proceedings and post-proceedings represents the core of LNNS.

Volumes published in LNNS embrace all aspects and subfields of, as well as new challenges in, Networks and Systems.

The series contains proceedings and edited volumes in systems and networks, spanning the areas of Cyber-Physical Systems, Autonomous Systems, Sensor Networks, Control Systems, Energy Systems, Automotive Systems, Biological Systems, Vehicular Networking and Connected Vehicles, Aerospace Systems, Automation, Manufacturing, Smart Grids, Nonlinear Systems, Power Systems, Robotics, Social Systems, Economic Systems and other. Of particular value to both the contributors and the readership are the short publication timeframe and the world-wide distribution and exposure which enable both a wide and rapid dissemination of research output.

The series covers the theory, applications, and perspectives on the state of the art and future developments relevant to systems and networks, decision making, control, complex processes and related areas, as embedded in the fields of interdisciplinary and applied sciences, engineering, computer science, physics, economics, social, and life sciences, as well as the paradigms and methodologies behind them.

Indexed by SCOPUS, INSPEC, WTI Frankfurt eG, zbMATH, SCImago.

All books published in the series are submitted for consideration in Web of Science.

More information about this series at http://www.springer.com/series/15179

Prasant Kumar Pattnaik · Mangal Sain ·
Ahmed A. Al-Absi · Pardeep Kumar
Editors

Proceedings of International Conference on Smart Computing and Cyber Security

Strategic Foresight, Security Challenges and Innovation (SMARTCYBER 2020)

Springer

Editors
Prasant Kumar Pattnaik
School of Computer Engineering
Kalinga Institute of Industrial Technology
KIIT Deemed to be University
Bhubaneswar, India

Mangal Sain
Division of Information and Communication Engineering
Dongseo University
Busan, Korea (Republic of)

Ahmed A. Al-Absi
Department of Smart Computing
Kyungdong University Global Campus
Gangwondo, Korea (Republic of)

Pardeep Kumar
Department of Computer Science
Swansea University, Bay Campus
Swansea, UK

ISSN 2367-3370 ISSN 2367-3389 (electronic)
Lecture Notes in Networks and Systems
ISBN 978-981-15-7989-9 ISBN 978-981-15-7990-5 (eBook)
https://doi.org/10.1007/978-981-15-7990-5

This Springer imprint is published by the registered company Springer Nature Singapore Pte Ltd.
The registered company address is: 152 Beach Road, #21-01/04 Gateway East, Singapore 189721, Singapore

Prasant Kumar Pattnaik · Mangal Sain ·
Ahmed A. Al-Absi · Pardeep Kumar
Editors

Proceedings of International Conference on Smart Computing and Cyber Security

Strategic Foresight, Security Challenges and Innovation (SMARTCYBER 2020)

Springer

Editors
Prasant Kumar Pattnaik
School of Computer Engineering
Kalinga Institute of Industrial Technology
KIIT Deemed to be University
Bhubaneswar, India

Mangal Sain
Division of Information and Communication Engineering
Dongseo University
Busan, Korea (Republic of)

Ahmed A. Al-Absi
Department of Smart Computing
Kyungdong University Global Campus
Gangwondo, Korea (Republic of)

Pardeep Kumar
Department of Computer Science
Swansea University, Bay Campus
Swansea, UK

ISSN 2367-3370 ISSN 2367-3389 (electronic)
Lecture Notes in Networks and Systems
ISBN 978-981-15-7989-9 ISBN 978-981-15-7990-5 (eBook)
https://doi.org/10.1007/978-981-15-7990-5

This Springer imprint is published by the registered company Springer Nature Singapore Pte Ltd.
The registered company address is: 152 Beach Road, #21-01/04 Gateway East, Singapore 189721, Singapore

Preface

The 1st International Conference on Smart Computing and Cyber Security—Strategic Foresight, Security Challenges and Innovation (SMARTCYBER 2020), took place in Kyungdong University Global Campus, Gosung, Gangwondo, South Korea, during July 7–8, 2020. It was hosted by the Department of Smart Computing, Kyungdong University, Global Campus, South Korea.

The SMARTCYBER is a premier international open forum for scientists, researchers and technocrats in academia as well as in industries from different parts of the world to present, interact and exchange the state of the art of concepts, prototypes, innovative research ideas in several diversified fields. The primary focus of the conference is to foster new and original research ideas and results in the five board tracks: smart computing concepts, models, algorithms, and applications, smart embedded systems, bio-Inspired models in information processing, technology, and security. This is an exciting and emerging interdisciplinary area in which a wide range of theory and methodologies are being investigated and developed to tackle complex and challenging real-world problems. The conference includes invited keynote talks and oral paper presentations from both academia and industry to initiate and ignite our young minds in the meadow of momentous research and thereby enrich their existing knowledge.

SMARTCYBER 2020 received a total of 143 submissions. Each submission was reviewed by at least three Program Committee members. The committee decided to accept 37 full papers. Papers were accepted on the basis of technical merit, presentation and relevance to the conference. SMARTCYBER 2020 was enriched by the lectures and insights given by the following seven distinguished invited speakers: Prof. Prasant Kumar Pattnaik, School of Computer Engineering, Kalinga Institute of Industrial Technology; Professor Ana Hol, Western Sydney University, Australia; Professor Aninda Bose, Senior Editor Springer India; Prof. Evizal Abdul Kadir, UIR, Indonesia; Dr. James Aich S, CEO Terenz Co. Ltd, South Korea; Prof. Mangal Sain, Dongseo University, South Korea; and Prof. Ahmed A. Al-Absi, Kyungdong University Global Campus, South Korea. We thank the invited speakers for sharing the enthusiasm for research and accepting our invitation to share their expertise as well as contributing papers for inclusion in the proceedings.

SMARTCYBER 2020 has been able to maintain standards in terms of the quality of papers due to the contribution made by many stakeholders.

We are thankful to the General Chairs, Prasant Kumar Pattnaik, KIIT Deemed to be University, India; Ahmed A. Al-Absi, Kyungdong University, South Korea; Mangal Sain, Dongseo University. We futher thank the Program Chairs, Baseem Al-athwari, Kyungdong University Global Campus, South Korea; Pardeep Kumar, Swansea University, UK; Deepanjali Mishra, KIIT Deemed to be University, India, for their guidance and valuable inputs.

We are grateful to Prof. John Lee, President of Kyungdong University (KDU) Global Campus, South Korea, and Honorary General Chair, SMARTCYBER 2020, for his constant support and for providing the infrastructure and resources to organize the conference. We are thankful to Prof. Sasmita Rani Samanta, Pro-Vice-Chancellor, KIIT Deemed to be University, India, Honorary General Chair, SMARTCYBER 2020, for providing all the support for the conference.

Thanks are due to the Program and Technical committee members for their guidance related to the conference. We would also like to thank the Session Management Chairs, Publications Chairs, Publicity Chairs, Organizing Chairs, Finance Chairs and Web Management Chair who have made an invaluable contribution to the conference. We acknowledge the contribution of EasyChair in enabling an efficient and effective way in the management of paper submissions, reviews and preparation of proceedings. Finally, we thank all the authors and participants for their enthusiastic support. We are very much thankful to entire team of Springer Nature for timely support and help. We sincerely hope that you find the book to be of value in the pursuit of academic and professional excellence.

Bhubaneswar, India — Prasant Kumar Pattnaik
Gangwondo, Korea (Republic of) — Ahmed A. Al-Absi
Busan, Korea (Republic of) — Mangal Sain
Swansea, UK — Pardeep Kumar

Contents

Editors and Contributors

About the Editors

Prasant Kumar Pattnaik Ph.D. (Computer Science), Fellow IETE, Senior Member IEEE, is a Professor at the School of Computer Engineering, KIIT Deemed University, Bhubaneswar. He has more than a decade of teaching and research experience and awarded half dozen of Ph.D. Dr. Pattnaik has published numbers of research papers in peer-reviewed international journals and conferences and filed many patents. He also edited book volumes in Springer and IGI Global Publication. His areas of interest include mobile computing, cloud computing, cyber security, intelligent systems, and brain–computer interface. He is one of the Associate Editors of Journal of Intelligent & Fuzzy Systems, IOS Press, and Intelligent Systems Book Series Editor of CRC Press, Taylor Francis Group.

Mangal Sain received the Master of Application degree from India in 2003 and the Ph.D. degree in Computer Science from Dongseo University, Busan, South Korea, in 2011. Since 2011, he has been an Assistant Professor with the Department of Information and Communication Engineering, Dongseo University, Busan, South Korea. He has published over 40 international publications. His current research interests include wireless sensor network, middleware, cloud computing, embedded system, and the Internet of Things. He is a member of TIIS and has participated as a TPC member in several international conferences.

Ahmed A. Al-Absi Ph.D (Computer Science), is an Associate Professor at the Smart Computing Department, Kyungdong University Global Campus, South Korea. He is currently Dean of International Faculty and Director of Global Academic Collaboration Centers at Kyungdong University Global. He has more than ten years of experience in teaching and university lecturing in the areas of database design and computer algorithms. Dr. Al-Absi has published numbers of research papers in peer-reviewed international journals and conferences. His research areas are Big Data, Large Scale Data Process Systems, Cloud Computing,

IoT, VANET, Deep Learning, Parallel Computing, Security, and Bioinformatics. His professional experience includes being a speaker at a number of renowned research conferences and technical meetings such as IEEE, Korea ICT leaders forum, and reviewer for refereed journals and conferences on data-intensive computing as well as an examiner for postgraduate scholars in his research areas.

Pardeep Kumar received the B.E. degree in Computer Science from Maharishi Dayanand University, Haryana (India), in 2002, the M.Tech. degree in Computer Science from Chaudhary Devi Lal University, Haryana (India), in 2006, and the Ph.D. degree in Ubiquitous Computing from Dongseo University, Busan (South Korea) in 2012. He is currently a Lecturer/Assistant Professor with the Department of Computer Science, Swansea University, Swansea, UK. From 2012 to 2018, he had held postdoc positions at the Department of Computer Science, Oxford University, Oxford UK (08/2016–09/2018), at the Department of Computer Science, The Arctic University of Norway, Tromso, Norway (08/2015–08/2016), and at Centre for Wireless Communications and the Department of Communications Engineering, University of Oulu, Finland (04/2012 to 08/2015).

Contributors

Satyabrata Aich Terenz Co., Ltd., Busan, Republic of Korea

Ahmed A. Al-Absi International Faculty (Academic), Global Academic Collaboration Centers, Kyungdong University-Global Campus, Wonju-si, South Korea

Ahmed Abdulhakim Al-Absi Dean of International Faculty (Academic), Director of Global Academic Collaboration Centers, Kyungdong University, Yangju, South Korea;
Department of Smart Computing, Kyungdong University, Bongpo, Gosung, Gangwondo, Republic of Korea

Mohammed Abdulhakim Al-Absi Department of Computer Engineering, Dongseo University, Sasang-gu, Busan, Republic of Korea

Baseem Al-athwari Smart Computing Department, Kyungdong University, Goseong-gun, Gangwon-do, South Korea;
Department of Computer Engineering, Kyungdong University, Global Campus (Goseong), Goseong, Gangwon-do, Republic of Korea

Amrizal Amrizal Department of Informatics, Universitas Islam Riau, Pekanbaru, Indonesia

S. Aravinthkumar SRM University, Sonepat, India

Kurnia Rina Ariani Muhammadiyah Surakarta University, Surakarta, Indonesia

Wafa Arsalane Jeonbuk National University, Jeonju, South Korea

Yudhi Arta Department of Informatics Engineering, Faculty of Engineering, Universitas Islam Riau, Pekanbaru, Indonesia

Hossain Md Azam Smart Computing Department, Kyungdong University, Goseong-gun, Gangwon-do, South Korea

Bimantara Department of Informatics Engineering, Faculty of Engineering, Universitas Islam Riau, Pekanbaru, Indonesia

Sabyasachi Chakraborty Department of Computer Engineering, Inje University, Gimhae, South Korea;
Terenz Co., Ltd., Busan, Republic of Korea

Ajayveer Singh Chandel VIT University, Vellore, India

Chiwoon Cho School of Industrial Engineering, University of Ulsan, Ulsan, Republic of Korea

Daehyoun Choi School of Industrial Engineering, University of Ulsan, Ulsan, Republic of Korea

Olivia Fachrunnisa Department of Management, Faculty of Economics, Universitas Islam Sultan Agung, Semarang, Indonesia

Muhamad Haddin Electrical Engineering, Universitas Islam Sultan Agung, Semarang, Indonesia

Ari Hanggara Department of Informatics, Universitas Islam Riau, Pekanbaru, Indonesia

Ana Hol School of Computer, Data and Mathematical Sciences, Western Sydney University, Rydalmere, NSW, Australia

Md Azam Hossain Department of Computer Engineering, Kyungdong University, Global Campus (Goseong), Goseong, Gangwon-do, Republic of Korea

Soonwook Hwang Korea Institute of Science and Technology Information (KISTI), Daejeon, Republic of Korea

Dian Indriani Informatics Engineering, Universitas Islam Riau, Riau, Indonesia

Eugene Istratova Department of Automated Control Systems, Novosibirsk State Technical University, Novosibirsk, Russia

Ju Jinquan Department of Computer Engineering, Dongseo University, Sasang-gu, Busan, Republic of Korea

Evizal Abdul Kadir Department of Informatics Engineering, Faculty of Engineering, Universitas Islam Riau, Pekanbaru, Indonesia

Ahmadhon Kamolov Department of Computer Engineering, Dongseo University, Busan, Republic of Korea

Sanghoon Kang Department of Electronic Engineering, Pukyong National University, Busan, South Korea

Jik-soo Kim Department of Computer Engineering, Myongji University, Yongin, Republic of Korea

Ki-Hwan Kim Department of Computer Engineering, Dongseo University, Busan, Republic of Korea;
Department of Ubiquitous IT, Dongseo University, Busan, Republic of Korea

Jeongseok Kim Department of Electrical and Computer Engineering, University of Seoul, Seoul, South Korea;
Security Labs, AIX Center, SK Telecom, Seoul, South Korea

Neeraj Kumar VIT University, Vellore, India

Sandra Kumi Department of Computer Engineering, Dongseo University, Busan, South Korea

Novendra Kurniadi Department of Informatics, Universitas Islam Riau, Pekanbaru, Indonesia

Hoon Jae Lee Division of Computer Engineering, Dongseo University, Busan, Republic of Korea;
Division of Information and Communication Engineering, Dongseo University, Sasang-gu, Busan, Republic of Korea

Jaeho Lee Department of Electrical and Computer Engineering, University of Seoul, Seoul, South Korea

Sang-Gon Lee Department of Computer Engineering, Dongseo University, Busan, South Korea

ChaeHo Lim BITSCAN Co., Ltd., Seoul, South Korea

Shreya Makkar VIT University, Vellore, India

P. Manjula SRM University, Sonepat, India

Winda Monika Library Science, Universitas Lancang Kuning, Riau, Indonesia

Ojasvita Muyal VIT University, Vellore, India

Akshay Nagpal VIT University, Vellore, India

Arbi Haza Nasution Informatics Engineering, Universitas Islam Riau, Riau, Indonesia

Salhazan Nasution Informatics Engineering, Universitas Riau, Riau, Indonesia

MD. Nur Alam Department of Smart Computing, Kyungdong University, Gosung, Gangwondo, South Korea

M. D. Nur Alam Department of Smart Computing, Kyungdong University, Bongpo, Gosung, Gangwondo, South Korea

Yuli Nurhasanah Department of Management, Faculty of Economics, Universitas Islam Sultan Agung, Semarang, Indonesia

Yustus Oko Oktian Department of Computer Engineering, Dongseo University, Busan, South Korea

Otabek Olimjonov Department of Smart Computing, Kyungdong University, Gosung, Gangwondo, South Korea

Hanhoon Park Department of Electronic Engineering, Pukyong National University, Busan, South Korea

Jinse Park Department of Neurology, Haeundae Paik Hospital, Inje University, Busan, Republic of Korea

Jong-Il Park Department of Computer Science, Hanyang University, Seoul, South Korea

Prasant Kumar Pattnaik School of Computer Engineering, KIIT University, Bhubaneswar, India

Dita Prameswari Department of Management, Faculty of Economics, Universitas Islam Sultan Agung, Semarang, Indonesia

Sri Arttini Dwi Prasetyowati Electrical Engineering, Universitas Islam Sultan Agung, Semarang, Indonesia

Gustita Arnawati Putri Veteran Bangun Nusantara Sukoharjo University, Kabupaten Sukoharjo, Indonesia

Tilak Rasaili Department of Smart Computing, Kyungdong University, Gosung, Gangwondo, South Korea

Happy Rhemananda Department of Management, Faculty of Economics, Universitas Islam Sultan Agung, Semarang, Indonesia

Fu Rui Department of Computer Engineering, Dongseo University, Busan, Republic of Korea

Hind R'bigui School of Industrial Engineering, University of Ulsan, Ulsan, Republic of Korea

Mangal Sain Division of Computer Engineering, Dongseo University, Busan, Republic of Korea;
Division of Information and Communication Engineering, Dongseo University, Sasang-gu, Busan, Republic of Korea

Dahit Santosh Department of Smart Computing, Kyungdong University, Gosung, Gangwondo, Republic of Korea

Mohammad Ibrahim Sarkar Department of Electronic Engineering, Chonbuk National University, Deokjin-gu, Jeonju-si, Jeollabuk-do, Republic of Korea

Jay Sarraf School of Computer Engineering, KIIT University, Bhubaneswar, India

Shahi Saugat Department of Smart Computing, Kyungdong University, Gosung, Gangwondo, Republic of Korea

Panji Rachmat Setiawan Department of Informatics, Universitas Islam Riau, Pekanbaru, Indonesia

Dima Roulina Simbolon Department of Management, Faculty of Economics, Universitas Islam Sultan Agung, Semarang, Indonesia

Dina Sin Department of Construction, Technical Oil and Gas Institute, Sakhalin State University, Yuzhno-Sakhalinsk, Russia

Apri Siswanto Department of Informatics Engineering, Faculty of Engineering, Universitas Islam Riau, Pekanbaru, Indonesia

Konstantin Strokin Department of Construction, Technical Oil and Gas Institute, Sakhalin State University, Yuzhno-Sakhalinsk, Russia

Des Suryani Department of Informatics, Universitas Islam Riau, Pekanbaru, Indonesia

Nesi Syafitri Department of Informatics, Universitas Islam Riau, Pekanbaru, Indonesia

Abdul Syukur Department of Informatics, Universitas Islam Riau, Pekanbaru, Indonesia

Rakesh Thapamagar Department of Smart Computing, Kyungdong University, Gosung, Gangwondo, South Korea

Obhi Thiessaputra Electrical Engineering, Universitas Islam Sultan Agung, Semarang, Indonesia

Gabit Tolendiyev Department of Computer Engineering, Dongseo University, Busan, Republic of Korea

Ihsan Ullah Department of Robotics Engineering, Daegu Gyeonbuk Institute of Science and Technology, Daegu, South Korea

Vaibhaw School of Computer Engineering, KIIT University, Bhubaneswar, India

Ariyani Wahyu Wijayanti Veteran Bangun Nusantara Sukoharjo University, Kabupaten Sukoharjo, Indonesia

Elizabeth Nathania Witanto Department of Computer Engineering, Dongseo University, Busan, South Korea

Proposal of Pseudo-Random Number Generators Using PingPong256 and Chaos Maps

Ki-Hwan Kim and Hoon Jae Lee

Abstract Internet of Things (IoT) devices are easily exposed to physical attacks; therefore, their design must consider authentication and encryption. Many authentication and encryption methods use algorithms such as advanced encryption standard (AES) and secure hash algorithm (SHA). A pseudo-random number generator (PRNG) can also be used for authentication and encryption, and linear feedback shift register (LFSR) provides an easy way to generate PRNGs. LFSR allows the mathematical generation of unique values proportional to a given length. However, as LFSR is mathematically predictable, it is not used alone for this purpose. PingPong256 uses a variable clock for LFSR that can generate very long periods. However, LFSRs are still potentially at risk of being attacked by correlation analysis attacks. There are several methods to account these security issues, including chaos maps (such as logistic maps), SHA, and AES. This paper proposes a method of using logistic maps corresponding to PingPong256 and chaos maps. For this purpose, various PingPong256 configurations are proposed and compared to verify the effectiveness of the proposed method. The method was tested using NIST SP800-22.

Keywords PRNG · PingPong256 · Logistic map · LFSR · IoT

1 Introduction

The demand for wearable equipment is rapidly increasing worldwide every year [1–4]. The cumulative sales volume of wearable equipment worn on the wrist was one of the world's highest [5]. Wearable equipment has several advantages. First, wearable equipment has high portability because it is designed to be worn on various parts

K.-H. Kim
Department of Ubiquitous IT, Dongseo University, Busan 47011, Republic of Korea
e-mail: ghksdl90@naver.com

H. J. Lee (✉)
Division of Computer Engineering, Dongseo University, Busan 47011, Republic of Korea
e-mail: hjlee@dongseo.ac.kr

P. K. Pattnaik et al. (eds.), *Proceedings of International Conference on Smart Computing and Cyber Security*, Lecture Notes in Networks and Systems 149,
https://doi.org/10.1007/978-981-15-7990-5_1

of the body and has low power requirements. Second, biosensors can continuously monitor health status such as heart rate, electrocardiography (ECG) [6], and blood pressure. Third, data transfer can be achieved conveniently using built-in wireless communication sensors (Wi-Fi, Bluetooth, etc.). Various types of data measured by wearable devices allow the observation of different biometric information such as stress and a user's mental state [7–9]. These devices can also incorporate artificial intelligence (AI) and a dedicated analytics system can monitor and alert a person regarding health readings [10–12]. However, various cyber-attacks on these devices have been occurring, including the use of robots in distributed denial of service (DDoS) attacks, remote control of medical devices, and the capture of privileges and control of security cameras and automobiles [13]. IoT devices have extremely low computing power and storage space and are often placed in environments that are vulnerable to attack due to limited communicability and physical exposure of the equipment. Therefore, lightweight hardware devices require a proper balance between performance and security.

The main purpose of the PingPong algorithm is to generate a pseudo-random noise sequence (PN code) with a long period. PN codes have basic requirements including randomness, unpredictability, and incapability of reproducing/repeating the same sequence (non-reproducibility). A linear feedback shift register (LFSR) is used in this method for generating the maximum length sequence (MLS). The MLS feature of LFSR describes the condition that when there is a memory of size m bits, the period is $L = 2^m - 1$ bits and all values generated during the cycle are unique. LFSR is mainly used for generating pseudo-random bit sequences (PRBS), signal signatures, and footings of signal sets. The main advantage of the system is that the operation can be interpreted precisely by the algebraic principle, and the digital conversion is very easy to perform. PingPong256 is based on PingPong128 and is a hybrid generator, combining the nonlinear Lee Moon (NLM) generator with a highly secure clock-controlled generator [14, 15]. It consists of two different LFSRs ($LFSR_{255}$ and $LFSR_{257}$ bits), LM generators, and two memory bit store allocations.

Chaos theory is the discipline dealing with deterministic nonlinear dynamic systems [16, 17]. Fractals are differentiable and unlike regular Euclidean geometric bodies, they have irregularly separated structures [18]. Time-averaged fractals (escape-time fractals) are color images of the speed at which each point emanates, usually on a complex plane. Various encryption schemes using a chaotic system have been studied [19–21]. The logistic map is a discrete-time dynamic system given as a quadratic polynomial of nonlinear differential equations representing chaotic phenomena. According to logistic thought, the $n + 1$ generation constitutes the function of the population of the n generation, which is represented by multiplying the opposite value of the input by a specific coefficient value. The logistic map according to r is shown in Fig. 1.

The organization of this paper is structured as follows. First, we present the method to use the PingPong algorithm and logistic map to construct a pseudo-random number generator (PRNG). In Sect. 2, we study the PingPong256 and logistic map. In Sect. 3, we propose two PRNGs. Section 4 is the conclusion.

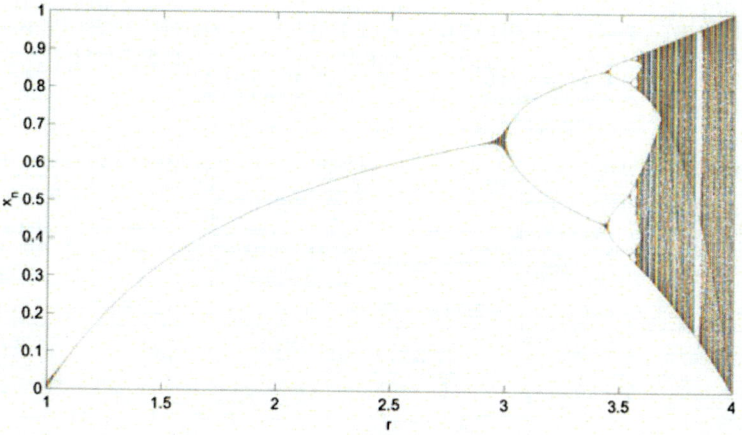

Fig. 1 Periodic change of the logistic map according to r value

2 Background

2.1 PingPong256

The overall structure of the PingPong256 algorithm is shown in Fig. 2. It is simple and easy to implement in hardware and software to ensure a long-term periodic output. The values of clock control called f_a and f_b are calculated by referring to some state values of different LFSRs. This method will provide different results over time even when initialized with the same value in the two LFSRs [22]. The linear structure of the LFSR can be extended with unpredictable results through the clock controller. This process can be viewed in detail in Fig. 3. Each LFSR in Fig. 3 has approximately 30 taps, making it harder to predict the cycle. The structure of the process can be changed according to the application. Even though two LFSRs with long periods of

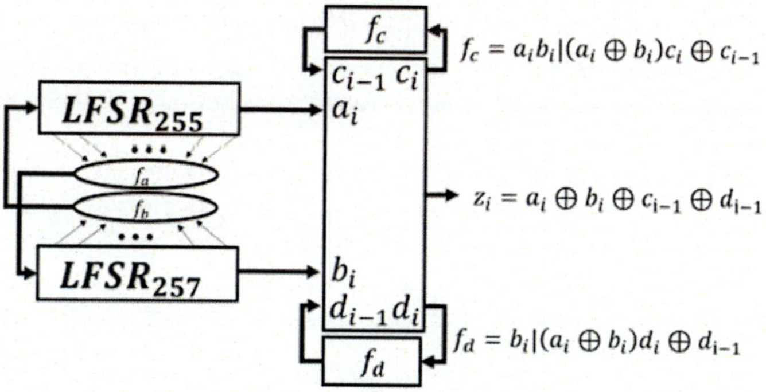

Fig. 2 PingPong256 generator structure

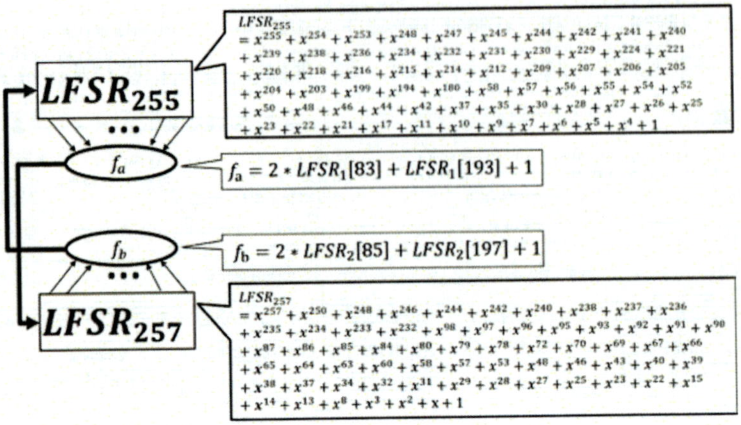

Fig. 3 Polynomial of PingPong256 function

a_i and b_i are influenced by an irregular clock controller, they will present an irregular periodic output unless the initial value is known. The output bits from the LFSR are inputs for the combined function (zi), carry function (fc), and memory function (fd) to produce the following memory states and key sequence bit: ci is a carry sequence, di is a memory sequence, and the initial values of the two variables are 0. The LM generator is an enhanced sums generator and must compute ci-1 and di-1 to compute zi. Therefore, PingPong256 acts as a random number generator in the LFSR core structure. LFSR has a structure that guarantees the maximum period of a defined size, but it also has a linear structure and can be predicted easily. Therefore, the PingPong256 has the ability to make the linearity of LFSR behave as a nonlinear structure through variable clocks and functions. In this paper, we attempt to improve the existing linearity of this system by replacing the linear disadvantage of LFSR LFSRwith the logistic map.

PingPong256 generates a period as defined in Eq. 1 that prevents the reuse of the same key sequence when encrypting long messages. PingPong256 also has linear complexity as shown in Eq. 2 which will withstand attacks using the Berlekamp–Massey algorithm. Finally, a statistical feature of the function is that the frequency of the sequence of keys "0" or "1" should have approximately the same ability to withstand an attack [9]. The calculation of the linear complexity and period of PingPong256 is shown below:

$$LC \geq 2^{4.6} \times 2^{\lceil \frac{512-11}{2} \rceil} = 2^{4.6} \times 2^{0.5} \times 2^{250} \approx 2^{256} \quad (1)$$

$$P \geq 2^{4.6} \times 2^{\lceil \frac{512-11}{2} \rceil} = 2^{4.6} \times 2^{0.5} \times 2^{250} \approx 2^{256} \quad (2)$$

2.2 Logistic Map

The experiment was conducted by setting the value of r to 3.999 in the logistic map function $xn + 1 = r * xn(1-xn)$. The r values were varied for each experimental iteration to produce a variety of results. The logistic map shows that to guarantee a random number, it is preferred to use a constant that cannot be predicted and has a decimal value of 20 bits or fewer [23]. Based on these criteria, and as shown in Fig. 4, the data structure corresponding to the double type variable of the C programming language is divided into four regions: sign, integer, prime number 0, and prime number 1. In this experiment, only the 32-bit value corresponding to the prime number 1 is used. A random number generator test investigates the properties and efficacy of a proposed random number generator. To complete the test, many samples are collected, and statistical analysis is used to determine whether random numbers are generated. If the generator does not pass the statistical test, it must not be recognized as a random number generator and should not pass any other complex tests. Generally, if the random number generator passes all statistical tests, it generates a random number. However, occasionally a generator which has passed the tests may not be able to generate a random number. There are two types of errors in statistical hypothesis testing. Let H0 be the hypothesis that a given sample output sequence is generated by a random bit generator. If the significance level (a) for H0 is too large, a test result of "reject" occurs even though the given sequence is random. This error is called a Type I error. However, if the significance level is too small, the test will "accept" the sequence even if the given sequence is not random. This error is called a Type II error. Generally, the significance level is chosen to be $0.001 \leq a \leq 0.05$. Essentially, when verifying the data set consisting of bits in a random number, the following method is used [24].

The logistic map is constructed as shown in Fig. 4, and the experimental setup is $r = 3.999$, $x = 0.300501$–0.300505. As shown in Table 1, most of the experimental results were satisfactory and passed the test. However, in three experiments, it was confirmed that values exceeding the limit range were found. These cases were primarily found in the poker test with $m = 5$. We have established a similar experimental environment and confirmed the need to extend the scope of these measurements based on this study [25]. Therefore, in this paper, the numerical values deviating from the allowable range are interpreted as inadequate for use as reference values due to the limitation of the measurement range. As a result, we confirmed that

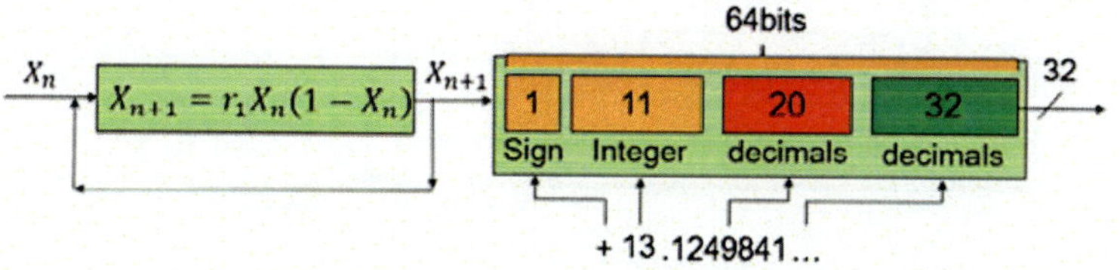

Fig. 4 Experiment 1: random number generation using the logistic map

Table 1 Experiment 1: randomness test result ($r = 3.999$, $x = 0.300501$–0.300505)

Test	Criterion	Case 1	Case 2	Case 3
Frequency	3.841	0.124	2.633	0.085
Serial	5.991	0.129	2.709	1.962
Generalize serial test				
$T = 3$	9.488	2.172	6.023	2.070
$T = 4$	15.507	6.192	16.063	5.520
$T = 5$	26.296	11.360	21.979	12.104
Poker test				
$M = 3$	14.067	2.007	6.558	6.842
$M = 4$	24.996	12.063	22.175	9.476
$M = 5$	44.654	46.279	52.804	46.298
Auto-collimation	Max ≤ 0.05	0.010404	0.009974	0.008802

using a decimal value corresponding to the lower 32 bits provides a result approaching a random number.

3 Proposal main model

The LFSR outputs 1 bit per round of calculation, but the structure using the logistic map outputs 32 bits per round. Figure 5 shows the structure in which all LFSRs of the PingPong256 are changed to a logistic map to unify the components which have different outputs. Thus, it has a 32-bit register output per round and has a 32-bit output per round compared to the existing PingPong algorithm. The diagram detailing the PingPong256 algorithm using the logistic map is shown in Fig. 5.

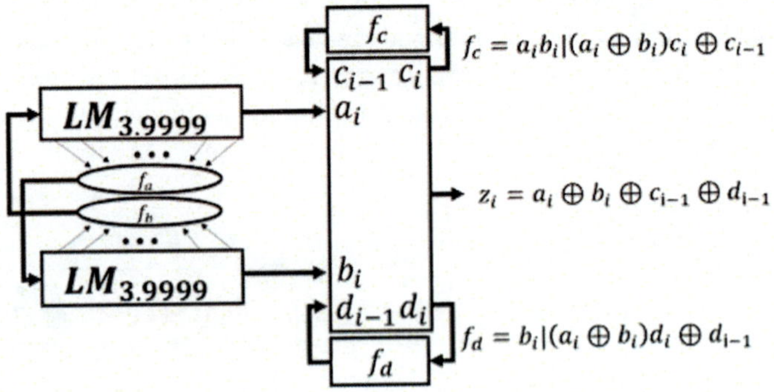

Fig. 5 Two logistic map PingPong256

3.1 Two logistic Maps and XOR Operations.

We used a random number generator in a logistic map experiment which used decimals to reduce the 32 bits of the result to 11 bit. This experiment induced chaos and created a random number generator structure that uses the control group to avoid guessing the initial value.

The structure of Experiment 1 is shown in Fig. 6, where the original logistic map X and the comparative logistic map Y are generated [26]. The experimental values were $r_1 = 3.999$, $r_2 = 3.999$, and $X_0 = 0.300501$ and $Y_0 = 0.300503$. To confirm whether the structure is effective, Y was increased by 0.0002 to make eight control groups. Table 2 lists the results of setting the initial values as $r_1 = 3.999$, $r_2 = 3.999$, $X_0 = 0.300501$, and $Y_0 = 0.300503$ as shown in Fig. 4. However, the poker test did not satisfy the conditions for an allowable value, and it was confirmed that it

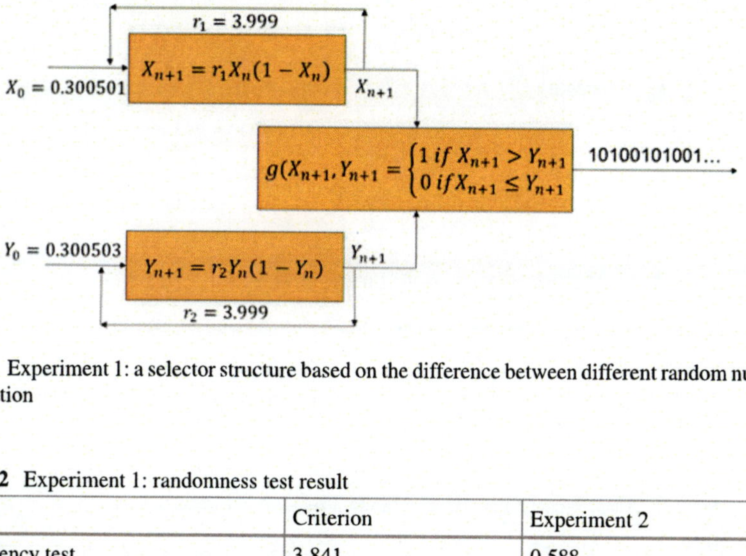

Fig. 6 Experiment 1: a selector structure based on the difference between different random number generation

Table 2 Experiment 1: randomness test result

Test	Criterion	Experiment 2
Frequency test	3.841	0.588
Serial test	5.991	7166.367
Generalize serial test		
$T = 3$	9.488	9440.613
$T = 4$	15.507	13982.530
$T = 5$	26.296	19077.599
Poker test		
$M = 3$	14.067	5112.210
$M = 4$	24.996	7617.505
$M = 5$	44.654	8358.768
Auto-collimation test	≤0.05	0.027487

was difficult to verify the randomness. In a similar paper [26], the opposite result was found, showing that the control structure using logistic map produced positive results. References to the paper [26] show that the above structure varies with input values, and it is uncommon to pass all the test results.

To account for this condition, we used an exclusive logical OR function as shown in Fig. 7. The setup for Experiment 2 was $r_1 = 3.999$, $r_2 = 3.999$, $X_0 = 0.300501$, and $Y_0 = 0.300503$. In this experiment, the r values of the different logistic maps were set with the same value, and the initial input values were varied. The results of the experiment are listed in Table 3. When $m = 5$ in the poker verification, which was the only portion in the previous experiment to fail, the result was acceptable. Thus, experimental results show that the logistic map with different initial values can be improved using the exclusive OR function. Experiments show that a random number generator using various chaotic functions can express desirable results. To overcome

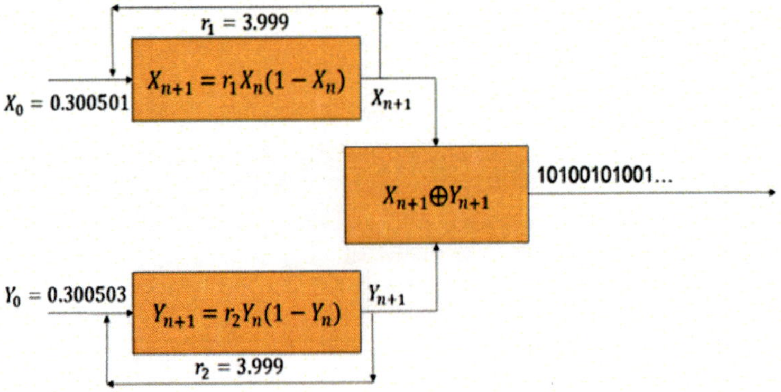

Fig. 7 Experiment 2: a random number generator using two logistic maps and exclusive OR function

Table 3 Experiment 2: logistic map random number verification result

Test	Criterion	Case 1	Case 2	Case 3
Frequency test	3.841	3.064	2.391	2.160
Serial test	5.991	4.937	2.417	2.222
Generalize serial test				
$T = 3$	9.488	5.932	2.687	8.144
$T = 4$	15.507	10.264	8.281	9.872
$T = 5$	26.296	13.946	20.171	22.652
Poker test				
$M = 3$	14.067	11.207	6.184	15.351
$M = 4$	24.996	17.78	16.492	20.957
$M = 5$	44.654	34.481	40.706	40.573
Auto-collimation test	≤ 0.05	0.010404	0.009974	0.008802

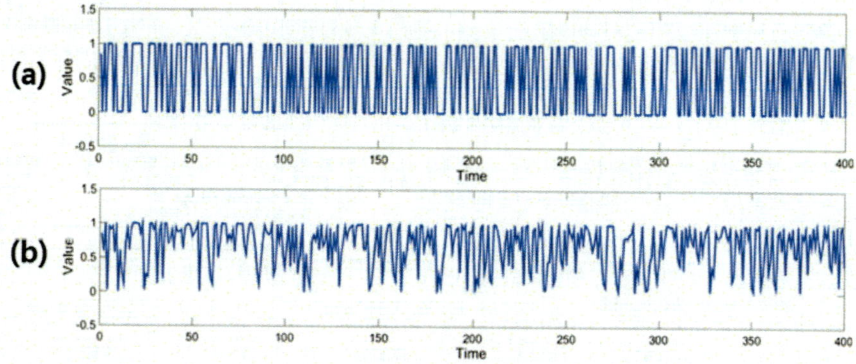

Fig. 8 400 outputs from each structure **a** Experiment 1, **b** Experiment 2

this problem, we added a PingPong256 structure corresponding to a stream cipher [27].

The results of Experiment 1 and Experiment 2 in graphic form are shown in Fig. 8. Experiment 1 produces results of 0 and 1, so the attacker has a 50% chance of choosing the correct answer each time. Experiment 2 produces real numbers from 0 to 1, so it is more sensitive to changes in value.

3.2 *Output Change According to Clock Control Function*

Logistic maps that do not use clock control can be deduced by continuous observation. The generated signal for this model is created using the XOR model proposed in Sect. 3.1. The initial values of the two different logistic maps are 0.5, and the r coefficient values, r_1 and r_2 are set to 3.999 and 3.9999, respectively. The blue line in Fig. 9 is the logistic map result using the x value, and the red-dotted line is the result using the y value. The functions f_a and f_b are configured to operate using an arbitrary value corresponding to each logistic map.

Figure 10 compares the results of the operation using the same initial value as Fig. 9 and referring f_a and f_b to different logistic maps. In this case, the blue line

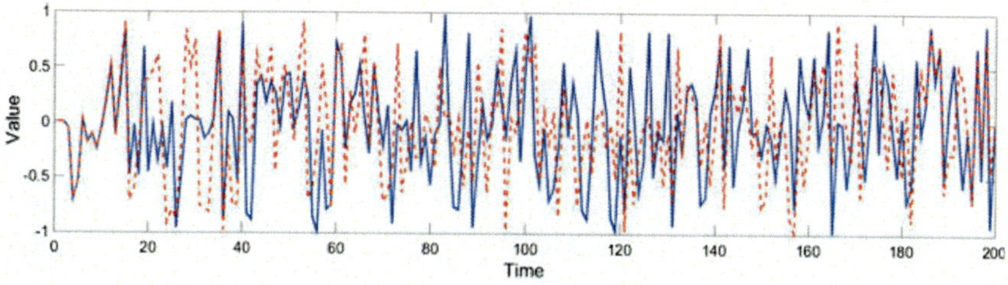

Fig. 9 Logistic map output without clock control function

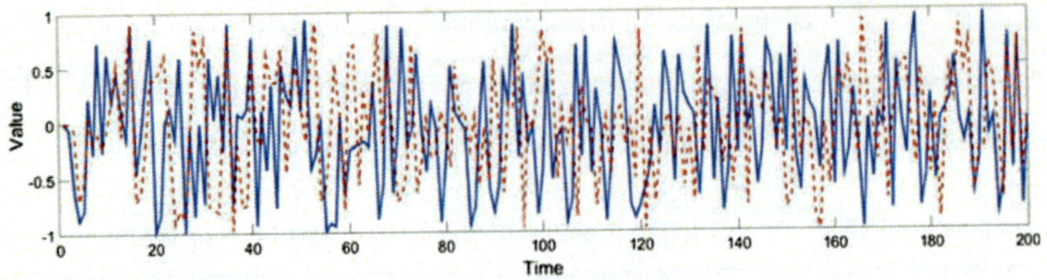

Fig. 10 Difference between the logistic map without clock control function and logistic map with clock control function

represents the difference value of the logistic map referring to the x value, and the red-dotted line represents the difference value of the logistic map referring to the y value.

Because the initial values of both sides are the same, the difference does not occur until approximately counting 13 on the x-axes. Subsequently, the difference between them can be observed, and the output is an actual value between 0 and 1, which makes it difficult to make accurate predictions.

3.3 R-Factor Relationship in Logistic Map

Table 4 lists the results of performing the PingPong256 procedure using two logistic maps three times with different initial values. A total of 2,257,600 bits were used for random number verification. The tests were varied by changing the initial value of the logistic map by increasing x_2 of x_1 and using another initial value. The initial values

Table 4 Two logistic map PingPong256 random verification results ($r_1 = 3.999$, $r_2 = 3.999$, total length = 2,457,600, $x_1 = 0.7861$, $x_2 = 0.859167$)

Test	Criterion	Case 1	Case 2	Case 3
Frequency	3.841	0.375	0.008	2.305
Serial	5.991	0.626	0.234	2.52
Generalize serial test				
$T = 3$	9.488	1.345	2.793	3.03
$T = 4$	15.507	14.599	6.079	5.391
$T = 5$	26.296	21.697	13.407	6.413
Poker test				
$M = 3$	14.067	3.631	3.996	5.972
$M = 4$	24.996	17.173	8.536	30.682
M=5	44.654	34.86	27.885	42.615
Auto-collimation	≥0.05	0.61717	0.635077	0.639495

Table 5 Two logistic map PingPong256 random number verification result ($r_1 = 3.999$, $r_2 = 3.9999$, total length = 2,457,600, x1=0.7861, $x_2 = 0.859167$)

Test	Criterion	Case 1	Case 2	Case 3
Frequency	3.841	1.818	0.095	0.549
Serial	5.991	5.089	0.412	4.462
Generalize serial test				
$T = 3$	9.488	10.665	0.75	6.425
$T = 4$	15.507	13.09	2.676	12.147
$T = 5$	26.296	24.726	9.242	14.489
Poker test				
$M = 3$	14.067	14.836	5.638	13.071
$M = 4$	24.996	13.638	7.025	14.36
$M = 5$	44.654	19.291	16.356	32.019
Auto-collimation	≥0.05	0.070461	0.573655	0.047886

used in the experiment passed all the tests listed in Table 4. In case 3, frequency verification showed that the number of 0s returned was 1,226,420 and the number of 1s returned was 1,231,180 which summed to a total of 2,457,600 bits. In the sequence verification process, the number of bits changed from 0-to-0 was 612,051, those changed from 0-to-1 was 614,369, those changed from 1-to-0 was 614,368, and those changed from 1-to-1 was 616,811. However, it was confirmed that the criterion was not passed in a similar manner for all cases and input values. The case results are negative in cases 1 and 2. Table 5 lists the nonlinear element by changing the value of r_2 to 3.9999. When this change is made, cases 1, 2, and 3 each passed all the tests.

With more acceptable results achieved, we re-implemented the PingPong256 of the logistic map using the changed r value above and confirmed the results as listed in Table 6. The characteristics of the system confirmed through this experiment are as follows. When generating a random number using multiple logistic maps, the values of r must be different. A logistic map has a wide range of random number generation ability according to r value. Using multiple logistic maps and nonlinear functions can help to ensure adequate randomness. It can be seen from the results of Tables 5 and 6 that the system is sensitive to initial values, and they are recommended to be set as asymmetrically as possible.

4 Conclusions

This paper proposes a new PRNG method using the PingPong256 algorithm and logistic maps corresponding to chaos maps. There are three primary benefits in using the proposed algorithm. First, the clock controller of PingPong256 expands

Table 6 Two logistic map PingPong256 random number verification results ($r_1 = 3.9999$, $r_2 = 3.9999$, total length = 2,457,600, $x_1 = 0.7861$, $x_2 = 0.859167$)

Test	Criterion	Case 1	Case 2	Case 3
Frequency	3.841	0.784	2.759	0.844
Serial	5.991	11.112	3.073	2.352
Generalize serial test				
$T = 3$	9.488	4.651	6.822	8.499
$T = 4$	15.507	11.112	7.454	12.462
$T = 5$	26.296	14.65	17.907	17.056
Poker test				
$M = 3$	14.067	6.368	5.638	10.088
$M = 4$	24.996	14.161	7.025	12.945
$M = 5$	44.654	31.945	32.757	22.624
Auto-collimation	$\geq$0.05	0.800961	0.576845	0.219434

the randomness by introducing misalignment over time even when the same initial value is used on both sides of the process. Second, the carry and memory functions make it easy to create a continuous signal by using the previous and current outputs simultaneously. Finally, the logistic map achieves superior randomness to the LFSR due to its irregular and continuous features. Therefore, the PingPong algorithm using logistic maps can generate random numbers using logistic maps corresponding to a chaotic function using the high sensitivity of the initial value. While some experiments have shown that one logistic map function works successfully, when we tested them we found problems in the poker test including that they were biased and returned specific output patterns. The proposed PRNG guarantees long periodicity of irregular output and can be created by an intuitive hardware and software structure.

Acknowledgments This work was supported by Dongseo University, "Dongseo Cluster Project" Research Fund of 2020 (DSU-20200008).

References

1. "Wearable device unit sales worldwide by region from 2015 to 2022 (in millions)," statistic
2. J.E. Mück, B. Ünal, H. Butt, A.K. Yetisen, Market and patent analyses of wearables in medicine. Trends Biotechnol. (2019)
3. IDC, IDC Forecasts Sustained Double-Digit Growth for Wearable Devices Led by Steady Adoption of Smartwatches, 17, Dec, 2018. https://www.idc.com/getdoc.jsp?containerId=prUS44553518
4. Marketer, Wearables 2019, 3, Jan, 2019. https://www.emarketer.com/content/wearables-2019
5. Forecast unit shipments of wearable devices worldwide from 2017 to 2019 and in 2022 (in million units), statistic, [Internet]

6. T.B. Garcia, D.J. Garcia, *Arrhythmia Recognition: The Art of Interpretations* (Jones & Bartlett Publishers, 2019)
7. C. Setz, et al., Discriminating stress from cognitive load using a wearable EDA device. IEEE Trans. Inf. Technol. Biomed. **14**(2), 410–417 (2009)
8. R. Jerauld, Wearable emotion detection and feedback system. U.S. Patent No. 9,019,174. 28 Apr. 2015
9. V.P. Rachim, W.-Y. Chung, Wearable noncontact armband for mobile ECG monitoring system. IEEE Trans. Biomed. Circ. Syst. **10**(6), 1112–1118 (2016)
10. CBINSIGHTS, From Virtual Nurses To Drug Discovery: 106 Artificial Intelligence Startups In Healthcare, 3, Feb, 2017. https://www.cbinsights.com/research/artificial-intelligence-startups-healthcare/
11. IBM, IBM Wation Health. https://www.ibm.com/watson-health/learn/artificial-intelligence-medicine
12. J. Bresnick, Top 5 Use Cases for Artificial Intelligence in Medical Imaging, Health it Analytics, October, 30, 2018. https://healthitanalytics.com/news/top-5-use-cases-for-artificial-intelligence-in-medical-imaging
13. Guest Writer, The 5 Worst Examples of IoT Hacking and Vulnerabilities in Recorded History, Iotforall, May, 10, 2017.
14. L.H. Jae, S.M. Sung, H.R. Kim, NLM-128, An Improved LM-type Summation Generator with 2-bit memories, in *2009 Fourth International Conference on Computer Sciences and Convergence Information Technology* (IEEE, 2009)
15. H.J. Lee, S.J. Moon, On an improved summation generator with 2-bit memory. Sig. Proc. **80**(1), 211–217 (2000)
16. B. Davies, *Exploring Chaos: Theory and Experiment* (CRC Press, 2018)
17. R.L. Devaney, *A First Course in Chaotic Dynamical Systems: Theory and Experiment* (CRC Press, 2018)
18. L. Méhauté, M.G. Alain, C. Tricot, *Fractal Geometry* (Carbon Black. Routledge, 2018), pp. 245–270
19. R.A. Elmanfaloty, E. Abou-Bakr. Random property enhancement of a 1D chaotic PRNG with finite precision implementation. Chaos, Solitons & Fractals **118**, 134–144 (2019)
20. Z. Lin et al., Security performance analysis of a chaotic stream cipher. Nonlinear Dyn. **94**(2), 1003–1017 (2018)
21. D. Eroglu, J.S.W. Lamb, T. Pereira, Synchronisation of chaos and its applications. Contemp. Phys. **58**(3), 207–243 (2017)
22. H.J. Lee, Chen, K. PingPong-128, a new stream cipher for ubiquitous application, in *2007 International Conference on Convergence Information Technology (ICCIT 2007)* (IEEE, 2007)
23. M. François, D. Defour, C. Negre, A fast chaos-based pseudo-random bit generator using binary64 floating-point arithmetic. Informatica **38**(3) (2014)
24. A. Rukhin, J. Soto, etc., *A Statistical Test Suite for Random and Pseudorandom Number Generators for Cryptographic Applications*. National Institute of Standards and Technology.
25. H. Choi, D. Won, On algorithm for finding primitive polynomials over GF(q). Korea Inst. Inf. Secur. Cryptology **11**(1), 35–42 (2001)
26. V. Patidar, K.K. Sud, N.K. Pareek, A pseudo random bit generator based on chaotic logistic map and its statistical testing. Informatica **33**(4) (2009)
27. K.H. Kim, T.Y. Kim, S.G. Lee, W.T. Jang, H.J. Lee, Proposal of parallelization structure for PingPong 256. J. Eng. Appl. Sci. **13**, 1124–1129 (2018)

Early Detection of Alzheimer's Disease from 1.5 T MRI Scans Using 3D Convolutional Neural Network

Sabyasachi Chakraborty, Mangal Sain, Jinse Park, and Satyabrata Aich

Abstract Alzheimer's disease is a neurodegenerative disease that affects the old age population and is affected by the neurofibrillary tangles and neurotic plagues as they impair the neuron's microtubule transport system. The onset of this disease leads to a decline in the normal cognitive functioning of a person. The commonly observed symptom of AD is the difficulty in remembering the latest events. Moreover, as the progression occurs in a person, it can include symptoms like issues with the language, mood swings, and behavioral issues. As for the particular disease, no cure has been found out yet, to completely eradicate the disease from the body, therefore detection in advance of the disease has proven to be effective in improving a person's life. In the study, 1.5 T T1 weighted MRI scans were acquired from the Alzheimer's disease neuroimaging initiative (ADNI) database of 910 patients, where 336 were healthy control, 307 were mild cognitive impairment(MCI), and 267 for Alzheimer's disease. The study leverages a 3D convolutional neural network (3D-CNN) for learning the intricate patterns in the magnetic resonance imaging (MRI) scans for the detection of Alzheimer's disease. The 3D-CNN model performed superiorly by plotting an accuracy of 95.88%, precision of 0.951, recall of 0.9601, and f1-score of 0.9538.

Keywords Deep learning · MRI · Alzheimer's · Disease · Neural network

S. Chakraborty · S. Aich (✉)
Terenz Co., Ltd., Busan, Republic of Korea
e-mail: james@sikonic.io

M. Sain
Division of Computer Engineering, Dongseo University, Busan, Republic of Korea

J. Park
Department of Neurology, Haeundae Paik Hospital, Inje University, Busan, Republic of Korea

P. K. Pattnaik et al. (eds.), *Proceedings of International Conference on Smart Computing and Cyber Security*, Lecture Notes in Networks and Systems 149,
https://doi.org/10.1007/978-981-15-7990-5_2

1 Introduction

The most common neurological disorder that pertains to the old age population in the world is considered to be Alzheimer's disease (AD). Alzheimer's disease is primarily caused due to the neurofibrillary tangles and neurotic plagues as they impair the neuron's microtubule transport system. Therefore, the cellular collapse happened in the brain cells which leads to shrinkage of the hippocampus and parts of the cerebral cortex and hence causes Alzheimer's disease. Moreover, Alzheimer's disease is progressive in nature, and therefore, early detection and monitoring of the disease lead to the improvement in the life of patients. Also, as the old age population is increasing rapidly, a requirement for the development of suitable methods for the detection of Alzheimer's disease at a very early stage is indeed very important.

For the early detection of Alzheimer's disease, the most widely used diagnostic paradigm is the analysis of magnetic resonance imaging (MRI) scans of the brain. The MRI scans provide anatomical details about the subcortical structures of the brain that are further analyzed to check for any aneurysms, which further deems helpful for the early diagnosis of a particular type of disease. However, as the MRI is a 3D structure, it becomes really difficult for the human eye to analyze the intrinsic details and heterogeneous properties of subcortical structures. Therefore, with the advancement of intelligent technologies, computer-aided detection systems have been proven to be very effective concerning the analysis and diagnosis of diseases by leveraging multidimensional health care data.

In the past, many studies have been performed and found out that the textural, morphological analysis of the tissue and cell imaging scans have provided some very astonishing results. The application of textural and morphological analysis was considered to be huge as it was able to perform the quantification of gray level patterns and derive the inter-pixel relationship within the regions of interest. Moreover, it was also observed that different areas in a scan or an image had different textural and morphological patterns which were difficult for human beings to calculate. Therefore, textural and morphological analysis of the imaging scans proved to be very much reliable for neurological studies and applications in the detection and diagnosis of progressive diseases. But with the advances in the field of computer applications and intelligent systems, the research community is now focusing more on data-driven feature representation rather than handcrafted feature engineering which requires domain-specific knowledge. Therefore, with the rapid development of deep learning architectures and technologies, it is proving to lay down some state-of-the-art methodologies for medical image applications.

2 Data Collection and Preprocessing

2.1 Data Collection

The data for the study was collected from the ADNI database (https://adni.loni.usc.edu/). ADNI database for neuroimages is considered to be a landmark, international, and multicenter study to research the biomarkers that are responsible for Alzheimer's disease progression. The MRI scans selected for the study were based on particular imaging protocols described in Table 1 and also correspond to the baseline visit. Further, all the scans that were considered in the study were obtained from a single type of scanner, i.e., GE medical systems. Moreover, all the acquired scans were based on magnetization prepared-rapid gradient echo (MP-RAGE) sequence. All the scans used in the study were acquired in a time range of 20–30 min field of view (FoV) of all the scans including vertex, cerebellum, and pons.

Post applying the filter based on the imaging protocol mentioned in Table 1, a total of 910 MRI scans were selected from the baseline visit of the patients. Out of 910 patients, 423 were female, and 487 were male. The scans that were considered for the study belonged to the subjects aged 68.26 ± 7.2. The scans primarily belonged to two research groups that are healthy control (HC), mild cognitive impairment (MCI), and Alzheimer's disease (AD). The scans were distributed into the respective research groups as 336 for healthy control, 307 for MCI, and 267 for Alzheimer's disease. The subjects who were considered for obtaining the scan were selected on certain criteria described in Table 2.

Table 3 plotted above shows the specifications of the scans that were obtained from the ADNI database.

Table 1 Parameters for choosing MRI scans from the ADNI study

Imaging protocol	Values
Modality	MRI
Research group	HC, MCI, and AD
Visit	Baseline
Acquisition plane	Sagittal
Acquisition type	3D
Field strength	1.5 T
Flip angle	8.0°
Scanner manufacturing	GE medical systems
Pixel spacing	0.9–1.5 mm (X&Y)
Slice thickness	1.0 mm
Weighting	T1

Table 2 Eligibility criteria for the subject to be included in a group

Research group	Criteria
Alzheimer's disease	1. Mini-mental state exam score between 20 and 26 2. Clinical dementia rating of 0.5 or 1.0 3. NINCDS/ADRDA criteria for probable AD
Control	1. Subjects must be free of memory complaints 2. Mini-mental state exam score between 24 and 30
Mild cognitive impairment	1. Clinical dementia rating of 0.5 2. The subject must have a subjective memory concern as reported by the subject, study partner, or clinician 3. Cognitively normal, based on an absence of significant impairment in cognitive functions or activities of daily living

Table 3 Specification of acquired scans from ADNI

Imaging parameters	Values
Dimensions	256 × 256 × 180 pixels
Interslice gap	0.0 mm
Slice thickness	1.0 mm
Spacing	0.9375 × 0.9375 × 0.9375 mm
Plane	Sagittal

2.2 Data Preprocessing

The dataset that was used in the study was obtained from the ADNI database, and as mentioned above that ADNI is a multicenter study, therefore, the imaging scans acquired in the study contained temporal and spatial differences. To solve this particular problem and to maintain a constant tendency between all the scans, it is required that all the scans need to be in the same space such as Montreal Neurological Institute (MNI) [1, 2] or Individual Brain Atlases using Statistical Parametric Mapping (IBASPM) [3]. Therefore, to bring all the scans to the same space, an image registration procedure is performed. Image registration is a process that mutates upon a fixed image to find the correct alignment parameters so that an unknown or unseen image can be aligned similarly to the fixed image. Trivially, image registration could be understood as the process of aligning two images to a particular space where one acts as the source image and the other as target image, and the source image is transformed in a method to align with the target image. In the specific study, the MRI scans obtained from the PPMI database are considered as the source image, and the atlas such as MNI or IBASPM are considered as the target image.

The registration of the MRI scans obtained from the ADNI database was performed using ICBM-152-T1w-Nonlinear-Symmetric atlas created by Fonov et al. [1, 2]. The specifications of the ICBM-152-T1w-Nonlinear-Symmetric atlas are described in Table 4. The registration of the MRI scans was performed using one

Table 4 Specification of the acquired scans from ICBM–152- T1w-Nonlinear-Symmetric atlas

Image parameters	Values
Dimensions	193 × 229 × 193 pixels
Interslice gap	0.0 mm
Slice thickness	1.0 mm
Spacing	1.0 × 1.0 × 1.0 mm
Plane	Sagittal

of the most effective normalization tools known as Advanced Normalization Tools Python (ANTsPy) [4]. ANTsPy is particularly used in the field of imaging research for extracting important information from complex imaging datasets to perform preprocessing on MRI, fMRI, and SPECT data. The registration of the acquired MRI scans with the ICBM-152-T1w-Nonlinear-Symmetric atlas was performed using symmetric normalization. Figure 1 shown below depicts a particular MRI scan before and after the registration process.

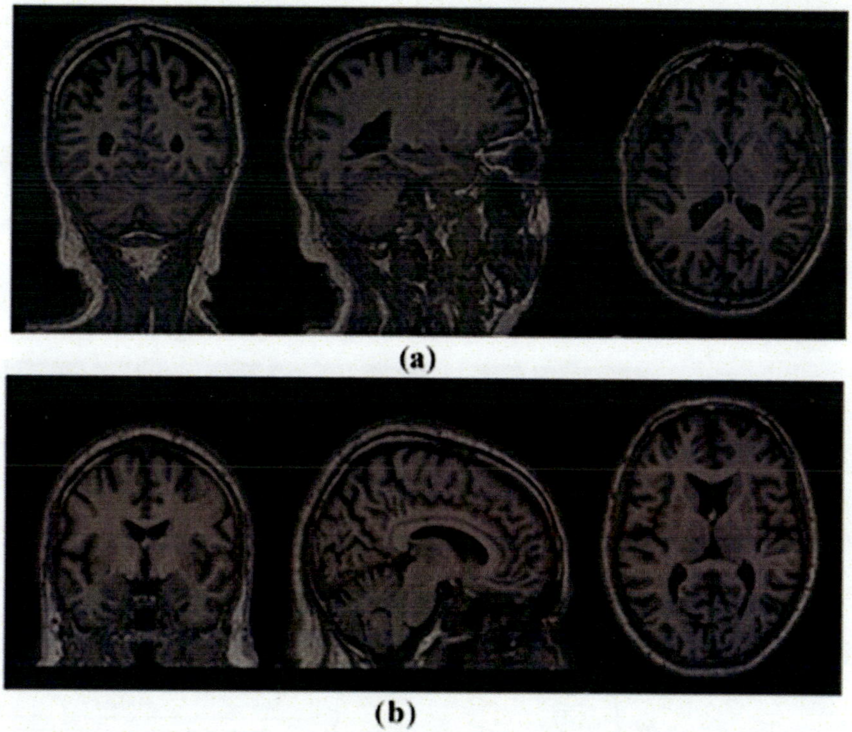

Fig. 1 Sample scan pre and post-registration. **a** MRI scan before registration **b** MRI scan after registration

3 Materials and Methods

The main premise of the study focusses on the detection of Alzheimer's disease and the classification of MRI scans as healthy control (HC), mild cognitive impairment (MCI), and Alzheimer's disease (AD) using 3D convolutional neural networks. The complete flow of process and methodology for the detection of Alzheimer's disease is been described in Fig. 2. The methodology is primarily divided into four (4) stages namely MRI scan acquisition from the ADNI database, data preprocessing, registration and transformation, 3D convolutional neural network architecture, and finally the results and performance evaluation of the CNN architecture based on particular performance metrics. The first two stages of the methodology have been thoroughly discussed in Sect. 2, and further, the third and the fourth stage will be discussed in the following sections.

3.1 3D Convolutional Neural Networks

In recent times, supervised learning techniques for solving problems have evolved massively. Moreover, the popularity and effectiveness of deep learning algorithms have also undergone a major paradigm shift in terms of architectural designs and optimizer functions [5]. Particularly, in the field of health care, the deep learning

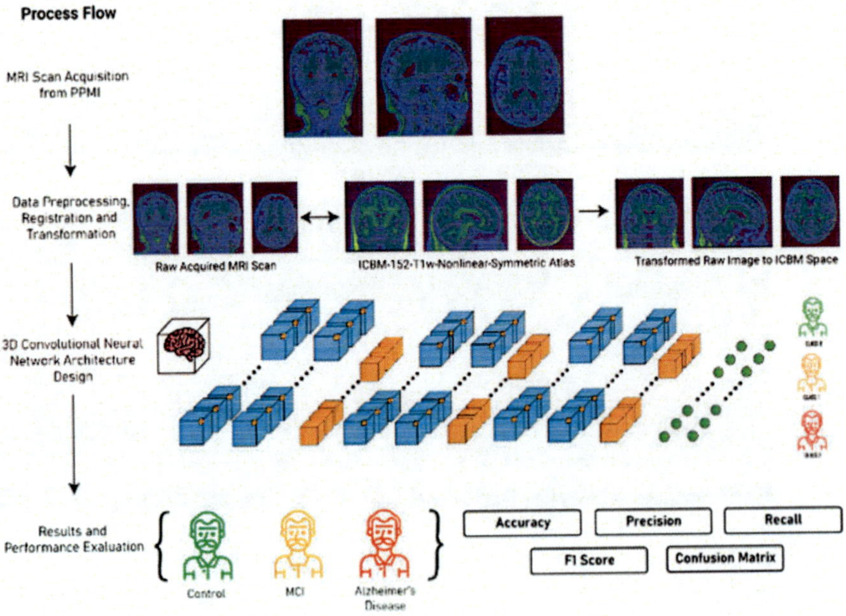

Fig. 2 Complete process flow

algorithms have shown much predominance over the previous techniques that were used for imaging analysis, aneurysms detection in images, biosignal analysis, etc.

In this study, a 3D convolutional neural network model has been developed for the detection of Alzheimer's disease from T1 weighted MRI scans. The primary proposition of the work presents a system that can be used to identify Alzheimer's disease from MRI or rather brain images. Additionally, the second proposition of the study is to determine the plausible regions of interest (ROIs) in the brain MRI images that are responsible for Alzheimer's disease. Therefore, to solve the primary proposition of the study, a 3D convolutional neural network has been developed as shown in Fig. 3. The CNN network developed in the work consists of 18 layers including the input and the output layer. Further, the network architecture consists of ten 3D convolution layers which allows the model to create the feature representations of the input brain MRI scans. Moreover, all the convolution layers are supported by activation functions. Further, all the feature representations are subjected to max-pooling layers which are responsible for downsampling the input feature matrix and provides an abstract form of the feature representation to avoid overfitting. After the complete process of feature learning, all the feature matrices are flattened so that it can be accepted by the dense layer or the fully connected layer. The representations from the dense layer are further subjected to the output dense layer with three neurons and SoftMax activation which corresponds to the three states that are healthy control, mild cognitive impairment (MCI), and Alzheimer's disease.

3.2 Hypothesis and Training Procedure

For developing statistical, machine learning, and deep learning model, the first step is considered to be the development of the hypothesis of the problem that needs to be solved. Therefore, the primary hypothesis that was devised for solving a particular problem is as follows:

1. The recall of Alzheimer's disease class must be 100%, and there should not be any mispredictions of the samples belonging to Alzheimer's disease class to any of the other two classes.
2. For the MCI class, there must not be any mispredictions of samples belonging to the MCI class to the healthy control class.
3. The recall of the healthy control class must be more than 85%.

Therefore, based on the above hypothesis, the performance of the 3D convolutional neural network model was evaluated. For the evaluation purpose, five different classification performance metrics were considered namely accuracy, precision, recall, f1-score, and confusion matrix. Also, for determining the generalizability of the model over unseen data, a five-split cross-validation was performed. The details regarding the evaluation of the performance metrics are described in the results section.

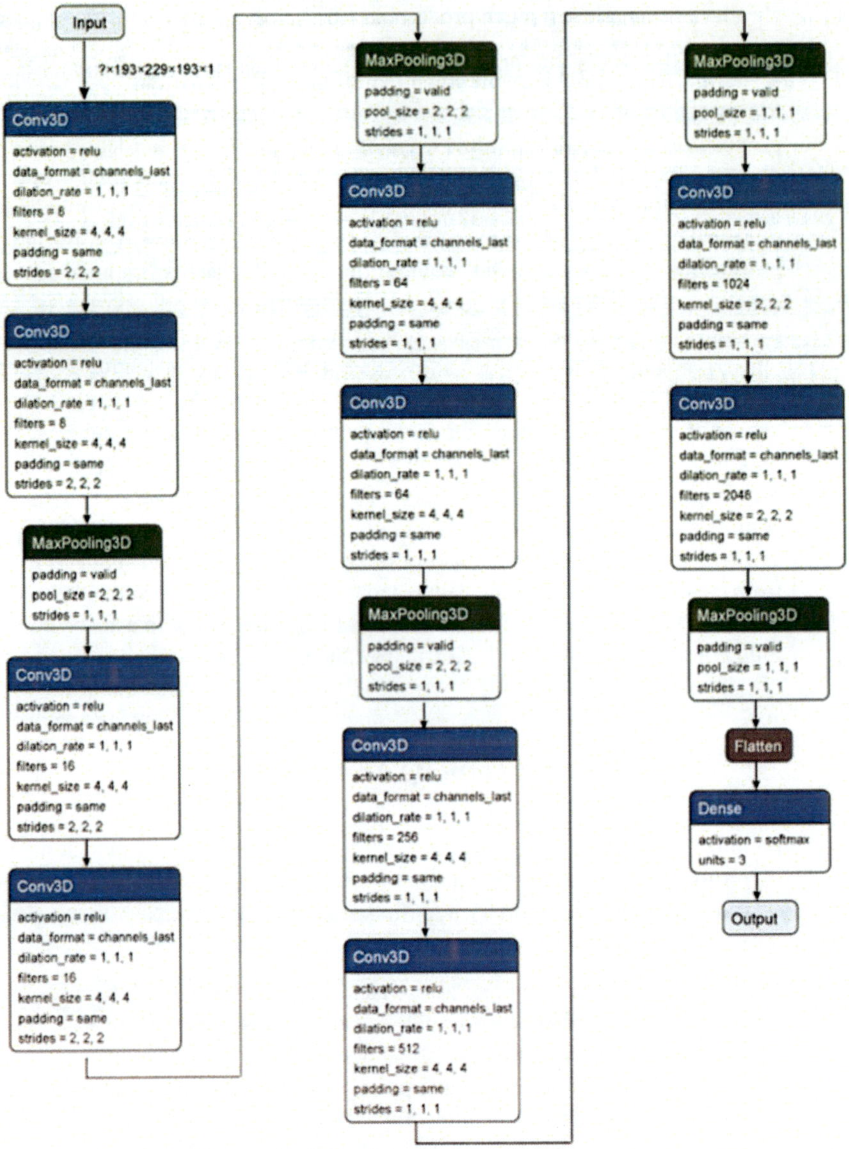

Fig. 3 3D-convolutional neural network architecture

3.3 *Model Optimization Hyperparameters and Loss*

The development of the 3D CNN architecture is indeed the most important aspect of the work. But more essentially, the component that needs to be considered carefully for creating the learning algorithm is choosing the right set of hyperparameters for optimizing the internal set of parameters of the network such as weights and biases, and the loss function.

The process of controlling the training process is considered very important while creating a deep learning model. The process is undertaken by the hyperparameters of the optimizer function that is responsible for tuning the optimizer algorithms. For the present study, the primary aspect that lies in the optimization algorithm is to minimize the validation and the testing error of the model. For performing the specific task, the hyperparameters that reside outside the primary deep learning model must be tuned in such a way we generate the perfect internal parameters of the model that are the weights and biases. But the challenge that is faced in the process is that the hyperparameters must be chosen in a particular way that it should be model-specific rather than training set specific to increase the generalizability of the model over unseen data. Therefore, for choosing the perfect set of hyperparameters to maintain the overall model generalizability and optimum objective score, Bayesian sequential model-based optimization (SMBO) is used.

Bayesian SMBO is an algorithm used for hyperparameter optimization the works on minimizing an objective function by creating a surrogate model (probability function) based on the evaluation results of the previous objective function. The basic objective function of the Bayesian SMBO is given by.

$$P(\text{score}|\text{hyperparameters}) = \frac{P(\text{hyperparameters}|\text{score})\,P(\text{score})}{P(\text{hyperparameters})}a. \tag{1}$$

The surrogate model that is developed by the Bayesian SMBO is considered to be less expensive than the main optimizer function [6]. Further, the next set of evaluation results is selected by using the expected improvement criterion [7]. The criterion is defined as

$$EI(x) = E\big(\max\big(f(x) - f^*, 0\big)\big) \tag{2}$$

where x belongs to the hyperparameter values and considered to be an improvement in the objective score of $f(x)$ and f^* is the maximum value of the objective score found in the process. Further, in the process, AdaDelta [8] is chosen as the optimizer algorithm for optimizing the weights and biases of the network. AdaDelta is considered to be a very robust algorithm relating to the gradient descent algorithm. The algorithm dynamically adapts over due course of the training process by leveraging only first-order information. Moreover, the algorithm does not require any manual tuning of the learning rate and is very robust towards noisy gradient information. Therefore, Bayesian SMBO was applied to the algorithm to generate the optimum hyperparameters and is mentioned below.

$$\text{Learning rate}: 0.07423;\ \rho : 0.751;\ \epsilon : 1.0$$

Another very integral part of the deep learning models is the loss functions. These functions are typically used to determine the variability between the prediction and the true value. The output of the loss functions is non-negative values that increase the generalizability of the model by decreasing the loss [9]. The loss function of a model is given by:

$$L(\theta) = \frac{1}{n} \sum_{i=1}^{n} L(y^{(i)}, f(x^{(i)}, \theta)) \tag{3}$$

where θ represents the parameters of the model, x represents the feature matrix and y represents the true values for a particular set of features.

The loss function used in the work is the categorical cross-entropy loss which is also known as the SoftMax loss. The categorical cross-entropy loss determines the performance of a model whose output is a probability. In the categorical cross-entropy loss function, each prediction is compared to the actual class value, and a score is calculated. The score is further used to penalize the probability of the prediction based on the difference from the actual value. The penalty that is offered to the predicted value is purely logarithmic in nature where a small score is been allotted to tiny differences and the huge score is allotted to larger differences [10]. The equation for the categorical cross-entropy loss is given by:

$$-\frac{1}{N} \sum_{i=1}^{N} \sum_{c=1}^{C} 1_{y_i \in C_c} \log(P_{\text{model}}[y_i \in C_c]) \tag{4}$$

where the double sum has been performed on the ith data samples ranging from 0 to N and the classes which range from 0 to C. The term in the equation,$1_{y_i \in C_c}$ acts as the indicator function for the ith observation for the Cth category. The term $P_{\text{model}}[y_i \in C_c]$ is the prediction probability for the ith observation in the Cth class.

4 Results

The 3D convolutional neural network model developed in the work presented decent results in terms of detecting Alzheimer's disease from brain MRI scans. The model developed in the work quantitatively presented effective results by prompting an average recall and precision of 0.9553 and 0.9411 for all three classes, respectively. Also, for the training procedure, a five-split cross-validation with the ratio of 80:20 was performed over the complete dataset, and it was found that all the data splits showcased a constant tendency toward the testing accuracy. Table 5 plots the results of the five-split cross-validation where were used to determine the generalizability of the 3D convolutional neural network model over unseen data.

Figure 4 demonstrates the confusion matrix that was generated based upon the results received from the best performing split of the 3D CNN architecture. Also, from the confusion matrix, it can be observed that the results completely align with the initial hypothesis which states the recall of the samples belonging to the Alzheimer's

Table 5 Performance evaluation of 3D CNN model

Metrics	Split 1	Split 2	Split 3	Split 4	Split 5
Accuracy (%)	90.36	94.65	95.88	95.37	92.70
Precision	0.906	0.9347	0.951	0.9577	0.928
Recall	0.9135	0.9481	0.9601	0.9436	0.9234
F1-score	0.9107	0.938	0.9538	0.95	0.9162

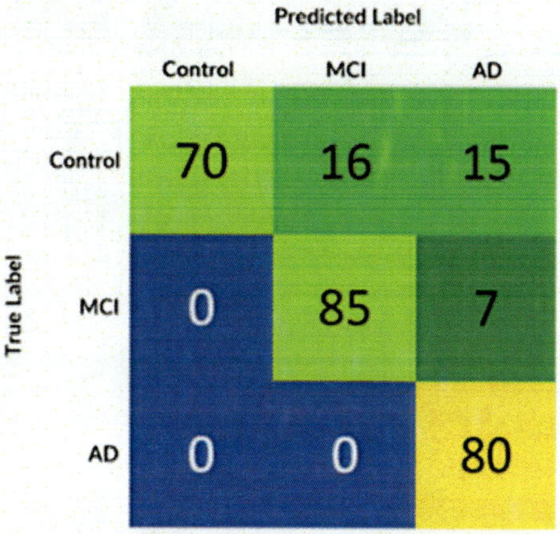

Fig. 4 Confusion matrix of third split cross-validation

disease class needs to be 100%, in MCI class, there should be no mispredictions in the control class, and recall of the healthy class should be more than 85%.

Another very important factor that needs to be measured for evaluating the performance of the deep learning models is the interpretability of the models. The field of health care is considered to be a critical field when it comes to the implementation of automated intelligent systems. So, the prime requirement that needs to be provided out of the model is the interpretation behind a particular prediction or causal-effect information that led to a particular prediction. Therefore, to interpret the predictions of the developed 3D CNN model, class activation map was used [11–13]. Figure 5 shows the class activation map on the sample MRI slices that has been predicted as Alzheimer's disease. The class activation map shows that the model paid much attention to the region of the hippocampus where the degradation took place.

5 Discussion

The study presented in the papers concerns the development of a 3D convolutional neural network architecture for the detection of Alzheimer'sdisease from 1.5T-T1 weighted MRI scans. For performing the study, MRI scans were collected from the ADNI database from three different research groups namely, healthy control, MCI,

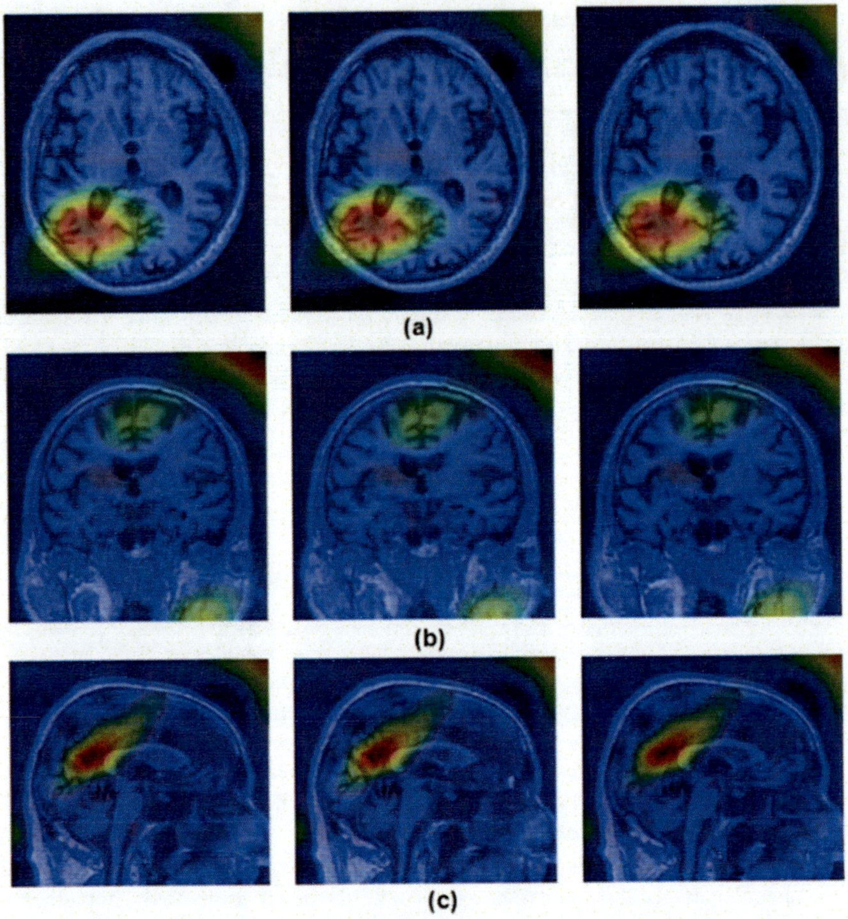

Fig. 5 Class activation maps for sample slices of MRI scan that has been predicted as Alzheimer's disease **a** Axial view **b** Coronal view **c** Saggital view

and Alzheimer's disease. Primarily as discussed that the ADNI is a multicenter study, therefore, the acquired MRI scans had spatial and temporal differences. So, to bring all the MRI scans to the same space, an image registration routine was performed over all the MRI scans. The registration of images was performed using ICBM-152-T1w-Nonlinear-Symmetric atlas. Post-registration of the brain MRI scans, a 3D convolutional neural network was developed for the learning intricate patterns in the MRI scans for the detection of Alzheimer's disease and classifying MRI scans into healthy control, MCI and Alzheimer's disease category, respectively.

Before the development of the model, a hypothesis was designed to evaluate the performance metrics of the model. The hypothesis stated that the recall of the samples belonging to the Alzheimer's disease class needs to be 100%, in MCI class, there should be no mispredictions in the control class, and recall of the healthy class should be more than 85%. Therefore, to satisfy the prior hypothesis, 5 (five) performance metrics namely confusion matrix, accuracy, precision, recall, and f1-score were evaluated. From the results, it can be observed that the model predicted superiorly by plotting a maximum accuracy of 95.88%, recall of 0.9601, the precision of 0.951, and an f1-score of 0.9538 for all the three classes, respectively. Also for understanding the generalizing capability of the model, five-split cross-validation was performed, and it was observed that the model performed constantly over all the cross-validation splits.

Before the development of the model, a hypothesis was designed to evaluate the performance metrics of the model. The hypothesis stated that the recall of the samples belonging to the Alzheimer's disease class needs to be 100%, in MCI class, there should be no mispredictions in the control class, and recall of the healthy class should be more than 85%. Therefore, to satisfy the prior hypothesis, 5 (five) performance metrics namely confusion matrix, accuracy, precision, recall, and f1-score were evaluated. From the results, it can be observed that the model predicted superiorly by plotting a maximum accuracy of 95.88%, recall of 0.9601, the precision of 0.951, and an f1-score of 0.9538 for all the three classes, respectively. Also for understanding the generalizing capability of the model, five-split cross-validation was performed, and it was observed that the model performed constantly over all the cross-validation splits.

Presently, as the research area in artificial intelligence, machine learning and deep learning are pretty focused on the interpretability of the black box models. Therefore, to determine and understand whether the model is choosing the correct areas or the regions that are responsible for the detection of Alzheimer's disease, a 3D class activation map (3D CAM) was developed. From the 3D-CAM plot depicted in Fig. 1, it can be observed that the model has predicted a particular MRI scans as Alzheimer's disease by paying much attention to the hippocampus. Therefore, it can be considered that the model understood the particular areas in the MRI scans which are responsible for Alzheimer's disease.

6 Conclusion

In the proposed study, a 3D MRI analysis was performed for the detection of Alzheimer's disease using 3D convolutional neural network. The study leveraged full brain 3D MRI scans to understand intricate patterns in all the subcortical structures of the brain for the detection of Alzheimer's disease. For the evaluation of the CNN model, certain performance metrics were considered, and to validate the values of the performance metrics, a prior hypothesis was designed. After the training of the 3D-CNN model, it was observed that the model performed superiorly by closely aligning with the prior hypothesis of the study and also demonstrated pretty astounding results. The model developed in the study plotted an overall accuracy of 95.88%, recall of 0.9601, the precision of 0.951, and an f1-score of 0.9538 for all the three classes, respectively. Moreover, the interpretation of the model over the MRI scans was also evaluated using 3D class activation maps, and it was found that the model paid maximum attention to the hippocampus for predicting a particular MRI scan as Alzheimer's disease.

To conclude, the outcome of the proposed study is very motivating. But there remains a huge scope of untouched study concerning the development of innovative architectures that can be leveraged for the detection of Alzheimer's disease using 3D-CNN. Moreover, presently the study focused on whole-brain MRI scans, but in the future, it is highly recommended to perform such research by considering specific subcortical structures and the development of more efficient architectures for the detection of Alzheimer's disease.

References

1. V. Fonov, A.C. Evans, K. Botteron, C.R. Almli, R.C. Mckinstry, D.L. Collins, Group TBDC. Unbiased average age-appropriate atlases for pediatric studies, **54**, 313–327 (2011).
2. V.S. Fonov, A.C. Evans, R.C. McKinstry, C.R. Almli, D.L. Collins, Unbiased nonlinear average age-appropriate brain templates from birth to adulthood. NeuroImage **47**, S102 (2009)
3. Y. Alemán-Gómez, IBASPM: Toolbox for automatic parcellation of brain structures, in *Proceedings of the 12th Annual Meeting of the Organization for Human Brain Mapping*, Florence, Italy, 11–15 June 2006.
4. C. Haaxma et al., Gender differences in Parkinson's disease. J. Neurol. Neurosurg. Psychiatry **78**(8), 819–824 (2007)
5. R. Miikkulainen, J. Liang, E. Meyerson, A. Rawal, D. Fink, O. Francon, B. Raju, H. Shahrzad, A. Navruzyan, N. Duffy, B. Hodjat, Evolving deep neural networks, in *Artificial Intelligence in the Age of Neural Networks and Brain Computing* (Academic Press 2019), pp. 293–312
6. P.I. Frazier, A Tutorial on Bayesian Optimization (2018). https://arxiv.org/abs/1807.02811.
7. M.A. Osborne, *Bayesian Gaussian Processes for Sequential Prediction, Optimisation and Quadrature* (Doctoral dissertation, Oxford University, UK, 2010).
8. M.D. Zeiler, *Adadelta: An Adaptive Learning Rate Method* (2012). ArXiv preprint arXiv: 1212.5701.
9. Z. Zhang, S. Mert, Generalized cross entropy loss for training deep neural networks with noisy labels, in *Advances in Neural Information Processing Systems* (2018), pp. 8778–8788
10. A. Björklund, S. Dunnett, Dopamine neuron systems in the brain: an update. Trends Neurosci. **30**(5), 194–202 (2007)
11. B. Zhou, A. Khosla, A. Lapedriza, A. Oliva, A. Torralba, *Learning Deep Features for Discriminative Localization.* CVPR'16 (2015). ArXiv: 1512.04150.
12. B.N. Patro, M. Lunayach, S. Patel, V.P. Namboodiri, U-cam: visual explanation using uncertainty based class activation maps, in *Proceedings of the IEEE International Conference on Computer Vision* (2019), pp. 7444–7453.
13. R.R. Selvaraju, M. Cogswell, A. Das, R. Vedantam, D. Parikh, D. Batra, Grad-cam: visual explanations from deep networks via gradient-based localization, in *Proceedings of the IEEE International Conference on Computer Vision* (2017), pp. 618–626.

Graph Theory-Based Numerical Algorithm to Secure WSAN Network with Low Delay and Energy Consumption

Ju Jinquan, Mohammed Abdulhakim Al-Absi, Ahmed Abdulhakim Al-Absi, Mangal Sain, and Hoon Jae Lee

Abstract Aiming at the defects of low speed and high consumption of traditional wireless sensor and actuator networks, this paper studies the wireless sensor and actuator network repair algorithm based on graph theory. This paper first analyzes the concept of the wireless sensor and sensor network, and then presents the cooperation algorithm in the wireless sensor and sensor network. The NS2 simulation model topology and truncation points are basically fixed. The simulation results show that the real-time performance and power consumption of the proposed algorithm are better than comparing the K-IDS algorithm and the RCR algorithm.

Keywords WSAN · Cut-point detection · Wireless sensor · Actuator network · Repair algorithm · Cooperative algorithm

---

J. Jinquan · M. A. Al-Absi
Department of Computer Engineering, Dongseo University, 47 Jurye-ro, Sasang-gu, Busan 47011, Republic of Korea
e-mail: Jujinquan351939917@gmail.com

M. A. Al-Absi
e-mail: Mohammed.a.absi@gmail.com

A. A. Al-Absi
Department of Smart Computing, Kyungdong University, 46 4-gil, Bongpo, Gosung, Gangwon-do 24764, Republic of Korea
e-mail: absiahmed@kduniv.ac.kr

M. Sain · H. J. Lee (✉)
Division of Information and Communication Engineering, Dongseo University, 47 Jurye-ro, Sasang-gu, Busan 47011, Republic of Korea
e-mail: hjlee@dongseo.ac.kr

M. Sain
e-mail: mangalsain1@gmail.com

P. K. Pattnaik et al. (eds.), *Proceedings of International Conference on Smart Computing and Cyber Security*, Lecture Notes in Networks and Systems 149,
https://doi.org/10.1007/978-981-15-7990-5_3

1 Introduction

Although the deployment of wireless sensors in many areas, including environmental monitoring, health care, etc., the requirements for combining actuators and sensors are increasing during the application process, and in this context, network systems must be used. The reactor interacts with the physical environment to be able to collect the corresponding information. From an engineering perspective, the operator is also a converter that accepts signals and converts corresponding signals. In the use of conventional sensors and engine networks, some mature research theories gradually emerged. Some small, low-cost wireless remote sensing implementers are gradually available, and in this development context, the scope of application has been extended to sensor operator networks.

WSAN belongs to the extension of WSN, and the difference between them is that WSAN can change the environment and the physical world, and at the same time, the wireless sensor network and engine can be applied to disaster detection, VANET [1–8] intelligent construction, biological and chemical detection fields and others in the future.

The wireless sensor and motor network consists of nodes of sensors and actuators. The sensor nodes are primarily responsible for collecting information, and the nodes in the engine are primarily responsible for changing the behavior of the environment. The sensor and the operator node are wireless, and the sensor node is primarily responsible for sensing and reporting environmental status information, and the operator node is primarily used to collect sensor location data and establish a basis for environmental status detection. The wireless sensor operator network can be self-organized and has the ability to work intelligently and independently, and users can connect to the network remotely and send some basic instructions.

In WSAN applications, the sensor is static and the operator is in motion, and in some specific environments, the sensor node is also free to move. For example, in the sensor-positioning problem, the sensor node can move to the location of the failed sensor node to ensure continuous signal coverage within the area. As a result, sensors and actuators can be integrated into the robot, allowing the robot to perform sensor while providing mobility.

1. Characteristics of wireless sensors and engine networks

The main advantages of wireless sensor networks and operating networks are synergy, as wireless sensors and operator networks require coordinated communication between sensors and actuators to improve overall application quality. Synergy between the sensor and the operator provides the WSAN path-building functionality to transfer data from the sensor to the operator. Some flow control work can also use this type of collaborative communication, where the reactor can help the sensor to obtain an accurate geographic location, and after receiving the corresponding data, triggers can be coordinated with each other to determine the best solution and complete the corresponding work tasks, in the collaborative process, including also design network standards, activity planning, fault tolerance and other issues.

In general, the engine is stronger than the sensor and the transmission radius is relatively large. In this case, the operator can deliver information directly to the sensor node, and the sensor can also transfer data to any operator, which improves the speed and quality of information transmission.

2 Wireless Sensor and Actuator Network Cooperation Algorithm

2.1 *Distribution Frame Algorithm*

Relevant scientists propose a distribution frame algorithm. In this type of algorithm, the event is mainly directed to different levels of problem models, the form of linear programming is adopted, and the multi-state distributed algorithm is mainly used to achieve the start of each sensor. Accelerate, idle and assemble any of the four switch types. After the original node sends the corresponding data, the original node can calculate the corresponding data arrival time on the package node. The data transfer rate accessed by r is indicated, and the corresponding data is propagated by the cyclical nature of the operator node. If the value of r decreases, the transmission power of the sensor node will also decrease, which reduces the power consumption of the transmission, thereby achieving efficient use of resources.

Since reliable upper and lower limits cannot be accurately calculated within this algorithm, it is impossible to summarize formulas for network changes, which is the main reason why these algorithms remain in the design phase. These algorithms can convert the AA-type collaboration problem into a nonlinear integer programming model. In the crash zone, the power of the operator node can be increased within task time constraints, so these algorithms focus on the cooperative power consumption of the operator node.

2.2 *Routing Framework Algorithms and Real-time Collaboration*

Shah G.A of the USA proposed an algorithm to implement the cooperation and guidance framework, which is mainly applied to collaborative communications (SA) within a semiautomatic mechanism structure. Its framework mainly includes Protocol for the Advancement of Women, Protocol for DEAR. DEWC belongs to the Chinese as a weighted dynamic aggregation algorithm, as shown in Fig. 1; the corresponding sensor nodes are configured in a dynamic hierarchical structure, and energy is the aggregation protocol. DEAR is an energy-aware routing algorithm that basically integrates two technologies: forward tracking and rollback, creating a route between nodcs and routers at a specific time.

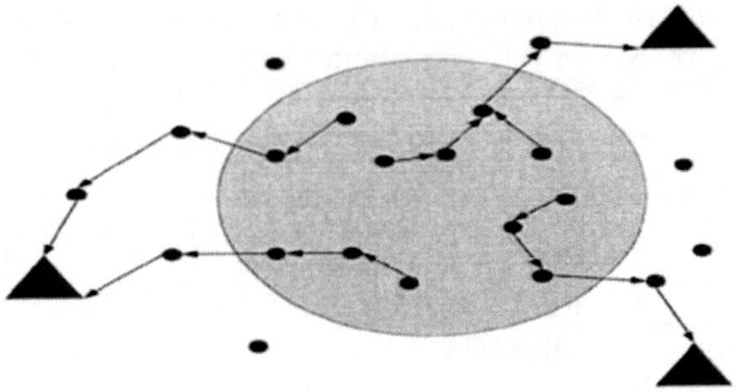

Fig. 1 Clustering schematic

Since there are SINK nodes, in a semi-automatic structure, sensor nodes and operator nodes must be coordinated by a central DEAR algorithm. In the absence of a SINK contract, SA collaboration is required by the distributed D-DEAR, and in the selection of multiple routes, the most effective route must be determined, and data transmission on this route is performed to meet the delay state.

2.3 Topology Maintenance Algorithms

The structure maintenance algorithm (shown in Fig. 2) essentially implements network repair by nodes adjacent to the aging node, which can effectively repair the network and ensure continuity of the network under the minimum information exchange and moving distance requirement. While fixing this algorithm, you must change the location of the nodes around the aging node, and you need to maintain a r/2 distance from the failed node. In this type of algorithm, r belongs to the radius of the node connection, and if the network suffers from a cohesion problem after the transmission of the adjacent node, it needs to repair the network by mobile cascading. The structure maintenance algorithm belongs to the network component in the sensor, and as long as the failed node has a problem, the neighboring node

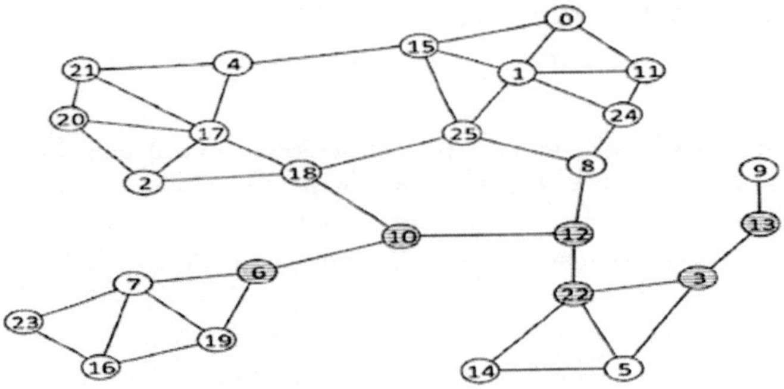

Fig. 2 Topology schematic

can repair the network. At the same time, repair of the structure is a prerequisite that does not require a cooperative connection between the nodes during operation.

2.4 Improved Network SA Collaboration Algorithms

Based on the above research, this paper proposes a low-latency collaborative algorithm, which is divided into four stages as follows.

Step 1: Based on the goal of optimizing the power in the network, according to the event area, the number of sensor nodes, factors such as the node that receives and consumes the transmission power, the frequency of the event occurring, and considering limitations such as network connectivity and delay, the optimal node is determined for the operator. Number of communication and communication radius between sensor nodes.

Step 2: Spread the sensor nodes and the engine nodes in the grid randomly and uniformly, and adopt the aggregation method to divide the grid into several groups, where the cluster can effectively reduce the load burden and avoid participating in excess engine nodes in the same event. There is only one operational node in each group, since the energy and computational energy of the port node are much higher than the sensor node, the port node is identified as the head of the block, and the other sensor nodes are used as members of the group, so that the mass is compared with the rotating mass. The first algorithm is simpler and more in real time. The sensor node within the mass detects the physical quantity and sends it only to the group head. The group head determines how to do it and takes appropriate action.

Step 3: After the assembly is completed, the port node needs to receive event information as quickly as possible and respond to the event in a timely manner. Reducing the distance between the sensor node farthest from the port node and the execution node in the cluster ensures that the operational node receives information from all cluster members in the cluster as soon as possible, finds the event as soon as possible and responds in a timely manner. Therefore, the port node needs to find the appropriate location in the cluster and redeploy it.

Step 4: After deploying the sensor node and the operator node, the connection between them begins. The sensor node determines the optimal path to transfer data information to the operator node to ensure low power consumption and substantial real time. In this paper, mixed one-hop and multi-hop connections are adopted, and within the radius of communication in the sensor node, the sensor node communicates directly with the operator node. When the distance exceeds this distance, the sensor node follows the shortest path algorithm (look for the jump relay node the following adopts multiple jump) to the port node.

It is assumed here that each executing node only executes the events that occur within the cluster in which it is located, so there is no need to consider coordination between the port and the port.

1. Establishment of network energy consumption model

Power grid consumption consists of two parts: the power consumption of the operator node and the power consumption of the sensor node. In this paper, the network structure is assumed to be mixed between both the league and the event-based. Under this assumption, the total energy consumption of an operator node can be derived as shown in Eq. 1:

$$E_{\text{actor}} = T_{\text{com}} E_{\text{rx}} \left(\frac{N_{\text{sensor}} * l}{N_{\text{actor}}} \right) + E_{\text{agger}} \left(\frac{N_{\text{sensor}} * l}{N_{\text{actor}}}, l \right) + \chi \left(\frac{N_{\text{sensor}}}{N_{\text{actor}}} \right) E_{\text{tx}}(d, l) + T_{\text{event}} E_{\text{action}}(t) t_{\text{event}} \tag{1}$$

Among them, the frequency of occurrence of the event is the number of rounds of data collected periodically, which is the power consumption of the operator node. The paper uses the Karush–Kuhn–Tucker (KKT) condition [9] as a first-class necessary condition for a local optimal solution.

2. Clustering algorithm

In wireless sensor networks, some cluster-based protocols have been proposed to perform multiple hop modes and integrate data into a specific group, effectively reducing the power consumption of the network, for example, LEACH is applied to WSN. \query{The sentence "A cluster-based routing..." seems to be incomplete. Kindly check.}A cluster-based routing protocol. In the LEACH algorithm, in order to save energy, sensor nodes are automatically grouped according to the received signal strength, and cluster members send data to the cluster header, then from the cluster header to the SINK node [10].

Because the port node and the sensor node are simultaneous, there are some obvious differences between WSAN and WSN node homogeneity: traditional, node deployment, node coordination and so on. Therefore, the proposed protocol for WSN is not used well in the application requirements and WSAN features.

In WSAN, some aggregate algorithms are proposed, such as the RECRP algorithm protocol [11], where the sensor node is first assembled for the first time, then a sensor node is selected as the cluster head, and the cluster head node is connected to the operator node. For the second cluster, the entire network forms a three-layer network structure of sensor nodes, cluster head nodes, operator nodes, operator nodes and operator nodes. In such a networked structure, the rotation of the cluster head is performed after each round of data transfer, which results in an increase in network delay and affects normal data collection.

This paper considers that the operator node is sufficiently dynamic, robust and has the task of accomplishing the task, so the operator node is usually fixed as the head of the block. The aggregation algorithm is proposed based on the Voronoi scheme of the operator node, and a detailed aggregation algorithm is described below.

The Russian mathematician m. C. Voronoi discovers the concept structure of Voronoi in 1908 which was named after him. It is essentially a global structure in which macro- and microentities in nature interact with distance. The basic definition is described below [12].

From the computational geometry obtained from the Voronoi structure, the V map is intended for separate points in the plane, which divide the plane into several regions, each area including a point and the area where the point is located is the distance to the point. Collect the last points.

Basic definition: Let P be a set of discrete points. Select the Voronoi region in P for all groups at the least distance:

$$V(P_i) = \{P|d(P, P_i) \leq d(P, P_j), j \neq i, j = 1, 2, \dots n\} \tag{2}$$

Let P be a set of discrete points set $P_1, P_2, \dots, P_n \in P$ definition P of V $V(P)$ map defining:

$$V(P) = \{V(P_1), V(P_2), \dots, V(P_n)\} \tag{3}$$

Among them, it is called a V map generator.

Geospatial (two-dimensional) extended definition of V-graph: two-dimensional geospatial space G is a collection of n points, lines and surface entities $g_1, g_2, \dots, g_n$ which is called a metric space that defines a dimension d.

Let G be a set of n entities $g_1, g_2, \dots, g_n \in G$, and the defined Voronoi area $V(g_i)$ is a collection of all the distances to the smallest point (grid):

$$V(g_i) = \{p|d(p, g_i) \leq d(p, g_j), i \neq j, j = 1, 2, \dots, n\} \tag{4}$$

Let G be a set of n entities $g_1, g_2, \cdots, g_n \in G$; define G the Voronoi diagram of $V(G)$ as:

$$V(g) = \{V(g_1), V(g_2), \dots, V(g_n)\} \tag{5}$$

After the Voronoi polygon map is expanded from point set generation to point, line and surface set, the Voronoi diagram has the following three characteristics:

1. There is a generator in each Voronoi polygon;
2. The distance from the inside of each Voronoi polygon to the generator is shorter than the distance from other generators;
3. The points on the polygon boundary are equal to the generator elements that generate this boundary;

Assume that both the sensor node and the actuator node have been deployed and enter the clustering phase. In this paper, the actuator node is used as the cluster head, the actuator node is equivalent to the generator element in the Voronoi diagram, and the Voronoi diagram of the actuator node is drawn. According to the Voronoi diagram characteristic, the sensor nodes in the Voronoi polygon of the actuator node are separated from this. The actuator nodes are closest to each other, and these sensor nodes automatically become cluster members, all sending data to this executor node. The clustering process is described in detail below.

First, assuming that the sensor node and the actuator node have been deployed, the sensor node and the actuator node know their own location (via a GPS receiver), and the entire network is divided into clusters based on the actuator node Voronoi diagram, each cluster including An actuator node and several sensor nodes. Each cluster is considered to be a different subnet. The clustering phase is as follows: As the cluster head of the subnet, the executor node broadcasts its own information (including position coordinates, node type, residual energy), and the sensor node receives all the signals sent by the executor node, and the sensor node automatically joins the distance. In the subnet of the nearest executor node, the sensor node sends an ACK confirmation message to the executor node.

The specific implementation steps of the clustering algorithm:

1. Initialization min $D(i) = \infty$
2. for each actuator A_j do
3. $A_j \rightarrow N_i$: Broadcasting your own information
4. end for
5. Each sensor node N_i; Receiving actor A_j Broadcast Information
6. for each sensor node N_i do
7. for each actuator node A_j do
8. if $d(N_i, A_j) <$ min $D(i)$ then
9. min $D(i) = d(N_i, A_j)$
10. min $DActor(i) = A_j$
11. End if
12. End for
13. End for
14. For each sensor node N_i do
15. $N_i \rightarrow$ min $DActor(i)$ send ACK
16. End for

In order to ensure the real-time nature of the network, the sensor node needs to monitor the event in time and send it to the actuator, and the actuator can perform the task as soon as possible. According to the analysis of real time in [13], once an event is detected by a sensor node, it usually needs to send information to the actuator node closest to it, thus ensuring real-time and low energy consumption.

3. Data transmission based on Dijkstra algorithm

The above analyzes the number of ideal actuators in the network and the communication radius of the sensor nodes and then clusters the network based on the Voronoi diagram of the actuator. The actuator node is the cluster head, and the sensor nodes in the Voronoi polygon of the actuator node are clusters. In order to improve the real-time performance of the network, the members analyzed the ideal position of the actuator node in the Voronoi polygon, redeployed the actuator, and then studied the data transmission path between the sensor node and the actuator node.

In order to improve the real-time performance of the network and reduce the network energy consumption, the data transmission mode in the cluster adopts a mixed mode of single hop and multi-hop, that is, if the distance between the sensor node and the actuator node is within the communication radius of the sensor node, then in the single-hop mode, the sensor node directly transmits data to the actuator node. If the distance between the sensor node and the actuator node is outside the communication radius of the sensor node, the multi-hop mode is adopted. The selection of the sensor node for the next hop is based on the Dijkstra algorithm, that is, the sensor node sends data along the shortest path to the executor node.

In each cluster in this paper, the optimal path within the cluster is searched based on the Dijkstra algorithm, the actuator node is used as the source point, the outer layer is extended until all the sensor nodes in the cluster are included, and then the optimal path is recorded. The executor node broadcasts this routing table to each sensor node, and the sensor node transmits according to this path.

From the above, the position of the sensor node and the actuator node can be known. First, the distance between the nodes is calculated. The node distance is used as the weight, and the actuator node is used as the source node. The shortest path tree is generated by the Dijkstra algorithm of this paper, and then, the actuator node will get the shortest path tree which is broadcast to the sensor nodes within the cluster, and finally the sensor node sends data along the shortest path to the executor node.

3 Simulation Result and Analysis of Cut-Point Detection Algorithm

1. Application of simulation model NS2

NS2 was originally developed by the University of California at Berkeley, and the functionality of NS2 is constantly expanding. NS2 uses C++ language development network simulation software (as shown in Fig. 3). The initial development purpose is to study the protocol interaction between large-scale networks and future networks. At present, NS2 has become the most widely used network in network research. In practical applications, NS2 can simulate multiple communication networks including routing algorithms, network transmission protocols, wireless networks, satellite communication networks.

NS2 is actually a simulator, mainly for discrete event simulation. In NS2, the Tcl class is a communication bridge between C++ code and front-end Tcl code, and is the base class of NS2. In addition to the Tcl class in the front-end and back-end of NS2, there are five other categories, including TclObject class, TclClass class, TclCommand class, InstVar class and EmbeddedTcl class, but all simulation processes are controlled by Tcl class. When using NS2 to carry out simulation work, some small tools are mainly used: dynamic graphic display tool, topology map generation tool, transmission event generation tool, simulation node moving tool, graphic display tool and so on.

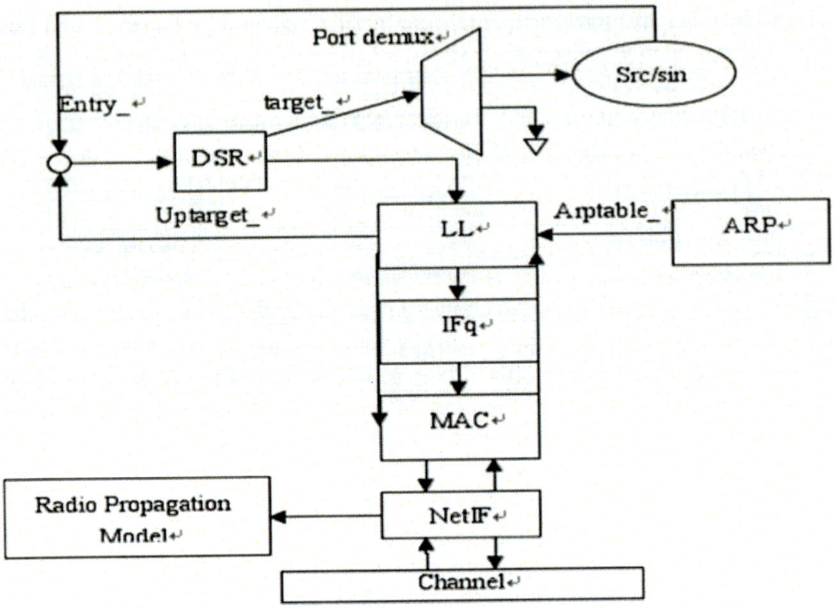

Fig. 3 Schematic diagram of the NS2 model structure

2. NS2 performance indicators

Because NS2 can support Windows, IOS, Linux and other operating systems, the Linux operating system is mainly used as an example. It is assumed that the simulation experiment is carried out in a square area of 200 m × 200 m, and there are 100 sensor nodes in the area, which are randomly and uniformly deployed. The temperature and humidity sensor is used to collect the physical signal, i.e., the temperature and humidity of the soil, and is powered by the battery. The sensor nodes are static and have a defined position. According to the analysis result of the ideal number of actuator nodes, there are nine actuator nodes, and the actuator nodes are mobile. We deploy these nine actuator nodes in the (50, 50), (100, 50), (150, 50), (50, 150), (100, 100), (150, 100), (50, 150), (100, 150) as well as (150, 150) position coordinates.

Based on the above proposed temperature and humidity control system, the clustering algorithm is based on the actuator node Voronoi diagram, the actuator node redeployment and routing model simulation.

3. Clustering model

The network is divided into nine clusters based on the Voronoi diagram of the actuator node in the region of 200 m * 200 m by the clustering algorithm. The obtained clustering model is shown in Fig. 4. In Fig. 4, the triangle represents the actuator node and the circle represents the sensor node. There is only one actuator node and several sensor nodes in each cluster. The cluster head is an actuator node. The distance

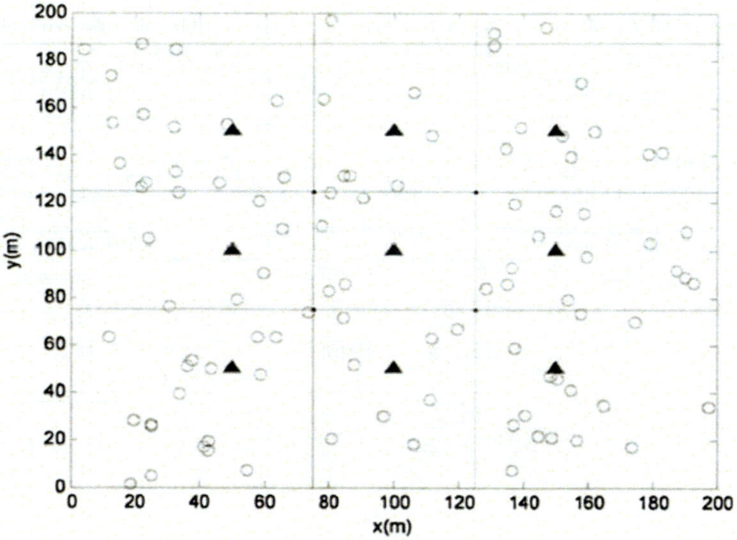

Fig. 4 Clustering model based on Voronoi diagram

between the sensor nodes in the cluster and the actuator nodes in the cluster is the closest to the distance from other actuator nodes.

In order to facilitate the analysis of the nodes, the actuator nodes and sensor nodes are displayed in the cluster. The triangle represents the actuator and the black dot represents the sensor. As shown in Fig. 4.

4. Network model after actuator node redeployment

It is found in Fig. 5 that after the entire node is clustered based on the Voronoi diagram of the actuator node, the network segmentation is not uniform, and the actuator node is not in the center of the cluster, which makes the sensor node in the cluster and the cluster head actuator node the farthest distance cannot be optimal. When the sensor node communicates with the actuator node, the intra-cluster communication delay is not optimal. Through the above research and analysis, the position of the actuator node is redeployed so that the actuator node can the minimax point in the cluster improves the real-time performance of the entire network. Figure 5 is a clustering model after the actuator node is redeployed.

It can be seen from the comparison of Figs. 5 and 6 that after the actuator node is redeployed, the size of each cluster in the entire area is uniform, and the actuator node is centered in the cluster, and the expected effect is achieved.

5. Data transmission model

After the sensor node and the actuator node are deployed, the sensor node monitors the environment and sends the monitored temperature and humidity physical quantity to the actuator node, and the actuator performs corresponding actions according to the

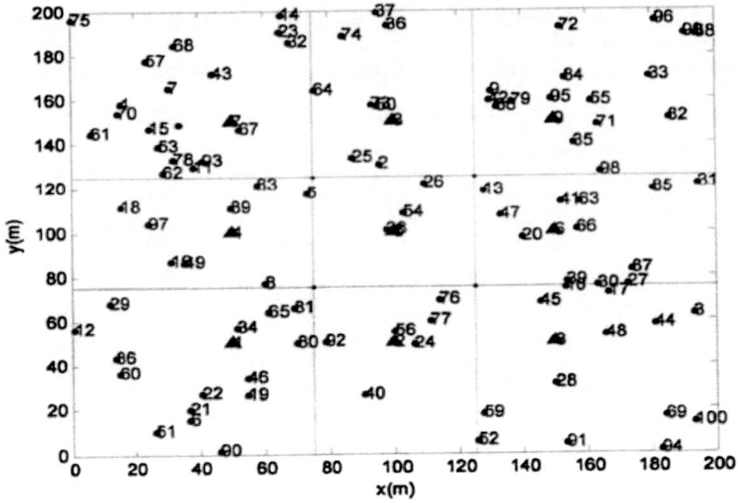

Fig. 5 Clustered model after labeling

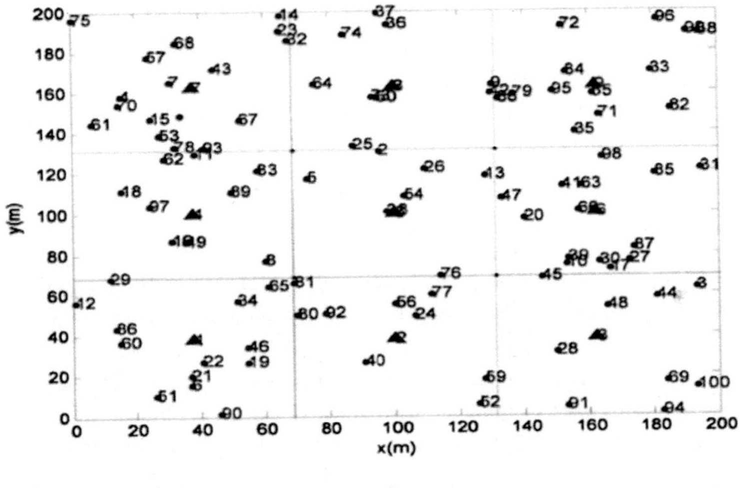

Fig. 6 Clustering model after actuator node redeployment

received physical quantity. When the distance between the sensor node and the actuator node is smaller than the communication node communication radius, the sensor node directly communicates with the actuator node. When the distance between the sensor node and the actuator node is greater than the sensor node communication radius, the sensor node searches for the optimal path. The next hop transmits data to the executor node through multi-hop communication. The entire network uses a multi-hop and single-hop hybrid communication mode. This paper simulates the data transmission path model based on Dijkstra algorithm and obtains the path of the

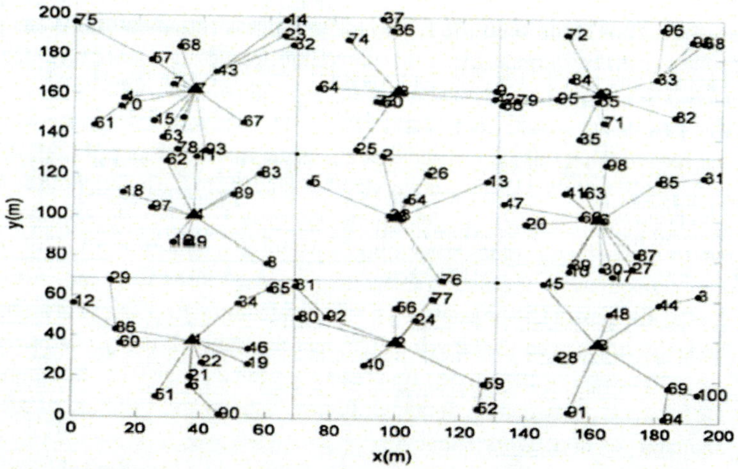

Fig. 7 Data transmission model

sensor node to transmit data to the actuator node. Figure 7 is a data transfer model between a sensor node and an actuator node, where a triangle represents an actuator node and a dot represents a sensor node.

6. Network performance analysis

Based on the above design low-latency low-energy SA cooperative algorithm (low-energy low-latency cooperative algorithm), the performance of the algorithm is tested from the following two aspects, namely network energy consumption and average delay.

First, let's study the average delay. The network's delay formula (6) has been proposed based on the ideal sensor node communication radius and the ideal number of actuator nodes.

$$t_{i-\mathrm{hop}}(R, N_{\mathrm{actor}}) = \frac{\sqrt{A}}{\sqrt{\pi N_{\mathrm{actor}}} \times R}(t_{\mathrm{tx}}(l) + t_{\mathrm{que}}(l) + t_{\mathrm{tx}}(l)) \tag{6}$$

When redeploying the actuator node, the delay formula (7) of the process from event occurrence to execution event in the network is presented in more detail:

$$\Gamma^{e2a} = \left|\frac{\mathrm{dist}(s, a)}{\lambda}\right| \times \varepsilon + f_s \times p + k \times \mathrm{dist}(s, a) + R \tag{7}$$

According to the definition of formula (7):

$$\left|\frac{\mathrm{dist}(s, a)}{\lambda}\right| \times \varepsilon = t_{i-\mathrm{hop}}(R, N_{\mathrm{actor}}) = \frac{\sqrt{A}}{\sqrt{\pi N_{\mathrm{actor}}} \times R}(t_{\mathrm{tx}}(l) + t_{\mathrm{que}}(l) + t_{tx}(l)) \tag{8}$$

Equation (8) represents the delay of the process of transmitting data to the actuator node by the sensor node. Since this part of the delay is the main delay in the network, it affects the real-time performance of the network, and the algorithm mainly improves this part of the transmission delay. Verify this part of the delay. It can be seen from the formula that this part of the delay is equal to the number of hops in the network multiplied by the single-hop delay time. According to the number of sensor nodes, the ideal number of actuator nodes can be obtained. Based on the deployment of sensor nodes and actuator nodes, clustering is performed. Based on the Dijkstra algorithm data transmission, the data transmission model can be obtained. According to the model, the sensor can be known. The maximum number of hops between the executor nodes is substituted into the formula, and the network delay can be approximated. Since the delay of the single hop is assumed to be consistent, the estimated value of the network delay can be obtained. Figure 7 shows the variation of the sensor node and changes in network latency.

Figure 8 shows the average delay in the process of transmitting data to the actuator node by the sensor nodes in the cluster at different scales. As can be seen from the figure, as the network scale becomes larger, the delay of the K-IDS and the RCR becomes larger. When the network changes dynamically, there is a lack of corresponding adjustment. However, in the low-energy low-latency cooperative algorithm, different scales have different actuator numbers and corresponding sensor communication radii, and each time the dynamic change occurs, the actuator node also finds the optimal position, so that it has the longest communication distance. The distance between the sensor nodes can be the closest, thus ensuring that the maximum hop count of the network is not too large and the delay is too large. As can be seen from Fig. 8, in the low-energy low-latency cooperative algorithm, the hop count is generally maintained. At around 2 hops, the guaranteed delay does not increase as the size increases.

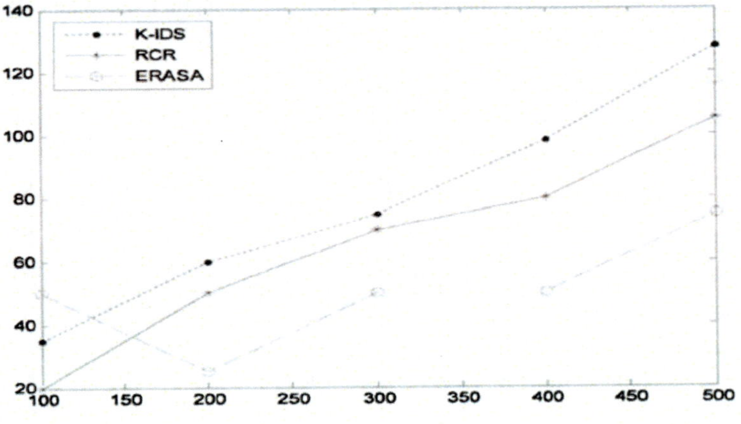

Fig. 8 Network average delays with sensor number curve

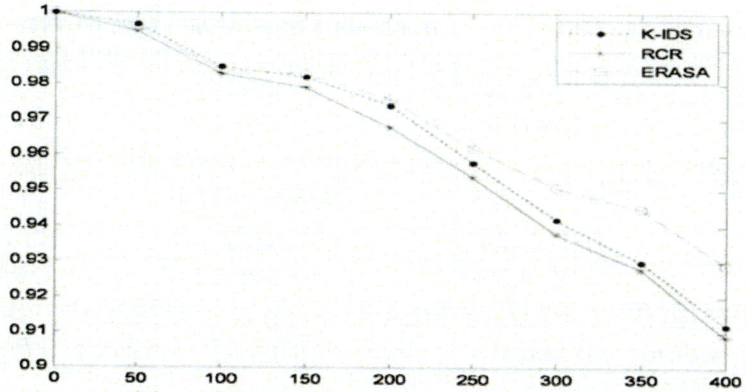

Fig. 9 Relationship between the average energy balance index and the number of periodic report rounds

The energy consumption of the low-energy low-latency cooperative algorithm is analyzed below. The sensor nodes are in the area 200 m × 200 m. Hypothesis $EB_i = \frac{E_{i-\text{rest}}}{E_{i-\text{init}}}(1 \leq i \leq N_{\text{actor}})$, among them EB_i for the energy balance index, $E_{i-\text{rest}}$ Actuator node residual energy, $E_{i-\text{init}}$ Indicates the initial energy of the actuator node, $N_{\text{actor}}1!$ The number of actuator nodes. Assume that the initial energy of the actuator is $10J$, Figure 8 reflects the change in the average energy balance index of the actuator nodes in the network as the monitoring time progresses, which reflects the overall trend of network energy consumption.

It can be seen from Fig. 9 that the average energy of the actuator nodes in the network using the low-energy low-latency cooperative algorithm is slower than that of the K-IDS and the RCR, the residual energy of the actuator node is more, and the energy consumption of the network is relatively smaller.

4 Conclusion

This paper proposes a WSAN (Wireless Sensor and Actuator Network) repair algorithm based on graph theory for the low speed and high consumption of traditional wireless sensor and actuator networks. Firstly, the wireless sensor and actuator network concepts are analyzed. Then, the wireless sensor and actuator network cooperation algorithm are described. At the same time, a low-energy low-latency SA cooperative algorithm is proposed. Finally, the cut-point detection algorithm is summarized. The algorithm is the topology of the simulation model NS2, and cut points are mainly repaired. The simulation analysis results show that the real-time performance and energy consumption of the proposed algorithm are better than other algorithms. This paper does not consider the reliability of the network. Future work is to consider the network from the perspective of reliability; (2) the actuator is not considered in this paper. In coordination between actuators, only the cooperation between the wireless sensor node and the actuator network is considered.

References

1. M.A. Al-Absi, A.A. Al-Absi, T. Kim, H.J. Lee, An Environmental channel throughput and radio propagation modeling for vehicle-to-vehicle communication. Int. J. Distrib. Sens. Network (IJDSN), SAGE Publisher **14**(4), 1–10 (2018)
2. M.A. Al-Absi, A.A. Al-Absi, H.J. Lee, V2V communication modeling for environmental channel throughput and radio propagation, in *8th IEEE International Conference on ICTC Convergence* (IEEE Publisher, Jeju Island, Korea, October 18-20, 2017), pp. 507–512
3. M.A. Al-Absi, A.A. Al-Absi, Y.J. Kang, H.J. Lee, Obstacles Effects on signal attenuation in line of sight for different environments in V2V, in *20th International Conference on Advanced Communication Technology (ICACT)* (IEEE Publisher, Chuncheon-si Gangwon-do, Korea, 11-14 Feb. 2018), pp.17–20
4. M.A. Al-Absi, A.A. Al-Absi, H.J. Lee, Performance Analysis for City, Highway, and Rural Area in Vehicle-to-Vehicle Network, in *8th IEEE International Conference on ICTC Convergence* (IEEE Publisher, Jeju Island, Korea, October 17–19, 2018)
5. M.A. Al-Absi, A.A. Al-Absi, H.J. Lee, Performance enriching channel allocation algorithm for vehicle-to-vehicle city highway and rural network. Sensors **19**, 3283 (2019)
6. M.A. Al-Absi, A.A. Al-Absi, H.J. Lee, A Secure Enhanced Non-Cooperative Cognitive Division Multiple Access for Vehicle-to-Vehicle Communication. Sensors **20**, 1000 (2020)
7. M.A. Al-Absi, A.A. Al-Absi, H.J. Lee, Varied density of vehicles under city, highway and rural environments in V2V communication. Int J Sens Netw **33**(3), 148–158 (2020)
8. A.A. Mohammed, A.A. Ahmed, H.J. Lee, Comparison between DSRC and other Short Range Wireless Communication Technologies. In Proceedings of the 2020 22nd international conference on Advanced Communication Technology (ICACT) Phoenix Park, PyeongChang, Korea, 16–19 Feb. 2020 pp.1–5
9. M. Fukushima, *Basis of Nonlinear Optimization* (Science Press, Beijing, 2011), pp. 76–78
10. Q. Tianshuang, T. Hong, L. Ting, J.Y. Yanghua, *Wireless Sensor Network Protocol and Architecture* (Electronic Industry Press, Beijing, 2018), pp. 67–68
11. Z.C. Dai, Z. Li, B.W. Wang, An energy-aware cluster-based routing protocol for wireless sensor and actor network. Inf. Technol. **8**(7), 1044–1048 (2009)
12. C. Qiang, *The Theory and Application of Limited Voronoi Mesh Generation* (Beijing University of Posts and Telecommunications Press, Beijing, June-7, 2010)
13. G. Nan, L. Zhijun, J. Shouxu, Actor deployment strategies in wsans. First international conference on pervasive computing. Sig. Proc. Appl. **8**(10), 150–154 (2010).

Decentralized Privacy Protection Approach for Video Surveillance Service

Jeongseok Kim and Jaeho Lee

Abstract Cloud-based video surveillance service is being deployed ubiquitously, and the service normally stores video streams in the cloud over the Internet. The main problem of this service is that the user never knows how secure the video stream is transferred and stored. Although the video contains sensitive data that is related to personal or area-specific information, there is no way to objectively check whether the video is securely managed. Therefore, a user must trust that the service provider will keep their video secure without any evidence. Ensuring confidential information and privacy is protected, the sensitive data should be encrypted and traceable when they are transferred. In this paper, an architecture to build a key management system based on blockchain and public-key cryptography is proposed. The video encryption is possible by a block-cipher algorithm with a given symmetric key. In the proposed system, the symmetric key is also encrypted by a public key and stored as a ledger in the blockchain. Thus, when the symmetric key is requested, every operation is traceable, so a user can trust how a video stream is transferred and stored securely because of the nature of blockchain, decentralization, transparency, and the impossibility of being tempered. In addition, the user can monitor the request flow of the key, and then, the key is selectively distributable by the user's intention. Therefore, the system meets the requirement of a fully secure method to manage the whole life cycle of video from creation to deletion. In this paper, we propose to apply this architecture to video surveillance service and also extend them to general-purpose systems.

Keywords Video surveillance · Blockchain · Privacy protection

---

J. Kim (✉) · J. Lee
Department of Electrical and Computer Engineering, University of Seoul, Seoul, South Korea
e-mail: jeongseok.kim@sk.com; justinkim@uos.ac.kr

J. Lee
e-mail: jaeho@uos.ac.kr

J. Kim
Security Labs, AIX Center, SK Telecom, Seoul, South Korea

P. K. Pattnaik et al. (eds.), *Proceedings of International Conference on Smart Computing and Cyber Security*, Lecture Notes in Networks and Systems 149,
https://doi.org/10.1007/978-981-15-7990-5_4

1 Introduction

This paper presents a specific approach to protect privacy or data in a video surveillance service from a user perspective. To protect privacy, video encryption method is taken, but a more important point is that the proposed method can provide a trustful environment for the user. In the existing service, there are few vulnerable points while capturing and recording video from the surveillance camera, but from the point of transmission over the Internet and storage on the service, it is required to protect against video leakage that can lead to privacy infringement.

The protocol discussed in this paper illustrates how the proposed method protects user privacy with common public-key cryptography and video encryption on the blockchain network. Away from the bidirectional encryption concept of cryptography, this paper proposed a unidirectional method to reduce architecture complexity. In addition, any operation to the encrypted video is traceable through blockchain memory, and encryption technology can be used to manage the encrypted video owned by the user. In particular, a user can invalidate the encryption key of video, and then, it has the same effect as permanent deletion.

2 The Privacy Problem in Video Surveillance Service

This paper addresses the privacy problems in video surveillance service. In nature of the service, video is streamed and recorded continuously and automatically without user intervention. For this reason, users need a way to authorize every piece of the system-generated content before moving them to cloud storage or have a special type of media storage that encrypts video on behalf of the user.

2.1 Related Works

In previous research, privacy protection is achieved by a strong encryption mechanism for personal data protection based on blockchain architecture. With regard to video encryption, the audio and video data itself can be protected in compliance with the MPEG-CENC standard [1] using a single AES key.

Most of media encryption is used for digital right management (DRM) systems like Widevine, Marlin, Playready, etc. Vishwa et al. [2] showed how to expand personal data protection method to DRM with a blockchain approach. In addition, public-key infrastructure (PKI) has been shown to play an essential role in identifying each participant and authenticating to establish secured communication channels among users and services in a distributed environment such as a blockchain network [3]. Another significant issue to share sensitive data with stakeholders was raised when handling electronic medical records (EMRs) [4]. Unlike the previous study, the research emphasized how patients can manage the status of medical information generated by third parties who are normally hospital. It is similar to the nature of video surveillance services and in that the creation and ownership of data must be completely separated; all medical data is owned by the user, but the data is created and used only in hospitals that are granted permissions from the user.

Although the blockchain-based approach provides tamper-proof provenance and user privacy protection, it is necessary to design the platform considering that the performance of the system is more affected by the number of records than the size of the file [5].

2.2 *Accessibility to Video Stream*

When considering privacy protection, it is devising a way to share data to authorized users or services by using the property of tamper prevention. However, like the situation of handling EMRs, it is more emphasized sharing sensitive data only with authorized stakeholders in video surveillance service. Since surveillance camera operates all day once installation is done, it is impossible for user to approve sharing recorded video stream with the others every time. In another case, mobile video surveillance devices like unmanned vehicles could be considered. If they become inoperable by any reason while performing a covert mission in an important area, sensitive information is still stored inside the device. Regardless the mobility of the surveillance camera, the device records video inside of its storage first. Then, the data is potentially unsafe from a malicious user if they can physically access into device. Therefore, video encryption should be applied across platforms.

2.3 *Complexity of Compound Identity*

Asymmetric encryption relies on pair of public and private keys. In essence, assuming that encryption and decryption are performed in two or more parties, since public key exchange is indispensable, mutual identification procedures by Compound are used to secure each other's public keys. Normally, *Compound* operation generates 2-tuple for public-key identity, or 5-tuple for entire identity between two party [6]. The number of tuples is a natural part of asymmetric encryption and is widely used to provide strong encryption between the two ends without much consideration of service complexity. However, if user and multiple services need a procedure to identify at the same time, Compound identity has $O(n!)$ complexity to secure each other's public keys.

$$\begin{aligned}\text{Compound}^{\text{public}}_{u,s_1,\ldots,s_n} = &\ \text{Compound}^{\text{public}}_{u,s_1} + \cdots + \text{Compound}^{\text{public}}_{u,s_n} + \\ &\ \text{Compound}^{\text{public}}_{s_1,u} + \cdots + \text{Compound}^{\text{public}}_{s_1,s_n} + \\ &\ \cdots \\ &\ \text{Compound}^{\text{public}}_{s_n,u} + \cdots + \text{Compound}^{\text{public}}_{s_n,s_n-1}\end{aligned} \tag{1}$$

The complexity is problematic in video surveillance system. Since user manages and supervises surveillance cameras installed in multiple locations, with general Compound identity approach, it limits the services scalability.

3 The Privacy Problem in Video Surveillance Service

The complexity of Compound identity comes from establishing an encrypted channel by sharing public keys. If the video surveillance camera is applied to establish a unidirectional encrypted channel for storing data, the complexity is remarkably reduced.

3.1 Asset Identity

In this paper, we introduce a concept of asset (camera or encrypted video) which belongs to user only for handling video stream encryption (Fig. 1). Procedure 1

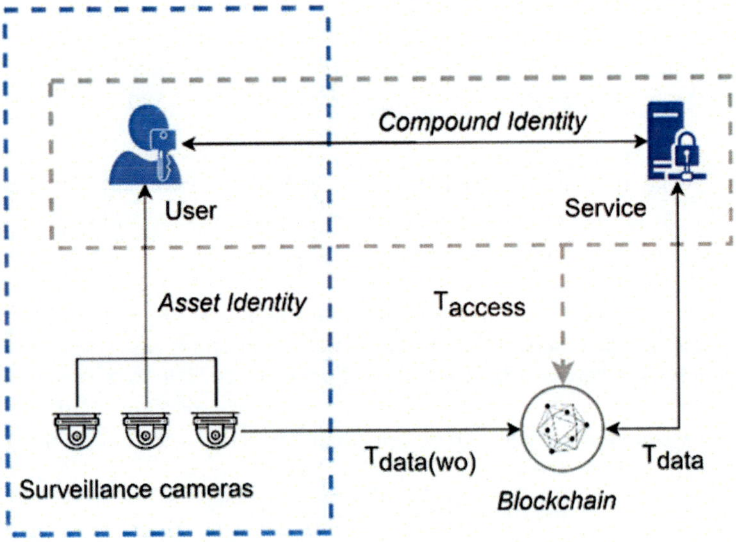

Fig. 1 Overview of privacy protection platform for surveillance service

illustrates the implementation for user and multiple assets. As illustrated, the identity consists of public key of user and nonce which can be used a secret key for symmetric encryption.

$$\begin{aligned} \text{Asset}_{u,a} &= \left(pk^{u}_{\text{sig}}, \text{Nonce}^{u}_{\text{enc}}\right) \\ \text{Asset}_{u,a_1,a_2,\ldots,a_n} &= \left(pk^{u}_{\text{sig}}, \text{Nonce}^{A}_{\text{enc}}\right) \end{aligned} \tag{2}$$

The nonce encrypted by public key is added to encrypted video stream file so that anyone can access to it. However, since the public key associated with the encrypted nonce cannot be found without blockchain, the video is protected from all other users and services as long as the user does not grant access permission. The nonce is able to be generated by asset or third party such as key management system. The advantage of introducing asset with nonce comparing to the general Compound approach is that, even as the number of camera or video files grow, the complexity stays $O(n)$ because it requires only user's public key.

Procedure 1. Generating asset identity

```
Require: A ≠ ∅
    procedure AssetIdentity(U, A)
        if (pk^U_sig, sk^U_sig) = ∅ then                // U executes
            (pk^U_sig, sk^U_sig) ← G_sig()
            sk^U_enc ← G_enc()                          // only U keeps
        end if
        for each a_k ∈ A do
            Nonce ← G_nounce()                          // only a_k keeps
            Nonce^{a_k}_enc ← Encrypt(pk^U_sig, Nonce)
        end for
        return (pk^U_sig, Nonce_enc)
    end procedure
```

3.2 Protocol

All the information used for privacy protection are stored in the blockchain memory, $\mathcal{L}$. In case of video stream, the actual storage is ds which is mapped with a cryptographic hash function, $\mathcal{H}$, because it is normally large set of byte octets, and even if it is sufficiently small, the stream should be able to be permanently removed when its life cycle is done.

In terms of privacy, the most important thing when managing a video is not only controlling access to the files once created, but also tracking them after transferring to the other users or services.

3.3 Data Transaction

Since a camera (asset) has already been acquired nonce by asset identity process, the asset can create encrypted media, M_{enc} with metadata of asset, a, and $\mathrm{Nonce}_{\mathrm{enc}}$ to consider off-chain access by owner. Now, Procedure 2 explains how *StoreSecure-DataTX* interacts with the blockchain memory and the data store. Since anyone can extract the metadata of asset and encrypted nonce from the encrypted media file, the store transaction can be done without performing decryption of the media file. This design lets our platform start the store transaction in different system from camera or cloud data storing service, so that it helps implement flexible service architecture.

ValidateAsset in *StoreSecureDataTX* is an auxiliary function for business logic to intervene in the transaction. Although M_{enc} has all information, $\mathrm{Nonce}_{\mathrm{enc}}$ must be decrypted by owner's public key to get plain M. In addition, the blockchain memory has only hashed key of the public key, so actual asset management should be handled in service platform.

Procedure 2. Storing secure data

Require: $M_{enc} \neq \emptyset$
  procedure $StoreSecureDataTX(pk^{k}_{sig}, M_{enc})$
    $(\mathrm{a_p}, \mathrm{Nonce}^{\mathrm{p}}_{\mathrm{enc}}) \leftarrow \mathrm{Parse}(\mathrm{M_{enc}})$
    if $\mathrm{ValidateAsset}(pk^{U}_{sig}, a_p, Nonce^{p}_{enc}) \neq \mathrm{True}$ then
      return $\emptyset$
    end if
    $h_{M_{enc}} \leftarrow \mathcal{H}(M_{enc})$
    $\mathcal{L}[\mathcal{H}(pk^{k}_{sig})] \leftarrow \mathcal{L}[\mathcal{H}(pk^{k}_{sig})] \cup h_{M_{enc}}$
    $\mathrm{ds}[h_{M_{enc}}] \leftarrow M_{enc}$
    return $h_{M_{enc}}$
  end procedure

After storing by *StoreSecureDataTX* in the blockchain memory, Procedure 3 shows how to extract the information to decrypt video and grant access permission to requestor. *CheckPolicy* is common function to confirm that the requestor has right permission through blockchain transaction. In this case, the permission is more high-level definition such as *downloadable and readable*. Hence, *CheckPolicy* is proper to check if the requestor can access M_{enc} from ds. After acquiring M_{enc}, the requestor should be verified by service platform by *ValidateAsset* with his public key to get decryptable $\mathrm{Nonce}_{\mathrm{enc}}$.

Procedure 3. Loading secure data

```
Require: m ≠ 0
    procedure LoadSecureDataTX(pk^k_sig, m)
        (h_Menc, x_p) ← Parse(m)
        if CheckPolicy(pk^k_sig, x_p) ≠ True then
            return Error
        end if
        if h_Menc ∈ L[H(pk^k_sig)] then
            M_enc ← ds[h_Menc]
            (a_p, Nonce^p_enc) ← Parse(M_enc)
                if ValidateAsset(pk^U_sig, a_p, Nonce^p_enc) ≠ True
then
                    return ∅
                end if
            end if
        return (pk^k_sig, h_Menc, a_p, Nonce^p_enc)
    end procedure
```

3.4 Tracing Transaction

The rest transactions are related to trace who or which service accesses to the encrypted media through blockchain.

The line 6 of Procedure 4 illustrates how to record the transfer process to support off-chain decryption. Since M^t_{enc} has both $\text{Nonce}^k_{\text{enc}}$ and $\text{Nonce}^t_{\text{enc}}$, users, k, and t, can decrypt the media file. In addition, this process does not require to decrypt M_{enc} for transferring.

Procedure 4. Transferring secure data

```
Require: T ≠ ∅ ∨ m ≠ 0
    procedure TransferSecureDataTX(pk^k_sig, T, m)
        (pk^k_sig, h_Menc, a_p, Nonce^p_enc) ← LoadSecureDataTX(pk^k_sig, m)
        M_enc ← ds[h_Menc]
        (a_p, Nonce^p_enc) ← Parse(M_enc)
                if M_enc ≠ ∅ then
            (pk^t_sig, Nonce^t_enc) ← AssetIdentity(T, a_p)
            M^t_enc ← M_enc ∪ Nonce^t_enc
            h_Menc ← StoreSecureDataTX(pk^t_sig, M^t_enc)
        end if
        return (pk^k_sig, h_Menc, a_p, Nonce^p_enc)
    end procedure
```

Another important point of video surveillance service is removing actual video data permanently. The Procedure 5 approach is inspired by the fact that the encrypted media will be no longer decryptable if the symmetric key is not available. The invalidating process removes encrypted Nonce which is generated by a specific user, t, from the data store and blockchain memory. Since no decryptable information of $\text{Nonce}_{\text{enc}}$ remains on M_{enc}, the only way to decrypt M_{enc} is to have plain Nonce in somewhere outside of the blockchain memory. However, $Nonce$ is normally generated by *random number generator*, and the number of M in the surveillance service is unpredictable, so maintaining Nonce in different system is virtually impossible.

Procedure 5. Invalidating secure data

Require: $h_{M_{enc}} \neq 0$
procedure InvalidateSecureDataTX$\left(pk_{sig}^{k}, t, h_{M_{enc}}\right)$
$M_{enc} \leftarrow ds\left[h_{M_{enc}}\right]$
$\left(a_p, Nonce_{enc}^{p}\right) \leftarrow Parse(M_{enc})$
$M_{enc} \leftarrow M_{enc} - \text{Nonce}_{enc}^{t}$
$StoreSecureDataTX\left(pk_{sig}^{k}, M_{enc}\right)$
if M_{enc} has no Nonce then
$ds\left[h_{M_{enc}}\right] \leftarrow \emptyset$
end if
end procedure

4 Conclusion

Video surveillance service is widely deployed, and it necessarily gathers sensitive data from user site. In situations where a large number of users are shared after an incident, privacy protection problem can arise regardless of intentional or mistake. The proposed platform in this paper enables that all sharing situation can be under the owner's control. User can be aware of how the file is used by combination of on- and off-blockchain storage. The approach in this paper focuses on video encryption. However, we expect that similar approaches can be applied to other platforms that separate data creation and distribution.

References

1. ISO/IEC 23001-7:2016, Part 7: Common encryption in ISO base media file format files, in *Information technology – MPEG systems technologies* Retrieved from https://www.iso.org/standard/68042.html
2. A. Vishwa, F. Hussain, A blockchain based approach for multimedia privacy protection and provenance, in *2018 IEEE Symposium Series on Computational Intelligence (SSCI)*, (Nov 2018), pp. 1941–1945

3. R. Wang, J. He, C. Liu, Q. Li, W. Tsai, E. Deng, A privacy- aware pki system based on permissioned blockchains, in *2018 IEEE 9th International Conference on Software Engineering and Service Science (ICSESS)*, (Nov 2018), pp. 928–931
4. A. Azaria, A. Ekblaw, T. Vieira, A. Lippman, Medrec: using blockchain for medical data access and permission management, in *2016 2nd International Conference on Open and Big Data (OBD)*, (Aug 2016), pp 25–30
5. X. Liang, S. Shetty, D. Tosh, C. Kamhoua, K. Kwiat, L. Njilla, Provchain: a blockchain-based data provenance architecture in cloud environment with enhanced privacy and availability, in *2017 17th IEEE/ACM International Symposium on Cluster, Cloud and Grid Computing (CCGRID)* (May 2017), pp. 468–477
6. G. Zyskind, O. Nathan, A. Pentland, Decentralizing privacy: using blockchain to protect personal data, in *2015 IEEE Security and Privacy Workshops* (May 2015), pp. 180–184

Exploring Generative Adversarial Networks for Entity Search and Retrieval

Wafa Arsalane

Abstract With the continuous growth of Web users and the variety of information needs, search engines face new challenges to satisfy users and provide them with accurate answers. Yet, it is remarkable that the majority of queries is related to real-world entities. For example, in Web search, queries such as: Microsoft founder or Bohemian Rhapsody are more frequent types of requests. This has lead to increase interest in entity search and retrieval. In its early steps, entity retrieval (ER) systems are benefited from the advances in ad hoc document retrieval and used existing models to retrieve entities considering similarities to documents. However, with the availability of structured data represented in knowledge bases, recent researches aim at building ER-oriented systems and studying entity representation and its characteristics. In this paper, we apply generative adversarial networks (GANs) to ER. Recently, GANs have shown outstanding results in some information retrieval tasks including Web search and question answering. We apply two variants of GANs to data from the benchmark test collection based on DBpedia-Entity (v.2). We use features from standard retrieval models such as language model, bm25 model, and the recent fielded sequential dependence model (FSDM). Within the experiments, we also explore performances on different types of queries and compare our results to some of the successful learning-to-rank (LTR) models.

Keywords Entity search · Entity retrieval · Wasserstein generative adversarial networks · Generative adversarial networks · DBpedia-Entity · Learning to rank

1 Introduction

For search and retrieval systems to satisfy the growing user's information need, understanding the nature of the query and intent of a user is basic requirement. Hence, the increasing number of Internet users and their interests upgrades the challenge. A

W. Arsalane (✉)
Jeonbuk National University, Jeonju, South Korea
e-mail: wafaearsalane@gmail.com

P. K. Pattnaik et al. (eds.), *Proceedings of International Conference on Smart Computing and Cyber Security*, Lecture Notes in Networks and Systems 149,
https://doi.org/10.1007/978-981-15-7990-5_5

	Searches
1	World Cup
2	Avicii
3	Mac Miller
4	Stan Lee
5	Black Panther
6	Meghan Markle
7	Anthony Bourdain
8	XXXTentacion
9	Stephen Hawking
10	Kate Spade

Fig. 1 Google trends top 10 searches for 2018

large part of a user's target information is or is related to entities. For instance, in Web search, most of queries are to do with real-world or abstract entities: locations, people, organizations, and so on. Figure 1 shows the top 10 global search queries on Google for the year 2018.[1] There so, the process of retrieval, in such cases, should be considered effective only when the desired entity, or list of entities, is provided as candidate answer. Entity retrieval (ER) [1] handles the tasks of extracting and presenting an entity, or entity-related properties, as an answer to a user's query. x' beyond entity retrieval for retrieval itself, it is an important task that supports intelligent Web technologies. Tasks like Web search question answering and recommendation systems can largely benefit from improvements in entity retrieval performances.

Despite being relatively a recent field of research, ER has taken advantages of the advances made in IR tasks, specifically ad hoc document retrieval. Standard document retrieval methods like BM25 [2] and language model [3] were widely used and extended to rank entities. This implicitly considers that the entity is flattened to a document-like description ignoring any structural forms or characteristic. However, with semantic search becoming a hot topic in information retrieval (IR), more structured data is being provided making it possible to build more effective representations of an entity.

Since several IR tasks are ranking-related problems, ranking techniques would obviously be a potential approach to enhance search and retrieval systems. Learning-to-rank models (LTR) were introduced to IR through document retrieval [4] and

[1] *Source* Google Trends https://trends.google.com/trends/yis/2018/GLOBAL/

recently applied to ER [5], achieving the state-of-the-art performances. LTR models for ER rely on extracting features for each query–entity pair using different scoring algorithms and then apply a reranking model on the candidate entities for a final ranking list (result).

In this work, we follow the same idea of LTR models and apply GANs to the task of reranking entities. GANs were previously introduced for IR systems leading to IRGAN model [6], which showed high effectiveness for Web search recommendation systems and question answering. We use two types of GANs: the IRGAN and Wasserstein generative adversarial networks (WGAN) [7]. Our experiments are conducted on dataset based on DBpedia-Entity v2 [8]. We build the query–entity features from the standard retrieval algorithms and use the FSDM [9] model as baseline retrieval. The evaluation shows improvements and promising results compared to some of the recently proposed approaches, including LTR models. We further make experiments to explore the effectiveness of these approaches on each type of query sets.

2 Related Work

For decades, search engines have focused on extracting probable related information as a response to user needs. However, with the rise of question answering systems, researchers start to notice that a document-based answer does not directly satisfy users' needs. In early stages, entities were related to their corresponding Wikipedia articles. It was assumed that the structured elements of an entity description can be merged in a bag-of-word document representation and eventually standard document retrieval approaches could work for the task. Language model [10], sequential dependence model, BM25 [2], etc., were successfully applied to ranking entities. Nonetheless, entity is a unique identifiable concept with special characteristics, types, attributes, and relationships to other entities. Additionally, entity representation separately became a task related to representation learning [11].

Recently, the emergence of semantic Web search put knowledge bases at the center of IR research, which helped data collections to become more organized and structured. Each entity in the knowledge base is represented by a unique identifier and has properties and relationships with other entities. The RDF triple was firstly used for entity representation TREC tracks [12]. The increasing demand for efficient search and retrieval system and open access to knowledge graphs gave another perspective to ER, the specific task of ad hoc entity retrieval from the Web of data (ERWD), consisting of retrieval entities straightforward from KBs [13] instead of using KB for the retrieval process [14]. A major KB exploited for ERWD is the DBpedia KB [10, 15] which is a large database extracted from Wikipedia[2] pages, is the backbone of the DBpedia-Entity[3] the current standard test collection for entity search and retrieval.

[2] https://wiki.dbpedia.org/about.

[3] https://wiki.dbpedia.org/projects/dbpedia-entity.

Availability of such benchmark has largely empowered the new researches and most of the recent works related to ERWD are based on data from the DBpedia-Entity test collection v1 and v2 [8, 16].

In ERWD, there are two major challenges, building effective entity representations and developing efficient retrieval models. The first challenge is as important to ER as it is to other tasks in interaction with entities, like entity linking [17], while building a representation is becoming a neural based task, such as in TextEnt model [11] and [18]. The second challenge, started from standard document retrieval models, which were applied to DBpedia-Entity reported in the test collections [8, 16], and then moved to more entity-oriented models. FSDM [9], for a recent instance, is an extension of the SDM model for the fielded entity representation, and it has scored the state-of-the-art results for ER along with LTR models [5].

Generative models are getting increasing attention. GANs [19], since their introduction in 2014, empowered several tasks in computer vision and image processing. However, they showed some training constraints and weaknesses in training. A new approach replaces the loss in GANs by another Wasserstein distance [7]-based function to ease the process. WGAN was applied to tasks in image processing, including image generation task, and showed better smoothness in training. GANs inspired some preliminary works in IR, IRGAN [6] which is an application of GANs to three IR tasks: Web search, question answering, and item recommendation. Following the concept of generator and discriminator, the generative model gets the probability of selecting a candidate document given the relevance, while discriminative model estimates the likelihood, whether the user approves the retrieved document. This scenario has similar settings to ER task, where users select a candidate entity as answer and judge if it is relevant to their query or not. Therefore, we explore these models, WGAN in addition to IRGAN, for reranking entities in DBpedia-Entity dataset and compare our results to the previous approaches cited above.

3 Generative Adversarial Networks for Entity Retrieval

3.1 IRGAN

IRGAN for Web search architecture was first introduced by Wang et al. in [6] with inspiration from GANs [19] that unify both generative and discriminative approaches in IR. IRGAN is composed of two main components: a generative network and a discriminative network. The underlying idea is that these two models play a game where each tries to compete against the other. The generative network generates fake data, and the discriminative tries to distinguish between real data and the fake data. See illustration in Fig. 2.

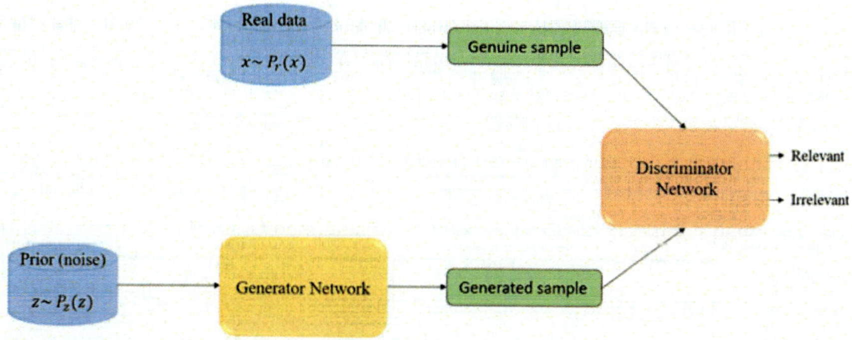

Fig. 2 GANs architecture and process description

Table 1 Formal notations related to GANs

Symbol	Meaning
G	The generator model
D	The discriminator model
P_z	Prior distribution (noise) over random data z
P_r	Distribution over real data sample x
P_g	The generators distribution over data z

In other words, using the notation in Table 1, D and G play a mini-max game where D tries to maximize a function F while G tries to minimize it, following (Table 1):

$$\min_G \max_D F(D, G) \tag{1}$$

where

$$F(D, G) = \mathbb{E}_{x \sim p_r(x)}[\log D(x)] + \mathbb{E}_{z \sim p_z(z)}[\log(1 - D(G(z)))] \tag{2}$$

$$= \mathbb{E}_{x \sim p_r(x)}[\log D(x)] + \mathbb{E}^*_{x \sim p_g(x)}[\log(1 - D(x))] \tag{3}$$

*Since $\mathbb{E}_{x \sim p_r(x)}[\log D(x)]$ does not affect G in the process of updating.

Formally, in ER scenario, the generative retrieval model G is supposed to generate a fake answer entity with its P_g distribution, and the discriminative retrieval model D measures how relevant this answer is by comparing it to real data and approves or rejects the answer. Both models improve while training until G gets to generate a real-like answer and D accepts it ($P_g = P_r$). For a query q, a candidate entity e and the given relevance criteria r, the two models can be defined as:

- **The discriminative retrieval model** D with the distribution $p(r|q, e)$ outputs a decision measuring how relevant the retrieved entity e is. It is first trained to get the real data distribution and then learns to score every sampling from G. It aims at maximizing the objective function with the gradient:

$$\nabla_\theta \frac{1}{n} \sum_{i=1}^{n} \left[\log\left(D\left(x^i\right)\right) + \log\left(1 - \left(D\left(G\left(z^i\right)\right)\right)\right)\right] \quad (4)$$

where x is real data and z is sampled data from G.

In other words, D estimates the probability of an entity e being relevant to the query q, and this is computed with a sigmoid function of the discriminator score:

$$D(e|q) = \sigma(D(q, e)) \quad (5)$$

In practice, D is a scoring function $s(q, e)$ that can be defined by a binary classifier, with score 1 meaning the pair query–entity matches and 0 for fake pair (not matching).

$$s(e|q) = w_2^T \tanh\left(W_1 x_{q,e} + b_1\right) + \mathrm{w}_0 \quad (5)$$

where xq, e is the vector representation of the query–entity pair (concatenation of query representation and entity representation which is basically scores from standard retrieval model, see next section), $W_1 \in R^{l,K}$ is the fully connected matrix for the first layer. $b_1 \in R^l$ is the bias vector for the hidden layer, and w_0 and w_2 are the weights for the output layer.

- **The generative retrieval model**: gets the conditional probability $p(e|q, r)$ of selecting (rather than generating) a candidate entity e given the query and relevance. It aims at selecting data approximately close to the real distribution Pr. Formally, G minimizes the objective of the model. Its gradient is defined as:

$$\nabla_\theta \frac{1}{n} \sum_{i=1}^{n} \log\left(D\left(G\left(z^i\right)\right)\right) \quad (7)$$

Since data in this work, and in IR in general, is discrete tokens (e.g., terms, text, IDs), computing the gradient cannot be done as in original GAN architecture. IRGAN uses instead the policy gradient-based reinforcement learning, the REINFORCE algorithm with the gradient defined as:

$$\nabla_\theta J^G(q) = \frac{1}{K} \sum_{k=1}^{K} \nabla_\theta \log P_g(e_k|q, r)\ \log(1 + \exp(s(e_k, q))) \quad (8)$$

where $\log(1 + \exp(s(e_k, q)))$ is defined as the reward for Pg, e_k the action and q the environment. With e_k refers to the kth document sampled from the current step of generator $Pg(e|q, r)$.

Putting all the formulas above together, we redefine the objective in Eq 3:

$$J^{G,D} = \min_{G} \max_{D} \sum_{i=1}^{N} \mathbb{E}_{e \sim p_r(e|q_i,r)} [\log D(e|q_i)] + \mathbb{E}_{e \sim p_g(e|q_i,r)} [\log(1 - D(e|q_i))] \quad (9)$$

Following this equation, the full process is divided into two steps turning in a loop until convergence:

- For the generative model *G*.
- For each query *q*, generate *K* documents with the distribution $P_g(e|q, r)$. Update parameters using the REINFORCE policy.
- For the discriminative model D: Sample from $G\ P_g(e|q, r)$ and get real data *x* and train *D* to score using $s(e, q)$.

3.2 WGAN

GAN usually uses JS divergence, and this has the problem of non-overlapping, leading to mode collapse and convergence difficulty. To overcome these two problems, WGAN extends GAN by using EM distance or Wasserstein distance [7].

The Wasserstein objective function in WGAN is redefined as the following:

$$L(\mathrm{P_r},\ \mathrm{P_g}) = \max_{\mathrm{w} \in W} \mathbb{E}_{x \sim P_r}[f_{\mathrm{w}}(x)] - \mathbb{E}_{z \sim P_z}[f_{\mathrm{w}}(G(z))] \quad (10)$$

where f is a K-Lipschitz continuous functions, $f_{\mathrm{w}}\mathrm{w} \in W$, parameterized by w.

The main part in the full process of WGAN is to find the f function minimizing the distance. WGAN has a similar architecture to GAN, two main models: generative model and discriminative model. The generator in WGAN has similar functionality as in GANs, and however, the discriminator changes goal to learn the f function and thus, measuring the Wasserstein distance to compute the loss. The discriminator role is no longer distinguishing the fake from the real data, instead it is used to learn parameters w to find a better f_w. During the process of minimizing the loss function, the Wasserstein distances between the distributions P_r and P_g decrease allowing the improvements of the generator output, getting closer to the real distribution (Fig. 3).

- Generative model: In our setting, since the role of the generator is rather selecting entities, *G* counts on the f function to minimize the distance to *Pr* and thus, by comparing the distance to given relevant entities, it selects closer entities as relevant ones. Its gradient is defined as:

$$\nabla_{\theta} \frac{1}{n} \sum_{i=1}^{n} f\left(G\left(z^{i}\right)\right) \quad (11)$$

- Critic model: initialized with random w parameters, it gets the sample from real data and from *G*, and trained to update w to _nd f with the following gradient.

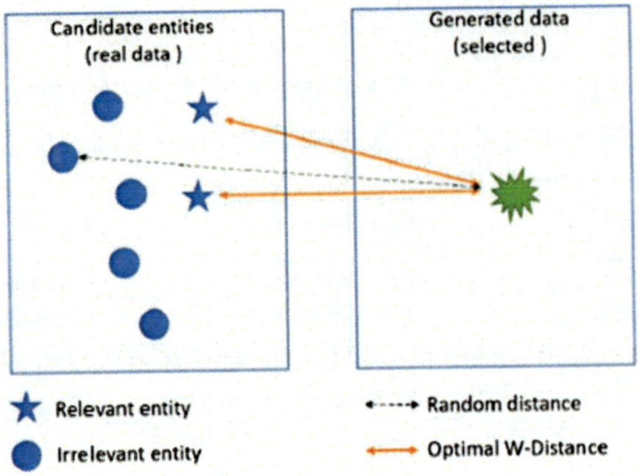

Fig. 3 WGAN process to compute the W distance to highly relevant entities and irrelevant entities

$$\nabla_\theta \frac{1}{n}\sum_{i=1}^{n}[f(x^i) + f(G(z^i))] \tag{12}$$

4 Experiment

4.1 DBpedia-Entity Test Collection

The DBpedia-Entity collection was first introduced by Balog and Neumayer [16], generated by assembling search queries from a number of entity-oriented benchmark campaigns and mapping relevant results to DBpedia knowledge base. The first dataset known as DBpedia-Entity v1 [16] was the first test collection designed for further research and development in Entity Search. The second version DBpedia-Entity v2 [8] came after and has become a benchmark test collection for entity retrieval. In this experiment, we use a novel five-field entity representation scheme, which was proposed by Zhiltsov et al. [9], that suggests to group an entity description into five main fields: Name: the full name of the entity, Cat: contains the entity categories, Attr: entity attributes (except the name), RelEn: the names of related entities (neighbors), and SimEn: entity aliases.

Queries in DBpedia are collected from competition campaigns, and both versions of the dataset have a total number of 485 queries grouped into four categories based on their search task, see Table 2.

A total of 13,090 judgments is provided for these queries. Most of the recent works adopt the binary judgment (relevant 1, non-relevant 0), and we do the same in this work following Zhiltsov et al. [9] setting (Fig. 4).

Table 2 Queries groups in DBpedia-Entity test collection

Query group	Number of queries	Type
SemSearch ES	130	Entity (Name)
ListSearch	115	Type (List)
INEX-LD	110	Entity, type, attribute, relation
QALD-2	140	Entity, type, attribute, relation

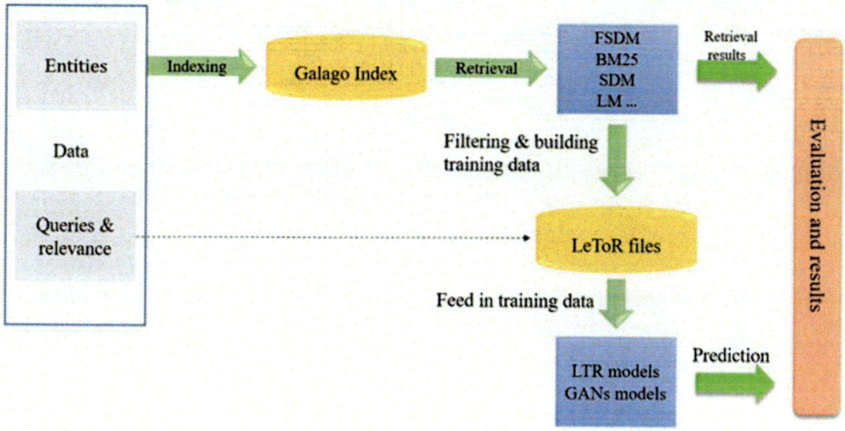

Fig. 4 Full process presented in this work

4.2 Indexing and Retrieval

We use the galago search engine toolkit,[4] for most processes in this part. We first build an index of the given entities with their five fields and another flattened description field (combine all the five field together) and then run baseline models over this index. We use the following approaches:

- LM: The standard language model used for document retrieval with mu = 1500.
- BM25: Okapi function with parameter values $k1 = 1.2$ $b = 0.75$ and $k_3 = 7$.
- BM25F: The fielded version of BM25.
- PMRS: Probabilistic retrieval model uses pseudo-relevance feedback model for query expansion.
- SDM: Sequential dependence model with component default weights, 0.8unigrams, and 0.15 ordered distance and 0.05 unordered window.
- FSDM: Fielded sequential dependence model

[4]https://sourceforge.net/p/lemur/galago/.

Table 3 Performances of different baseline retrieval models

Models	MAP@100	NDCG@10
BM25	0.1776	0.1606
LM	0.1701	0.1521
BM25F	0.1821	0.1790
PMRS	0.1973	0.1991
SDM	0.1724	0.176
PSDM	0.2330	0.3440

4.3 Reranking

The reranking process uses results from baseline models as features. We set the FSDM as a base for the ranking file. We first filter the top 100 entities and combine the other models to build a ranking file following the learning-to-rank (LeToR) format. The final LTR file is used as the input for all ranking models in this section.

We note that the LTR models in this experiments, LambdaRank, is based on the RankLib library.[5] We perform a fivefold cross-validation for all models, and we use Pytorch for all processes.

5 Results

In this section, we report our results on DBpedia-v2 test collection for both the baseline retrieval models and the GANs. We also report results of learning-to-rank models on our ranking data.

Table 3 shows performances of all referenced methods using the original relevance judgments from the test collection. We report results on different metrics, MAP which was the official metric of entity search introduced by the DBpedia-v1 along with the P@10, and NDCG@10the official metric for DBpedia-v2. Models in this section are manually trained with different parameters (including the default ones in galago search engine).

5.1 GANs and Ranking Models

Since both IRGAN and WGAN in these experiments proceed similarly to LTR methods, the results are directly comparable. The LTR models presented in Table 4 are trained with parameters from Ranklib library.

---

[5]https://sourceforge.net/p/lemur/wiki/RankLib/.

Table 4 Reporting performances of GANs and LTR models on DBpedia-v2 data

Models	MAP@100	NDCG@10	P@10	P@20
FSDM	0.233	0.344	0.233	0.189
LambdaRank	0.234	0.343	0.240	0.188
WGAN	0.238[b]	0.352[b]	0.234	0.189
IRGAN	0.272[a]	0.378[a]	0.251[a]	0.192[a]

[a] and [b] represent the highest results and improvements over the baselines, respectively

Results in Table 4 show how IRGAN and WGAN outperform LTR models on the full set of queries for both evaluation metrics.

We further investigate the effectiveness of our proposed models for each of the four types of queries. Previous works have also studied their models for different query sets and showed that some models perform better for a specific type than others. It is assumed that some models benefit from underlying queries characteristics. We report our results for both a mixed set of all 485 queries in Tables 3 and 4 while results for each query type are presented in Table 5. The IRGAN outperforms other models add set of all the four types (Fig. 5).

Another experiment consists of studying the contribution of entity fields in performances of GAN models. Figure 6 shows the results of IRGAN (left) and WGAN (right) with and without some field features. We can clearly notice that the field name has a powerful contribution and performances drop extremely to the lowest level when removing this field, given that a large number of queries is based on entity-name retrieval. Cat (Category) and Att (Attributes) fields are also important as features since the results when removing them are also among the lowest. However, for the last two fields (similar entities and related entities), the models perform relatively better. This analysis confirms the assumption that the fields name, category, and attributes hold more important information about the entity that is the reason why GANs perform poorly without them. While this can be different from query to query and query type.

Table 5 Performances of different models on each type of query set (where + indicates improved scores over baselines while * refers to the highest scores)

Models	MAP@100	NDCG@10	P@10	P@20
SemSearch ES				
FSDM	0.391	0.500	0.310	0.232
LambdaRank	0.437	0.538	0.315	0.231
WGAN	0.487+	0.561+	0.328+	0.252+
IRGAN	0.578*	0.624*	0.414*	0.491*
ListSearch				
FSDM	0.209	0.310	0.265	0.219
LambdaRank	0.224	0.332	0.315	0.243
WGAN	0.241+	0.338+	0.319+	0.250+
IRGAN	0.370*	0.412*	0.352*	0.392*
QALD-2				
FSDM	0.231	0.340	0.152	0.121
LambdaRank	0.249*	0.357*	0.154	0.129*
WGAN	0.245	0.341	0.159*	0.121
IRGAN	0.244	0.341	0.145	0.119
INEX-LD				
FSDM	0.201	0.262	0.281	0.24
LambdaRank	0.209	0.359	0.291	0.245
WGAN	0.213+	0.361+	0.329+	0.261+
IRGAN	0.242*	0.368*	0.351*	0.297*

+ indicates improved scores over baselines
* refers to the highest scores

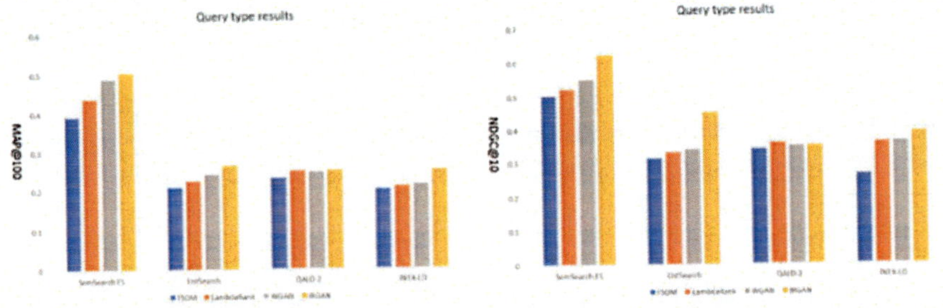

Fig. 5 Histograms showing performances of the four models and improvements of GAN-based models over FSDM and LambdaRank for each query set

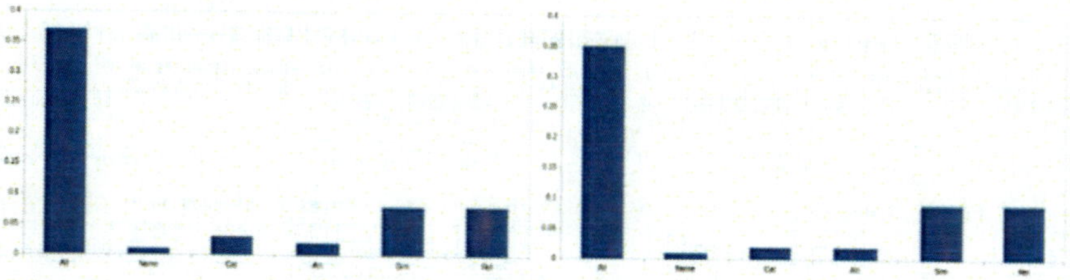

Fig. 6 Results of IRGAN (first) and WGAN (second) (in NDCG@10) using the set of all fields and without using a specific field. Each column name represents the results when removing the corresponding field name

6 Conclusion

Entity retrieval performances are still considered low compared to other IR tasks such as Web search and question answering. The main focus of the community is to improve the retrieval and ranking processes using available or new methods and approaches. Thus, any improvement over existing results can help the future works either in ERWD or other tasks handling entities. Our proposed approach gets insight from the LTR models and uses IRGAN and WGAN as rankers with features from standard baseline retrieval methods. While WGAN performances are, in some cases, approximately comparable to LTR models, it slightly outperforms them in some query sets, IRGAN outperforms LTR models on the DBpedia-v2 test collection recording improvements over up-to-date results. As for a future work, we might consider build representations using neural networks and train an end-to-end GAN model for entity retrieval.

References

1. K. Balog, *Entity Retrieval*. Springer Science + Business Media LLC (2017)
2. S. Jones, S.W. Karen, S.E. Robertson. A probabilistic model of information retrieval: development and comparative experiments. Inf. Proc. Manage. **36**(6), 779–808, 809–840 (2000)
3. P. Ogilvie, J. Callan, Language models and structured document retrieval (2003)
4. T.Y. Liu, Learning to rank for information retrieval. Found. Trends Inf. Retrieval Arch **3**(3), 225–331 (2009)
5. J. Chen, C. Xiong, J. Callan, An empirical study of learning to rank for entity search. SIGIR Proc. (2016)
6. J. Wang, L. Yu, W. Zhang, Y. Gong, Y. Xu, B. Wang, P. Zhang, D. Zhang, A minimax game for unifying generative and discriminative information retrieval models. SIGIR Proc. (2018)
7. M. Arjovsky, *Soumith Chintala* (Wasserstein gan, L.B., 2017)
8. F. Hasibi, F. Nikolaev, C. Xiong, K. Balog, S. Bratsberg, A. Kotov, J. Callan, Dbpedia-entity v2: a test collection for entity search (2017). https://doi.org/10.1145/3077136.3080751
9. N. Zhiltsov, A. Kotov, F. Nikolaev, Fielded sequential dependence model for ad-hoc entity retrieval in the web of data (2015). https://doi.org/10.1145/2766462.2767756

10. J. Lehmann, R. Isele, M. Jakob, A. Jentzsch, D. Kontokostas, P.N. Mendes, S. Hellmann, M. Morsey, P.V. Kleef, S. Auer, C, Bizer, Dbpedia a large- scale, multilingual knowledge base extracted from wikipedia. Semantic-web J. (2013)
11. I. Yamada, H. Shindo, Y. Takefuji, Representation learning of entities and documents from knowledge base descriptions. CoRR abs/1806.02960 (2018). http://arxiv.org/abs/1806.02960
12. K. Balog, A. de Vries, P. Serdyukov, P. Thomas, T. Westerveld, Overview of the TREC 2009 entity track (2009)
13. X. Lin, W. Lam, K.P. Lai, Entity retrieval in the knowledge graph with hierarchical entity type and content. In *Proceedings of the 2018 ACM SIGIR International Conference on Theory of Information Retrieval*. pp. 211–214. New York, NY, USA (2018). https://doi.org/10.1145/3234944.3234963
14. J. Dalton, L. Dietz, J. Allan, Entity query feature expansion using knowledge base links, p. 365374 (2014)
15. P. Mendes, M. Jakob, C. Bizer, Dbpedia: a multilingual cross-domain knowledge base, in *Semantic-web journal Proceedings of the Eighth International Conference on Language Resources and Evaluation (LREC-2012)*, 18131817 (2012)
16. K. Balog, R. Neumayer, A test collection for entity search in dbpedia, pp. 737–740 (2013). https://doi.org/10.1145/2484028.2484165
17. F. Hasibi, K. Balog, S.E. Bratsberg, Exploiting entity linking in queries for entity retrieval, in *Proceedings of the 2016 ACM International Conference on the Theory of Information Retrieval*, pp. 209–218. ICTIR '16, ACM, New York, NY, USA (2016). https://doi.org/10.1145/2970398.2970406
18. I. Yamada, H. Shindo, H. Takeda, Y. Takefuji, Learning distributed representations of texts and entities from knowledge base, p. 397411 (2017)
19. I. Goodfellow, J. Pouget-Abadie, M. Mirza, B. Xu, D. Warde-Farley, S. Ozair, A. Courville, Y. Bengio, Generative adversarial networks. Adv. Neural Inf. Proc. Syst. **3** (2014)

Secure Marine Communication Under Distributed Slotted MAC

Mohammed Abdulhakim Al-Absi, Ahmadhon Kamolov, Ki-Hwan Kim, Ahmed Abdulhakim Al-Absi, and Hoon Jae Lee

Abstract The automatic identification system is used by the ships, and the VHF data link (VDL) mode 4 system is used by the avionics industry. Based on ground infrastructure, the traditional surveillance applications for ships are built with radar support. The major drawback of radar is the inability to see behind large obstacles. This obstacle affects the attenuation signal at receiving ship. Therefore, we present an obstacle-based radio propagation model that considers the effect caused by the presence of obstructing ships in line of sight. This model is evaluated in terms of throughput, collision, and packet transmission under distributed slotted MAC.

Keywords Ship to ship · Obstacle · MAC · Wireless channel · Marine · SANET

1 Introduction

Different ad hoc networks have different characteristics, where the main medium between source and destination is the radio channel in all wireless communication. Wireless ad hoc networks can be classified through their application into general aggregation where ad hoc network include mobile ad hoc network (MANET), vehicular ad hoc network (VANET) [1–4], flying ad hoc network (FANET) [5], and sea

M. A. Al-Absi · A. Kamolov · K.-H. Kim
Department of Computer Engineering, Dongseo University, 47 Jurye-ro, Sasang-gu, Busan 47011, Republic of Korea
e-mail: Mohammed.a.absi@gmail.com

A. A. Al-Absi
Department of Smart Computing, Kyungdong University, 46 4-gil, Bongpo, Gosung, Gangwon-do 24764, Republic of Korea
e-mail: absiahmed@kduniv.ac.kr

H. J. Lee (✉)
Division of Information and Communication Engineering, Dongseo University, 47 Jurye-ro, Sasang-gu, Busan 47011, Republic of Korea
e-mail: hjlee@dongseo.ac.kr

P. K. Pattnaik et al. (eds.), *Proceedings of International Conference on Smart Computing and Cyber Security*, Lecture Notes in Networks and Systems 149,
https://doi.org/10.1007/978-981-15-7990-5_6

ad hoc network (SANET) [6]. Sea ad hoc network (SANET) is one of the popular types of ad hoc network, where ships they may can communicate using two-mode combinations. First one is they can communicate using satellite where this satellite can connect ship to shore, shore to ship, and ship to ship with the high bandwidth frequency. Second is ship to ship, base station to ship, and ship to base station communication where using the limited range for communication.

2 Sea Ad Hoc Network (SANET) Environment

There are two useful ship applications: First is parking availability notification and the spot locator. Ship parking is a technology that uses moving ships as nodes, allowing ships approximately 100–300 m of each other to communicate and create a wireless network with wide range. Figure 1 shows an illustration of the architecture composed of many ships and the base stations where this base station can supply the ships with difference of services. Ships are equipped with maritime very high-frequency connection system that is allowed it to be connected with the base stations and the other ships how is near the shore. With the use of the very high frequency, the ship can send and receive the data from and to other ships.

Ships they may can communicate using two-mode combinations. First one is they can communicate using satellite where this satellite can connect ship to shore, shore to ship, and ship to ship with the high bandwidth frequency. The satellite biggest downfall after bandwidth is latency. Distance from ship to satellite and from satellite to ship creates high delay communication which makes the real-time communication more difficult. Moreover, large communication satellite can deliver wide area coverage with high cost and huge amounts of money. Second is ship to ship, base

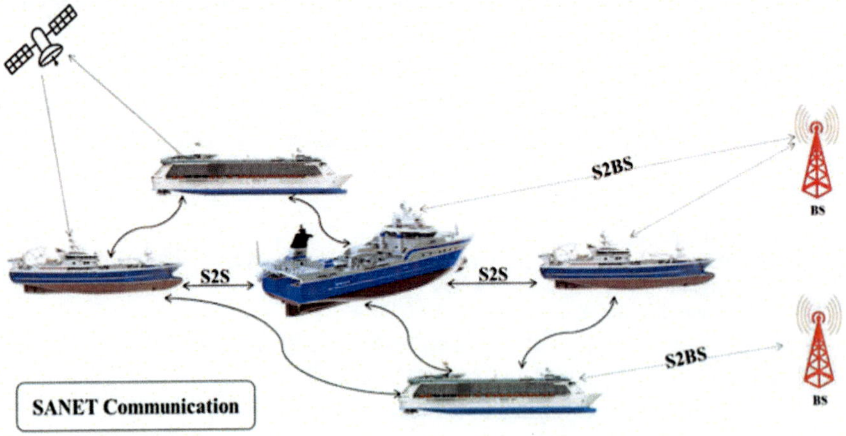

Fig. 1 Illustration of the architecture composed of many ships and the base station

station to ship, and ship to base station communication where using the limited range for communication. The main medium between transmitter and receiver is the radio channel in all wireless communication. The signal propagated either line of sight or non-line of sight where the network provides a low cost, low bandwidth, and low delay communication between the source and destination. The LOS is blocked if any part of any ships between the source and destination.

3 Maritime Wireless Communications

The ship communication system mainly refers to the GMDSS system, and the GMDSS is the global maritime distress and safety system (global maritime) [7]. GMDSS is based on modern radio communication technology maritime search and rescue and secure communications, distress and safety communication systems established to meet the needs of maritime communications, and the system also satisfies the regular communication service of the ship. The ship communication system has undergone many changes over the years. Due to the development of modern digital communication and navigation technology, including the development of satellite communications, satellite navigation, large-scale integrated circuits, and micro-processing technologies have enabled the establishment of new types of maritime communication systems necessary and possible. The International Maritime Organization (IMO) convened a meeting at its headquarters in London in November 1988, deliberating and adopting the amendments to the 1974 International Convention for the Safety of Life at Sea and the 1979 SOLAS Protocol, which are based on the law, i.e., 1988 amendment to the SOLAS Convention. The amendment introduced the GMDSS into the Convention and specified the GMDSS in the SOLAS Convention. GMDSS is a shore-based ship communication system. The basic concept of GMDSS is the search and rescue authorities on the shore ships in danger, and other vessels in the vicinity of the distressed personnel can quickly receive the alarm of the distress incident and promptly conduct search and rescue coordination assistance help. GMDSS can also provide emergency and secure communications and broadcast maritime safety information (sailing warnings, weather warnings, and weather) forecasts, and other emergency safety information. In other words, regardless of the sea area in which the ship sails, it is possible to complete the ship and navigation the safety of other ships in the same sea area is very important for all communication tasks.

In the International Maritime Organization (IMO), the global maritime distress safety system (GMDSS) [8] has divided the world into four subs communication coverage areas named A1, A2, A3, and A4 as showing in Table 1 where A1 and A2 are the big complicated areas of each country where small ships or boats are floating. However, the sea areas A3 and A4 are the oceans where there the ships are floating. The various radio communication systems are required by vessel to be carrying on board ships and rely on the area of operation of that specific vessel. Sea Area A1, it is about 37.04–55.56 Km (20–30 nautical miles) from shore within

Table 1 Definition of sea area

Sear Area	Device	Communication Coverage	Technology
Sea Area A1		(20- 30 nautical miles)	VHF-MF/DSC
Sea Area A2		(400 nautical miles)	MF/DSC
Sea Area A3		North latitude 70 degrees- South Latitude 70 degree	INMARSAT/HF
Sea Area A4		Worldwide	EPIRB/HF

area range of a shore-based of at least one very-high-frequency (VHF)-medium-frequency (MF)radio station where the digital selective calling (DSC) alerting is available. Digital selective calling (DSC) radio is an item of GMDSS that utilized by recreational boaters. Sea Area A2 should cover about (400 nautical miles) 740.8Km offshore but in practice it extends up to (100 nautical miles) 185.2Km offshore but this should exclude A1 areas. The digital selective calling (DSC) and MF radio range are required for A1 areas. Sea Area A3, in this area the coverage is within 70° north and 70° south and is within Inmarsat satellite coverage range, excluding Sea Areas A1 and A2 where continuous alerting is available. Sea Area A4, these are the regions outside sea zones of A1, A2, and A3. The polar regions North and South of 70° excluding Sea Areas A1, A2, and A3.

Sea Area A1 Distress Alerting:

- VHF DSC on Channel 70
- VHF radiotelephony on Channel 16
- MF DSC on 2187.5 KHz
- Inmarsat
- EPIRB.

Sea Area A2 Distress Alerting:

- VHF DSC on Channel 70
- VHF radiotelephony on Channel 16
- MF DSC 0n 2187.5 KHz

- Inmarsat
- EPIRB.

Sea Area A3 Distress Alerting:

- HF DSC on 8414.5 KHz and all other HF DSC frequencies
- F DSC on 2185.7 KHz
- Inmarsat
- EPIRB.

Sea Area A4 Distress Alerting:

- HF DSC on 8414.5 KHz and all other HF DSC frequencies
- HF DSC on other frequencies to alert coast stations
- MF DSC on 2187.5 KHz
- EPIRB (Cospas-Sarsat only).

4 Proposed Model (Sh-ANET)

In this section, we present the radio propagation model. First let us consider that the shore consists of set of ships which are in coverage range of each other as shown in Fig. 2. The parking area in the shore is divided into large and small zones which

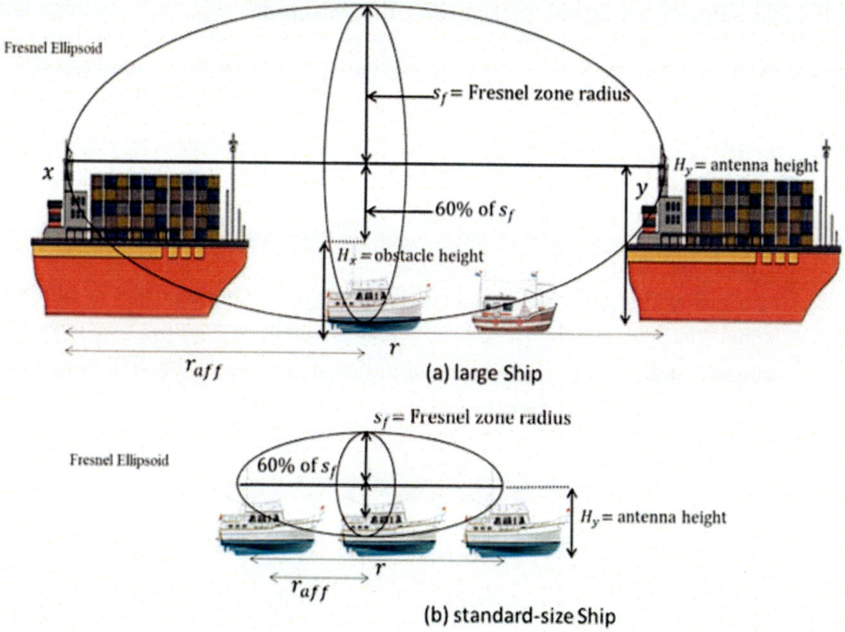

Fig. 2 Obstacle-based Sh-ANET model

are based on the height and the width of the ships. We assumed that each ship is equipped with sensors to protract it from any obstacles in the sea or in the shore within a small range. However, let us consider that there are ships moving one after another and communicating with each other's when they are in the same region. The hop-based communication is adopted for transmission (ship 1 sends the message to ship2, and then ship 2 sends the message to the base station) and must be done if one ship wants to communicate with the base station in the shore and informing them before it reaches to the shore.

Ships they may communicate using two-mode combinations. First one is they can communicate using satellite where this satellite can connect ship to shore, shore to ship, and ship to ship with the high bandwidth frequency. The satellite biggest downfall after bandwidth is latency. Distance from ship to satellite and from satellite to ship creates high delay communication which makes the real-time communication more difficult. Moreover, large communication satellite can deliver wide area coverage with high cost and huge amounts of money. Second is ship to ship, base station to ship, and ship to base station communication where using the limited range for communication. The main medium between transmitter and receiver is the radio channel in all wireless communication. The signal propagated either line of sight or non-line of sight where the network it provides a low cost, low bandwidth, and low delay communication between the source and destination. The LOS is blocked if any part of any ships is between the *Sx* and *Sy*. The transmission signal might get affected due to the blocked of Fresnel ellipsoid caused by the ship. The Fresnel zone is the ellipsoid concentric that locates the antenna aperture volumes that come from the antenna diffraction. In the radius, it computes the signal attenuation required for transmission among x and y with the presence of multiple obstacle in LOS between tx and rx first, find the set of probable obstructive SHIP in LOS between x and y. If any probable obstructing SHIP obtained, find the number of obstacle in LOS between x and y. If no obstacle is found, terminate the loop and no additional attenuation is required.

If any probable obstructing SHIP obtained, then compute LOS of probable obstacles and number of SHIPS obstructing LOS between x and y and cumulate the overall attenuation required for transmission.

Here we consider a distributed slotted MAC, where the time is divided into the equal length of slot $\delta 1$. The total amount of time that a device is in a range of adjacent yth ship is $L_y = \left\lceil \frac{2S_y}{u\delta 1} \right\rceil$. To compute Lth slot time when VANET device is in range of adjacent device is given in Eq. (1) as follows:

$$\mathcal{V}(y, L) = \sum\nolimits_{x=0}^{y-1} L_x + L, \forall \mathcal{N} \in \{1, \ldots, L_y\}, \tag{1}$$

where $L_0 = 0$. The collection of time slots in yth ship is shown in Eq. (2) as follows:

$$L_y = \{\mathcal{V}(y, 1), \ldots, \mathcal{V}(y, L_y)\}. \tag{2}$$

5 Comparison of Parameters and Propagation Environment

Table 2 [9] shows the channel model comparison of parameters and propagation. However, in our model we have used 5.9 GHz bandwidth.

6 Simulation Result and Analysis

We have used Windows 10, 16 GB RAM. The new obstacle-based radio propagation for Sh-ANET environment channel model and the distributed MAC are both incorporated into the simulator. For the evaluation, distributed slotted MAC is considered and the proposed model is evaluated in terms of throughput, collision, and packet transmission for Sh-ANET environment with the use of the vehicle parameters.

Figure 3 shows the throughput performance for Sh-ANET considering 40 ships which are moving at a speed of 10 m/s. The outcome indicates that the throughput achieved for the ship environment is high; Fig. 4 shows the collision performance considering 40 ships which are moving at a speed of 10 m/s. The outcome shows the collision which is low for Sh-ANET environment. Figure 5 shows that the successful packet transmission considering 40 ships are moving at a speed 10 m/s. The outcomes show that the packet transmission achieved for the Sh-ANET environment is high.

7 Conclusion

The proposed Sh-SANET obstacle-based radio propagation channel model is evaluated in terms of throughput, collision, and packet transmission under distributed slotted MAC. The experiments are carried out considering 10 km * 10 km for Sh-ANET network size, 40 ships which are moving at a speed of 10 m/s, 27 Mbps for Bandwidth, QAM-64 Modulation scheme.

Table 2 Channel model comparison of parameters and propagation

Channel model	Channel parameters	Duct considered atmospheric	Receiver mobility	Transmitter mobility	Propagation type	Propagation loss model
Sh-ANET (Ours)	Propagation loss, path loss, and received power	–	Yes	Yes	Line of sight	Log normal path loss distribution
Ref. [10]	Propagation loss, mean excess delay, and RMS delay spread	–	No	Yes	Line of sight	Basic path loss model
Ref. [11]	Propagation loss, time delay, and angle of arrival	–	Yes	Yes	–	Custom path loss model using
Ref. [12]	Mean excess delay, and RMS delay spread	–	Yes	Yes	Line of sight	–
Ref. [13]	Propagation loss	Evaporation duct	No	No	nLoS and LoS	Parabolic equation method
Ref. [14]	Propagation loss, RMS delay spread, path loss, and received power	Evaporation duct	No	Yes	Line of sight	Modified two-ray path loss model
Ref. [15]	Propagation loss, RMS delay spread, and mean excess delay	–	No	Yes	Line of sight (LoS) and non-line of sight (nLoS)	Long-distance path loss model

(continued)

Table 2 (continued)

Channel model	Channel parameters	Duct considered atmospheric	Receiver mobility	Transmitter mobility	Propagation type	Propagation loss model
Ref. [16]	Time delay, phase delay, and amplitude gain	–	–	–	–	–
Ref. [17]	Propagation loss, power delay profile, and inter-symbol interference	–	No	Yes	Line of sight	–
Ref. [18]	Propagation loss, channel fade margin, slow fading, and fast fading	Evaporation duct	–	–	Line of sight and non-line of sight	Log normal path loss distribution
Ref. [19]	Propagation loss, transmitted power, transmitter height, and receiver height	Evaporation duct	Yes	Yes	Line of sight	Three-ray path loss model
Ref. [20]	Propagation loss	–	Yes	Yes	Line of sight	–
Ref. [21]	Propagation loss	Evaporation duct	Yes	Yes	bLoS	–
Ref. [22]	Power delay profile, coherence time, bandwidth, and distance	–	Yes	Yes	Line of sight	–
Ref. [23]	Propagation loss, and large- and small-scale fading	–	No	Yes	Line of sight	Two-ray path loss model

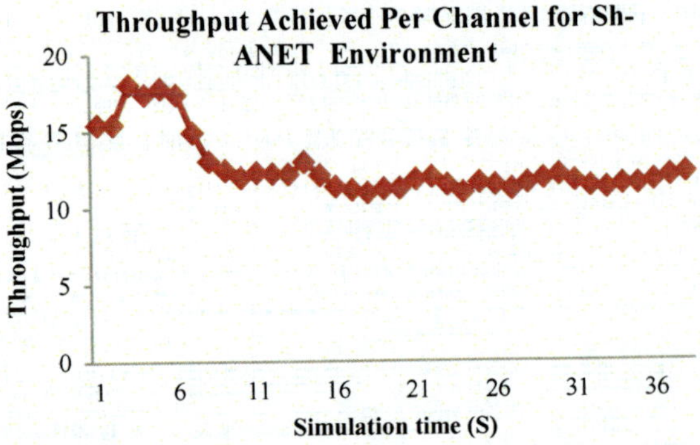

Fig. 3 Throughput performance for Sh-ANET environment

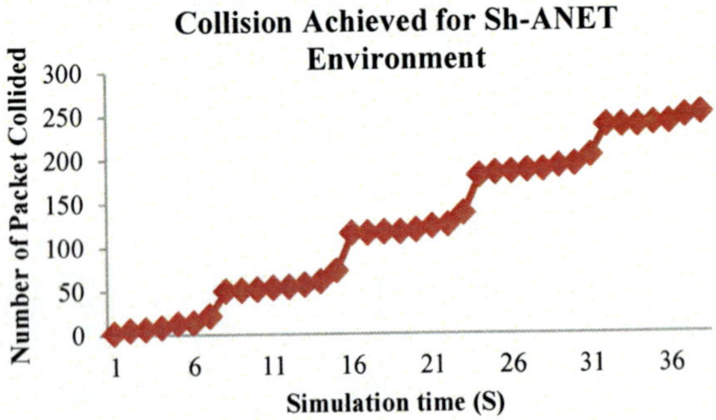

Fig. 4 Collision for Sh-ANET Environment

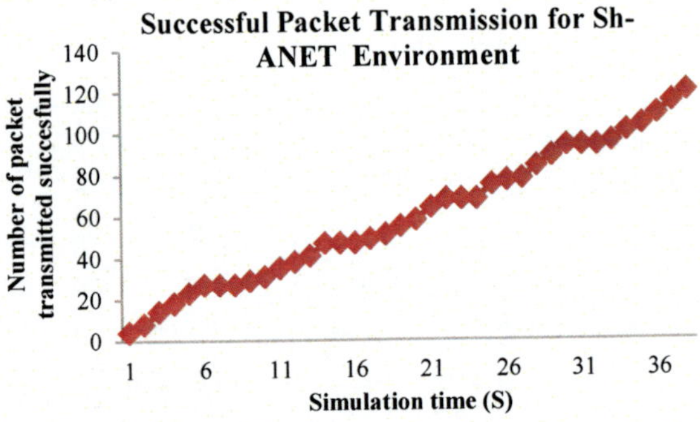

Fig. 5 Successful packet transmission for Sh-ANET environment

Acknowledgements This work was supported by Basic Science Research Program through the National Research Foundation of Korea(NRF) funded by the Ministry of Education, Science and Technology (grant number: NRF-2016R1D1A1B01011908).

References

1. M.A. Al-Absi, A.A. Al-Absi, T. Kim, H.J. Lee, An environmental channel throughput and radio propagation modeling for vehicle-to-vehicle communication. Int. J. Distrib. Sens. Netw. (IJDSN) **14**(4), 1–10 (2018) (SAGE Publisher)
2. M.A. Al-Absi, A.A. Al-Absi, H.J. Lee, V2V communication modeling for environmental channel throughput and radio propagation, in *8th IEEE International Conference on ICTC Convergence, Jeju Island, Korea, IEEE Publisher, pp.507–512*, 18–20 Oct 2017
3. M.A. Al-Absi, A.A. Al-Absi, Y.J. Kang, H.J. Lee, Obstacles effects on signal attenuation in line of sight for different environments in V2V, in *20th International Conference on Advanced Communication Technology (ICACT) Chuncheon-si Gangwon-do, Korea, IEEE Publisher, pp.17–20*, 11–14 Feb 2018
4. M.A. Al-Absi, A.A. Al-Absi, H.J. Lee, Performance Analysis for city, highway, and rural area in vehicle-to-vehicle network, in *8th IEEE International Conference on ICTC Convergence, Jeju Island, Korea, IEEE Publisher*, 17–19 Oct 2018
5. A. Kamolov, M.A. Al-Absi, H.J. Lee, S. Park, Smart flying umbrella drone on internet of things: AVUS, in *International Conference on Advanced Communications Technology (ICACT), pp. 191–195*, 17–20 Feb 2019
6. A. Kamolov, S. Park, An IoT based smart berthing (parking) system for vessels and ports international conference on mobile and wireless technology. *ICMWT: Mobile and Wireless Technology*, pp. 129–139 (2018)
7. A summary of IMO Conventions, International Maritime Organization, 4 Albert Embankment, London (2009). https://www.imo.org/en/KnowledgeCentre/ReferencesAndArchives/FocusOnIMO(Archives)/Documents/Focus%20on%20IMO%20-%20A%20summary%20of%20IMO%20Conventions%20(2009).pdf
8. GMDSS greatly streamlines ship-to-ship/ship-to-shore maritime distress communication. https://www.furuno.com/en/merchant/gmdss/
9. A. Habib, S. Moh, Wireless channel models for over-the-sea communication: a comparative study. Appl. Sci., 1–32 (2019)
10. I. Timmins, S. O'Young, Marine Communications channel modeling using the finite-difference time domain method. IEEE Trans. Veh. Technol. **58**, 2626–2637 (2009)
11. Y. Wu, Z. Gao, C. Canbin, L. Huang, H. Chiang, Y. Huang, H. Sun, Ray tracing based wireless channel modeling over the sea surface near Diaoyu Islands, in *Proceedings of the 2015 First International Conference on Computational Intelligence Theory, Systems and Applications, Yilan, Taiwan*, 10–12 Dec 2015
12. K. Yang, T. Roste, T. Ekman, Experimental multipath delay profile of mobile radio channels over sea at 2 GHz, in *Proceedings of the 2012 Loughborough Antennas and Propagations Conference, Loughborough, UK*, 12–13 Nov 2012
13. G. Woods, A. Ruxton, C. Huddlestone-Holmes, G. Gigan, High-capacity, long-range, over ocean microwave link using the evaporation duct. IEEE J. Ocean Eng. **34**, 323–330 (2009)
14. N. Mehrnia, M.K. Ozdemir, Novel maritime channel models for millimeter radiowaves, in *Proceedings of the 24th International Conference on Software, Telecommunications and Computer Networks (SoftCOM), Split, Croatia*, 21–23 Sep 2016
15. N.M. Konstantinos, P. Loulis, M. Chronopoulos, Measurements and wideband channel characterization for over-the-sea propagation, in *Proceedings of the 2006 IEEE International Conference on Wireless and Mobile Computing, Networking and Communications, Montreal, QC, Canada*, 19–21 June 2006
16. X. Cao, T. Jiang, Research on sea surface ka-band stochastic multipath channel modelling, in *Proceedings of the 2014 3rd Asia-Pacific Conference on Antennas and Propagation, Harbin, China*, 26–29 July 2014
17. F. Huang, X. Liao, Y. Bai, Multipath channel model for radio propagation over sea surface. Wirel. Pers. Commun. **90**, 245–257 (2016)
18. K. Zaidi, V. Jeoti, M. Drieberg, A. Awang, A. Iqbal, Fading characteristics in evaporation duct: fade margin for a wireless link in the South China sea. IEEE Access **6**, 11038–11045 (2018)

19. Y.H. Lee, F. Dong, Y.S. Meng, Near sea-surface mobile radiowave propagation at 5 GHz: measurements and modeling. Radio Eng. **24**, 824–830 (2014)
20. Z. Shi, P. Xia, Z. Gao, L. Huang, C. Chen, Modeling of wireless channel between UAV and vessel using the FDTD method, in *Proceedings of the 10th International Conference on Wireless Communications, Networking and Mobile Computing (WiCOM 2014), Beijing, China*, 19–20 June 2015
21. A. Iqbal, V. Jeoti, Feasibility study of radio links using evaporation duct over sea off Malaysian shores, in *Proceedings of the 2010 International Conference on Intelligent and Advanced Systems, Manilla, Philippines*, 15–17 June 2010
22. K. Mehrina, K.M. Ozdemir, Multipath delay profile and doppler spread of millimeter radiowaves over the sea channel, *in Proceedings of the 2017 IEEE International Black Sea Conference on Communications and Networking, Istanbul, Turkey*, 5–8 June 2017
23. J. Lee, J. Choi, W. Lee, J. Choi, S. Kim, Measurement and analysis on land-to-ship offshore wireless channel in 2.4 GHz. IEEE Wirel. Commun. Lett. **6**, 222–225 (2017)

IoT Technology with Marine Environment Protection and Monitoring

Mohammed Abdulhakim Al-Absi, Ahmadhon Kamolov, Ahmed Abdulhakim Al-Absi, Mangal Sain, and Hoon Jae Lee

Abstract The Internet of things is an important part of the new generation of information technology and represents the latest achievements in the development of the world's information industry. IoT technology makes ships intelligent. It is emerging technology that has been proposed in the past 20 years. Internet-based IoT integrates with many of its sensor processing and processing functions, drawing attention in all areas. Recently, it has been widely applied in many fields such as intelligent transportation, smart cities, smart industries, smart homes, and intelligent agriculture. Nowadays, it has also been used in shipping industry. The ship network is an important application of LAN communication network technology on ships, which can provide information exchange capabilities of various equipment, communication device, and personnel in ship. This paper presents a review of the Internet of things layer architecture, ship area network architecture in the field of marine environment protection and monitoring.

Keywords IoT · Ship architecture · Marine · Communication · Monitoring

M. A. Al-Absi · A. Kamolov
Department of Computer Engineering, Dongseo University, 47 Jurye-ro, Sasang-gu, Busan 47011, Republic of Korea
e-mail: Mohammed.a.absi@gmail.com

A. A. Al-Absi
Department of Smart Computing, Kyungdong University, 46 4-gil, Bongpo, Gosung, Gangwon-do 24764, Republic of Korea
e-mail: absiahmed@kduniv.ac.kr

M. Sain · H. J. Lee (✉)
Division of Information and Communication Engineering, Dongseo University, 47 Jurye-ro, Sasang-gu, Busan 47011, Republic of Korea
e-mail: hjlee@dongseo.ac.kr

M. Sain
e-mail: mangalsain1@gmail.com

P. K. Pattnaik et al. (eds.), *Proceedings of International Conference on Smart Computing and Cyber Security*, Lecture Notes in Networks and Systems 149,
https://doi.org/10.1007/978-981-15-7990-5_7

1 Introduction

The main medium between transmitter and receiver is the radio channel in all wireless communication. Ad hoc network has attracted a lot of attention because of many characteristics such as a networks, environment, topologies, and mobility and…act. Ad hoc network is collection of mobile stations that communicate with each other through broadcast radio transmissions, for example, the transmission signal that reach to all stations in the transmission range. So, because of this limitation of the radio range, it cannot cover all the terminals and the packets are based on the multihop to reach to the destination.

Different ad hoc networks have different characteristics, where the main medium between source and destination is the radio channel in all wireless communication. Wireless ad hoc networks can be classified through their application into general aggregation where ad hoc network includes mobile ad hoc network (MANET), vehicular ad hoc network (VANET) [1–4], flaying ad hoc network (FANET) [5], and sea ad hoc network (SANET) [6]. Sea ad hoc network (SANET) is one of the popular types of ad hoc network, where ships they may can communicate using two mode combinations. First one is they can communicate using satellite where this satellite can connect ship to shore, shore to ship and ship to ship with the high bandwidth frequency. Second is ship to ship, base station to ship, and ship to base station communication where using limited range for communication [7, 8].

The increasing volume of international and domestic trade poses new challenges to the efficiency, safety, and energy efficiency of transportation. As an important means of trade transportation, how to improve its transportation carrier-ship intelligence level is a key issue to be solved urgently. In recent years, under the guidance of sensing, communication, computer, information, automation, intelligent control and other technologies, ship intelligence has developed rapidly and achieved some new results. In the intelligent navigation of ships, the advancement of computer technology, sensor technology, communication technology, and information technology has promoted the updating, upgrading of ship navigation equipment, automation equipment, environment-aware equipment, the application of Internet of things technology, information physics system, and big data technology has accelerated. The development of information exchange between ships, ships, and shores provides the necessary foundation for the construction of a new generation integrated bridge system (IBS). The new generation of IBS has perfect integrated navigation, automatic ship handling, fault automatic diagnosis, automatic collision avoidance, and alarm functions. By monitoring the surrounding and own state of the ship, it not only ensures the safety of the ship but also saves it.

In terms of ship intelligent management and service, vessel traffic services (VTS) system still plays an important role in port ship management and inland river traffic management, but the current VTS system is limited to covering in important waters, mainly relying on radar. The ship's automatic identification system (AIS) and voice and other means of ship-shore interaction cannot achieve full monitoring and management of the navigation ship. Harmonized River Information Services (RIS) is an information collaboration service concept proposed by Europe to support inland navigation, traffic management, transportation management, and multimodal transport, providing users with electronic Static information such as river maps, laws and regulations, ship registration, and dynamic information such as ship location, cargo information, and estimated time of arrival.

2 Internet of Things

On the basis of communication networks such as the Internet and mobile communication networks, the Internet of things uses various intelligent objects with sensing, communication, and computing capabilities to automatically acquire various information of the physical world, and all of them can be independently addressed. Physical objects are interconnected and implemented comprehensive sensing, reliable transmission, intelligent analysis, processing, and construction of an intelligent information service system in which people and things, things and things are interconnected.

In the field of ships, global positioning system (GPS), automatic radar plotting aid (ARPA) radar, AIS, electricity Sub-hai chart display and information system, integrated bridge system, radio frequency identification (Radio Frequency Identification, RFID), video surveillance, etc. The application of monitoring and other technical means makes the ship develop rapidly in the direction of informationization and intelligence, but there is still a certain gap from the intelligent ship with true automatic perception, subjective analysis and intelligent operation.

The emergence of the Internet of things is the intelligentization of ships development provides new ideas. In [9] described based on Petri net method conflicts and redundancy between service components of the marine fishing vessel service system. The structure has improved the management efficiency of fishing vessels. In [10], use 3G, wireless technology such as RFID and ZigBee builds the Inland River Management Department System. In [11], the authors proposed "Smart Ship Autonomous Ship and Shore Architecture," where the information between the Smart Autonomous Ship and Data Center converges and is organically integrated and operated by applying these technologies to the Smart Autonomous Ship and Data Center and is converged and operated by implementing these intelligent data center technologies for ship and shore. In [12], the authors suggest Internet platform (IoT) for ocean observing buoys and check its operation in both outdoor and indoor environments. The proposed system consists of different sensor modules, a gateway, and a remote control location. In order to integrate sensor modules with different communication interfaces, they suggest a sensor-based network data area network (CAN) and gateway protocol. Based on previous studies and by conducting an interview with relevant field experts, this paper [13] authors applied IoT technologies for real-time remote control and monitoring of the navigation aids that were pointed out as the problems in navigation aids management.

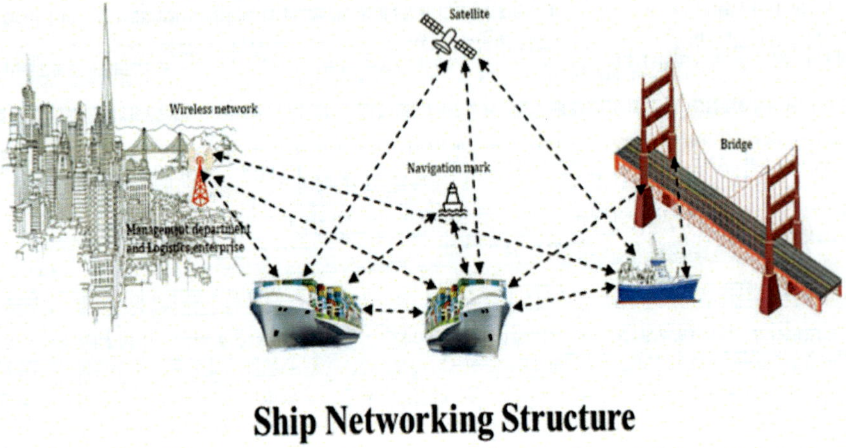

Fig. 1 Ship network structure

Based on the specific characteristics of the ship and the application environment, the Internet of vessels is built based on the Internet of things framework. Ship network generally refers to the purpose of integrating the logistics management, the comprehensive service of the industry, and the humanization of the travel experience. It covers the waterways, ship locks, bridges, ports and terminals with the enterprise, boat people, ships, and goods as the object, and integrates the core technology of the Internet of things. The data centered on the inland river intelligent shipping information integrated service network, which realizes the interconnection of people and ships, ship and ship interconnection, ship and cargo interconnection, and ship-shore interconnection, its structure is shown in Fig. 1.

3 Requirements for Maritime Wireless Communications

Detection, tracking, and inspection functions provided by the maritime communication system are necessary for preventive movement and data transmission between ships such as e-mail, SMS, and so on. Effective doses are discussed through the ventilation system. Considering the large distribution of ships, most vessels equipped with auxiliary vessels for vessels within 30 km are typically 50 km in length. For VHF modems, it can cover 30 km with a data transfer rate of 9.6–14.4 kbps. You can place multiple-hop links on a network or on a network if you have a ship that is 30 km from the ship. However, there is no way to build a dedicated network or network because the situation will catch up if the adjacent ship is not surrounded by 30 km.

3.1 Ship Ad Hoc Network Structural Design in Shore:

If the vessel is less than the base station, it can exchange information directly with BS. However, if the vessel is moving from the coverage area of the BS station, the network system must be configured with a previous ship or floating point because there is no direct link to the BS station. Figure 1 illustrates the structure of the maritime communications network.

3.2 Ship Ad Hoc Network Structural Design in Ocean:

In the case of NANET, the BS station lacks the wisdom of setting up the BS station on the ground and creating a BS at sea. As a result, it is desirable to apply an ad hoc network so that the BS does not have to communicate from peer to peer. Differentiate the grid on the beach without existing links to the BS. If there is no ship in the transmission coverage area of the broadband modem, an active HF range modem must be used for communication. The HF band modem has a transmission distance of approximately 40,000 km; therefore, it can be connected to the ground from a fixed BS line.

4 Ship Area Network Model

The entire network infrastructure, with the ship area network, has a layered architecture from a small field network to a global network. Ship area network model [14] consists of four-layer architecture as shows in Fig. 2.

- **Ship-to-Shore Network**:
  The top-level network includes ship LAN, satellite modem and antenna, satellite space segment, and network connected remote maintenance server and client. Ship companies and manufacturers can access the shipyard network via satellite links providing the ability to remotely maintain services.
- **Ship-to-LAN**:
  This network is a converged network that includes a gateway and a local Ethernet-based control network. The communication in the ship network is based on the IFc61162-450 standard protocol and uses User Datagram Protocol (UDP) transmission to provide a reliable/unreliable multicast transmission mode.
- **Local Control Network**:
  This network is a local network connected with various controllers, I/O modules, and various devices. It can independently realize various control targets, such as alarm monitoring system, ship maneuvering system, navigation data record and bridge information integration system. Network internal communication can use Ethernet or fieldbus technology such as PROFIBUS or CANopen. It should be

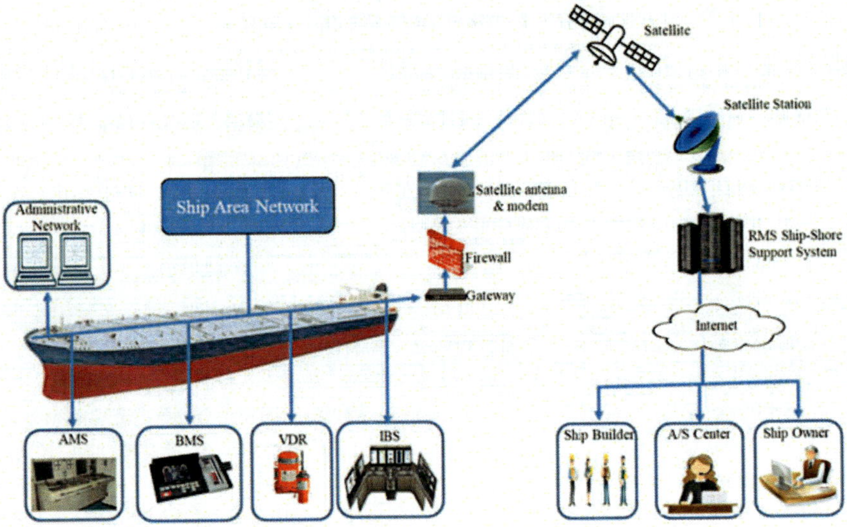

Fig. 2 Ship area network architicture

noted that the local control network can be used as a gateway for other network or devices that do not support the IEC62-450 standard.

- **Sensor/Actuator Network**:
  The network is connected to a console or LAN. This also combines fieldbus technology with wires or easier than local control networks. The wireless sensor/actuator network is a new type of wireless sensor network developed in recent years. It combines the advantages of wireless sensor networks and network control systems; so that the traditional wireless sensor network not only has monitoring functions but also has control and Execution function. Therefore, the wireless sensor/actuator network has broad application prospects in the fields of disaster relief work, precision agriculture, and smart home. Because wireless sensor/actuator networks can dramatically change the way people access information and interact with the environment.

5 IoT Common Later Architectures for Marine Environment Protection and Monitoring

The IoT generally can achieve controllable, thinkable, and knowing, to all around the world [15], however, IoT can be able to control the world by thinking, analysing the data, processing, and collecting data. As shown in Fig. 3, IoT common layer architectures for marine environment protection and monitoring system has five layers: business layer, application layer, data processing layer, data transmission layer, and perception and execution layer.

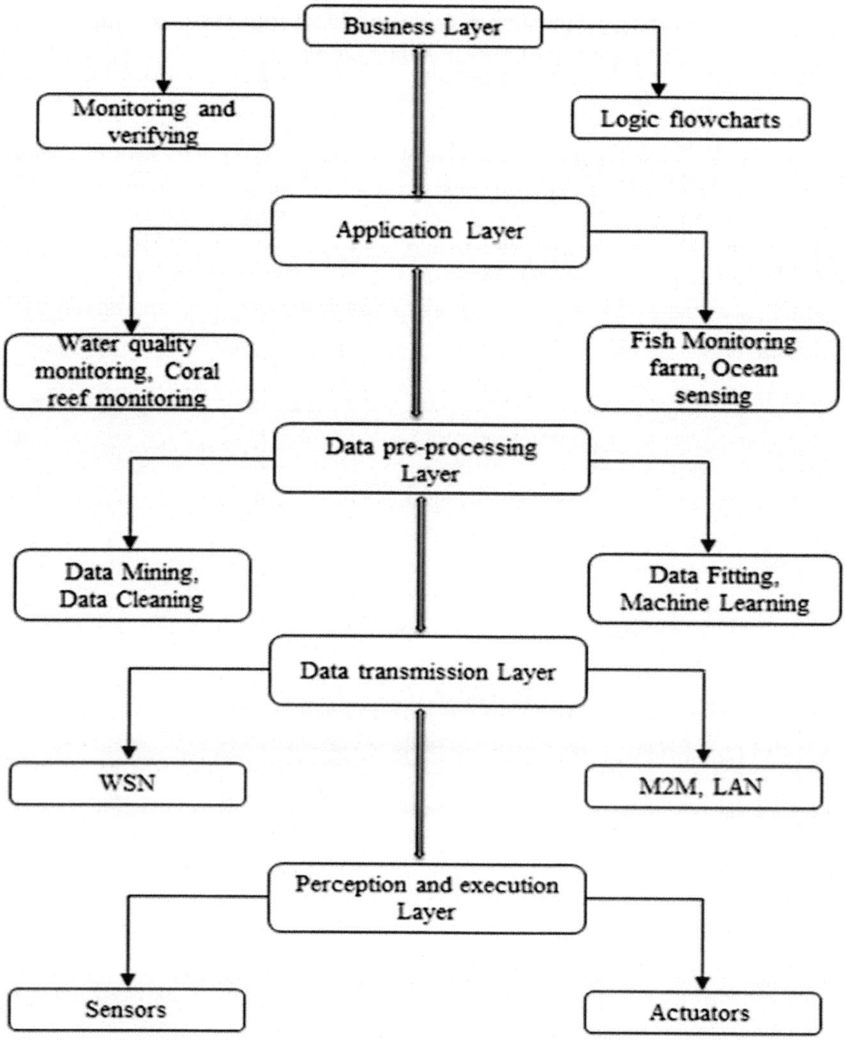

Fig. 3 IoT layer architecture for marine environment

- Execution and Perception Layer:
  This layer is the lowest layer in the architecture. Here, it includes actuators devices and sensors to collect data sensor information. In the Internet-based environment marine monitoring system and protection, this layer includes GPS sensors, power collection equipment, current water quality sensors, and quality control sensors. Current marine environment monitoring systems do not have execution functions.
- Data Transmission Layer:

The main task is to send from transmission layer to data processing layer different collected data through telecommunications networks, especially phones or wireless networks. The application layer sends to the execution and perception layer at the same time so the device or operator can perform the necessary actions, for instance, moving devices, lowering or increasing the temperature settings.

- Data Pre-Processing Layer:
  This layer is located in the middle of the Internet of things architecture, Here, advanced data extraction techniques can be used to store and process received raw data. You can also turn on alerts or alerts based on data collection or analysis, installation, filtering, sharing, and, in some cases, predefined rules.
- Application Layer:
  The main goal of the application layer is to support intelligent application services to meet the needs of users. This layer supply services based on the different applications that users require.
- Business Layer:
  This layer controls and manages the overall functionality and services of IoT platform.

6 Conclusion

With the rapid development of economy and the continuous progress of science and technology, Korea has more and more activities for the development and utilization of marine resources, and has also put forward higher requirements for marine environmental monitoring technology. The Internet of things is an important part of the new generation of information technology and represents the latest achievements in the development of the world's information industry. Exploring the construction of a marine environment monitoring system based on the Internet of things technology is of great significance for improving Korea's marine environment monitoring capacity and marine environment management level.

References

1. M.A. Al-Absi, A. A. Al-Absi, T. Kim, H.J. Lee, An environmental channel throughput and radio propagation modeling for vehicle-to-vehicle communication. Int. J. Distributed Sensor Netw. (IJDSN) **14**(4), 1–10 (2018) (SAGE Publisher)
2. M.A. Al-Absi, A.A. Al-Absi, H.J. Lee, V2V communication modeling for environmental channel throughput and radio propagation, in *8th IEEE International Conference on ICTC Convergence*, Jeju Island, Korea, IEEE Publisher, October 18–20, 2017, pp. 507–512
3. M.A. Al-Absi, A.A. Al-Absi, Y.J. Kang, H.J. Lee, Obstacles effects on signal attenuation in line of sight for different environments in V2V, in *20th International Conference on Advanced Communication Technology (ICACT)* Chuncheon-si Gangwon-do, Korea, IEEE Publisher, 11–14 Feb. 2018, pp. 17–20

4. M.A. Al-Absi, A.A. Al-Absi, H.J. Lee, Performance analysis for city, highway, and rural area in vehicle-to-vehicle network, in *8th IEEE International Conference on ICTC Convergence*, Jeju Island, Korea, IEEE Publisher, October 17–19, 2018
5. K. Ahmadhon, M.A. Al-Absi, H.J. Lee, S.H. Park, Smart flying umbrella drone on Internet of Things: AVUS, in *International Conference on Advanced Communications Technology (ICACT)*, February 17–20, 2019, pp. 191–195
6. A. Kamolov, S.H. Park, An IoT based smart berthing (parking) system for vessels and ports international conference on mobile and wireless technology, in *ICMWT: Mobile and Wireless Technology* (2018), pp. 129–139
7. D. Moon et al., A development of remote ship maintenance system based on ship area network. J. Soc. Naval Arch. Korea **47**(5), 751–756 (2010)
8. Y. Yoo, Domestic technology trends of standard ship networks. TTA J. **133**, 116–121 (2011)
9. C.M. Chen, L.J. Yu, P.L. Ling et al., The architecture of IOT smart service system of Ocean fishing vessel and its application based on Petri Net. Appl. Mech. Mater. **385**, 1771–1775 (2013)
10. Y. Zhuang, S. Song, Use of Internet of Things for ship management of Inland Rivers, in *Proceedings of ICTIS: Improving Multimodal Transportation Systems-Information, Safety, and Integration* (ASCE, Wuhan, 2013)
11. I. Im, D. Shin, J. Jeong, Components for smart autonomous ship architecture based on intelligent information technology, in *15th International Conference on Mobile Systems and Pervasive Computing* (2018), pp. 91–98
12. S.M. Kim, U.H. Lee, H.J. Kwon, J.-Y. Kim, J. Kim, Development of an IoT platform for Ocean observation buoys. IEIE Trans. Smart Process. Comput. **6**(2) (2017)
13. M. Cho, H.R. Choi, C.H. Kwak, A study on the navigation aids management based on IoT. Int. J. Control Autom. **8**(7), 193–204 (2015)
14. X. Zhang, Application of Internet of Things technology in agricultural production, in *Proceedings of the 2018 the International Symposium on Big Data and Artificial Intelligence, Hong Kong, China*, 29–30 December 2018, pp. 269–274
15. L. Antão, R. Pinto, J. Reis, G. Gonçalves, Requirements for testing and validating the industrial Internet of Things, in *Proceedings of the 2018 IEEE International Conference on Software Testing, Verification and Validation Workshops (ICSTW), Vasteras, Sweden*, 9–13 April 2018, pp. 110–115

Automatic Detection of Security Misconfigurations in Web Applications

Sandra Kumi, ChaeHo Lim, Sang-Gon Lee, Yustus Oko Oktian, and Elizabeth Nathania Witanto

Abstract Improper configuration of web applications or servers can lead to various security flaws. Security misconfiguration is ranked number 6 on the OWASP top 10 2017 list, meaning it is a critical risk in web applications that web developers need to focus on. The exploitation of this kind of vulnerabilities can lead to exploitation of other severe vulnerabilities and complete compromise of web applications. In this paper, we collaborate with security experts from a web security company to propose a tool to detect security misconfigurations in web applications. Our proposed tool, BitScanner, can effectively identify misconfiguration issues in all web applications regardless of the platform and technology they are built. The proposed tool is to enable web developers to fix any misconfiguration issues in applications before deployment in real development scenarios. Evaluation results show that our proposed tool has higher detection coverage and avoids false positives.

Keywords Web application security · Security misconfigurations · Vulnerability analysis · Penetration testing

---

S. Kumi · S.-G. Lee (✉) · Y. O. Oktian · E. N. Witanto
Department of Computer Engineering, Dongseo University, Busan, South Korea
e-mail: nok60@dongseo.ac.kr

S. Kumi
e-mail: kumisandra54@gmail.com

Y. O. Oktian
e-mail: yustus.oktian@gmail.com

E. N. Witanto
e-mail: elizabethnathaniaw93@gmail.com

C. Lim
BITSCAN Co., Ltd., Seoul, South Korea
e-mail: skscogh@naver.com

P. K. Pattnaik et al. (eds.), *Proceedings of International Conference on Smart Computing and Cyber Security*, Lecture Notes in Networks and Systems 149,
https://doi.org/10.1007/978-981-15-7990-5_8

1 Introduction

The exponential growth of web applications has led to an increased attack surface for organizations. According to OWASP's top 10 projects, security misconfigurations have been a persistent issue over the years. It ranked number 6 in the 2017 OWASP top 10 projects [1]. Security misconfigurations can happen at any level of an application stack. Attackers exploit misconfiguration vulnerabilities through unprotected files and directories, unused web pages, unpatched flaws and unauthorized access to default accounts. The exploitation of security misconfiguration vulnerabilities can lead attackers to exploit more critical vulnerabilities and also ultimately compromise an application.

As application security becomes sophisticated, refined techniques to prevent data breaches have been developed but simple human errors remain an issue. Insecure coding and errors from a developer can result in security misconfiguration vulnerabilities. The reliance on third-party components to develop web applications can cause misconfiguration issues. Attackers exploit configuration weakness in applications to gain knowledge about an application to exploit other critical vulnerabilities, which can pose a severe risk to organizations. Configuration flaws can lead an attacker to compromise an application entirely. Poorly maintaining or ignoring unused features in web applications can leave the application open to attackers. Improper handling of error messages in applications provides information for attackers to discover vulnerabilities.

According to Contrast Labs research [2] in October 2019, 36% of web applications are vulnerable to security misconfigurations. The researchers also discovered that 72% of.NET applications have security misconfigurations. In 2018, IBM [3] reported that data breaches related to improper configurations increased by 424%.

Despite the criticality of security misconfigurations, it has received little attention in the research field. A lot of research work has been done to detect injection vulnerabilities in web applications. The most recent researches in security misconfigurations are limited to particular technologies used in web applications.

In this paper, we present a tool that automatically scans web applications to detect security misconfigurations before deployment. BitScanner uses black-box testing (Dynamic Application Security Testing) to detect misconfiguration vulnerabilities in web applications. The main contributions of this paper are:

- We proposed BitScanner, which performs enhanced detection of security misconfigurations in web applications
- We evaluated our proposed approach on eight web applications running on different web servers.

The paper is organized as follows. Section 2 discusses the common vulnerabilities in web applications, background study on security misconfiguration and web application security testing. The proposed approach, BitScanner, is presented in Sect. 3. Evaluation results of the proposed approach and comparison with other tools are presented in Sect. 4. Section 5 describes related work, and we conclude in Sect. 6.

Table 1 OWASP top 10 web application security risks 2017

A1	Injection
A2	Broken Authentication
A3	Sensitive Data Exposure
A4	XML External Entities (XXE)
A5	Broken Access Control
A6	Security Misconfiguration
A7	Cross-Site Scripting (XSS)
A8	Insecure Deserialization
A9	Using Components with Known Vulnerabilities
A10	Insufficient Logging & Monitoring

2 Overview of Web Applications

This section discusses the common vulnerabilities and background on security misconfigurations of popular technologies in web applications. We also give a brief description of the security testing techniques used in web applications.

2.1 Common Web Application Vulnerabilities

The OWASP top 10 [4] identifies the most critical web application security risks. The goal of the project is to ensure that web applications built are secure against critical risks. The list is updated every three or four years. The 2017 OWASP top 10 [1] is shown in Table 1.

2.2 Security Misconfigurations in Web Applications

The security of a web application depends on the configuration of its web server and application server. According to OWASP [5], security misconfiguration vulnerabilities happen through overly informative error messages, unpatched security flaws, misconfigurations that permit directory listing, misconfigured SSL certificates and encryption settings and other varieties of server software flaws. In Netcraft's November 2019 [6] web server survey, Nginx and Apache have ranked the most popular web servers with 37.47% and 27.44%, respectively. According to the November 2019 W3Tech survey [7], WordPress and Joomla were the most popular content management systems. The most used server-side programming languages in developing web applications were PHP, ASP.NET and Java. In the following, we present some issues about the security configuration of the most popular web technologies mentioned above.

Nginx. nginx.conf [8] contains all the configuration settings about the Nginx server. server_tokens on; displays Nginx version on error pages and in server response header field which may allow hackers to gather information. Enabling unwanted HTTP methods to pose a critical risk on web applications such as information leakage, read and modify files and steal credentials of legitimate users. autoindex on will enable directory listing output when an HTTP GET request is made to a web application.

Apache. Option+ Indexes enable directory listings in Apache. ServerTokens Full and ServerSignature On will allow attackers to identify web server details.

PHP. php.ini configuration file contains directives you can set to configure PHP setup. Enabling expose_php will cause PHP information leakage. display_errors= On will expose error messages to all users of a web application.

ASP.NET. The behavior of an ASP.NET application is controlled by the machine.config and web.config configuration files. customErrors mode= "off" will expose internal error messages to end-users. debug= "true" allows attackers to see the line of code where errors occurred. trace enabled= "true" exposes information about the internal operations of a web page.

2.3 *Web Application Security Testing (AST)*

In the 2019 magic quadrant for AST [9], Gartner identifies three main techniques of AST.

- *Static Application Security Testing (SAST).* SAST is a set of technologies designed to analyze application source code, byte code, and binaries for coding and design conditions that are indicative of security vulnerabilities in a non-running state [10].
- *Dynamic Application Security Testing (DAST).* DAST technologies are designed to detect conditions indicative of a security vulnerability in an application in its running state. DAST tools run on operating code to detect issues with interfaces, requests, responses, scripting, sessions, data injection and authentication [11].
- *Interactive Application Security Testing (IAST).* IAST tools use a combination of static and dynamic analysis technologies. IAST tools use knowledge of application flow and dataflow to create advanced attack scenarios and perform a dynamic scan. IAST tools are proficient at reducing the number of false positives.

3 Our Proposed Approach

The proposed approach uses DAST to detect security misconfiguration vulnerabilities in web applications. BitScanner simulates an attack against target web applications through HTTP requests and analyzes responses from the web application's server to determine security misconfigurations. BitScanner detects nine misconfigurations as shown in Table 2. Our approach as shown in Fig. 1 uses four steps to detect security misconfigurations in web applications.

Step 1—Crawling web application. The *Crawler* module performs a search on the target application to identify all URLs before a scan. BitScanner leverages a full-depth crawling strategy to analyze the structure of a web application to identify all internal links excluding external links. A normalization feature is implemented to avoid crawling the same content more than once. BitScanner's crawler is configured to avoid crawling links that will cause it to log off.

Step 2—Identification of input parameters. *Structural analysis* is performed on collected pages to identify all vulnerable points within web forms. We parse and analyze each page to detect forms and extract all input parameters within forms. Also, we detect GET and POST parameters in JSON, XML and AJAX. The identified input parameters are used as starting points to generate attacks on web applications.

Table 2 Description of security misconfiguration vulnerabilities detected by BitScanner

Vulnerability	Description	Risk level
Internal server error 500 Page error	Application improperly handles errors that occur during processing and reveals sensitive information to allow an attacker to have access	External access (Medium)
Script error Exception error Validation error	Web application generates an error message that contains sensitive information about users and also display a line of source code where the error occurred	Information reveal (Low)
Directory listing	Web server is configured to display the list of files in a directory	
PHP information leaked	Injects code to reveal information about the current state of PHP	
Source code disclosure	Allow attackers to obtain server-side source code of web application	
POST method allowed	Attackers use the POST method to exploit vulnerabilities	

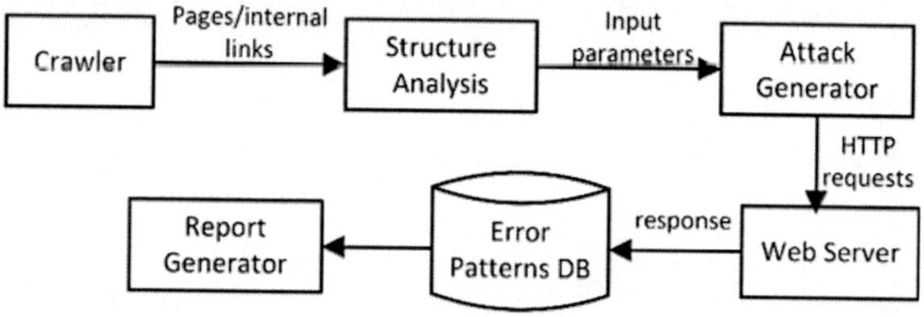

Fig. 1 Architecture of BitScanner

Step 3—Attack Generation. Input parameters detected are fuzzed with predefined attack patterns of security misconfigurations and sent as an HTTP request to the web application. Responses returned by web applications are matched against *Error Patterns DB* to determine vulnerabilities. The *Error Patterns DB* contains a wide range of error messages and error status, from different well-known DBMS.

Step 4—Report Generation. BitScanner generates a report of security misconfigurations detected in a scanned application in PDF and HTML form.

4 Experimental Evaluation

BitScanner is implemented in Perl and deployed on a laptop/desktop PC and a server. The server operates on the Intel Xeon processor and MySQL DB for the database. It uses a bandwidth of 1G and 10G for scanning. To evaluate the effectiveness of our tool, we selected eight real-life web applications running on different web servers such as Apache, Nginx and Cloudflare. For security reasons, we do not show domains of tested target web applications in this paper. The duration for complete scanning of web applications depends on the content and number of URLs it has. It takes 102 s to scan a web application with 61 URLs. Figure 2 shows the number of detected vulnerabilities and the vulnerability type of each target web application.

To compare BitScanner's performance, we use two well-known scanners, Acunetix [12] and Arachni v1.5.1 [13], since they operate in a similar way to our tool. We used three web applications to test the performance of each tool. Two applications use the Microsoft-IIS/8.5 web server, and the third application runs on the Nginx/1.4.1 web server. The comparative analysis with other tools is shown in Table 3. Acunetix has the highest number of detected vulnerabilities since it covers a wide range of vulnerabilities. Arachni detected few vulnerabilities using much time in execution. Some security misconfigurations detected by Acunetix which

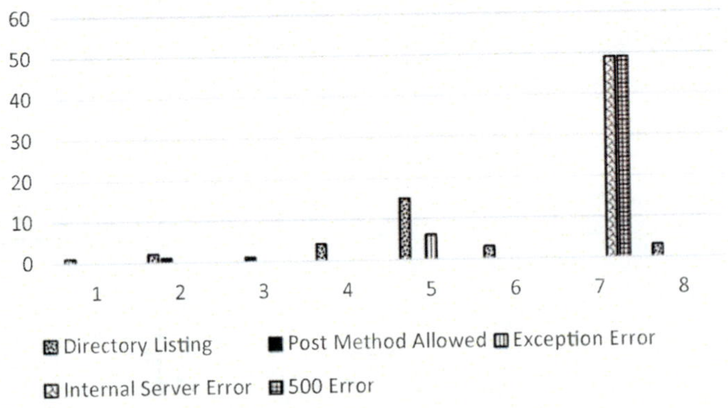

Fig. 2 Detected number of vulnerabilities per vulnerability type

Table 3 Comparative analysis with other tools

Application	BitScanner	Acunetix	Arachni
1	4	6	3
2	3	9	1
3	2	10	0

are not on our misconfiguration table include cookies without secure flagset, directory listing, weak password, Nginx SPDY heap buffer overflow, stack trace disclosure and unicode transformation issues. Currently, our tool only detects security misconfiguration vulnerabilities shown in Table 2.

5 Related Work

The authors of [14] proposed a tool that automatically audits security configuration settings of web server environments and was evaluated on Apache, MySQL and PHP (AMP) server environments. This approach does not address the detection of security misconfigurations at the application level, and it is limited to only AMP server packages.

In [15, 16], an automated scanner of web application configurations at the environment and application level to detect, quantify and fix misconfiguration is presented. Common Configuration Scoring System (CCSS) is used to determine the severity of configuration vulnerabilities detected. However, these tools can only detect misconfigurations in PHP web applications.

OWASP Orizon[17] is a static analysis tool for the detection of vulnerabilities in J2EE web applications. Pixy[18] scans the source code of web applications developed in PHP to detect XSS vulnerabilities.

In [19–21], dynamic analysis technique was used to detect SQL injections and XSS vulnerabilities in web applications. Although similar to our approach, these tools cannot detect security misconfigurations in web applications.

Our tool, motivated by discussed tools above, can detect security misconfigurations. Although in this work, we focus on the detection of security misconfigurations, vulnerabilities such as SQL injections, XSS, Web Shell, and XXE injections can be detected by our tool.

6 Conclusion

Security misconfigurations are identified to be part of OWASP's top 10 critical web application security risks but have less attention in the research area. In this paper, we presented a tool, BitScanner, to detect security misconfigurations in web applications.

We implemented and evaluated BitScanner on eight real-life web applications. A comparative analysis with other tools shows that our tool detects web configuration issues at a faster rate but limited to only misconfiguration vulnerabilities shown in Table 2.

In the future, we wish to improve on our work by including more security misconfiguration vulnerabilities to our framework.

Acknowledgements We thank BitScan, Inc., Ltd., and BNST Inc., Ltd., for their collaboration in this study. This work was supported by Basic Science Research Program through the National Research Foundation of Korea (NRF) funded by the Ministry of Education (Grant Number: 2018R1D1A1B07047601).

References

1. OWASP Top 10—2017. https://www.owasp.org/images/7/72/OWASP_Top_10-2017_%28en%29.pdf.pdf. Accessed 14 Aug 2019
2. Contrast Security News. https://www.contrastsecurity.com/security-influencers/two-years-after-the-release-of-the-2017-owasp-top-ten-limited-improvements-shown. Accessed 04 Dec 2019
3. IBM X-Force Report. https://newsroom.ibm.com/2018-04-04-IBM-X-Force-Report-Fewer-Records-Breached-In-2017-As-Cybercriminals-Focused-On-Ransomware-And-Destructive-Attacks. Accessed 04 Dec 2019
4. OWASP top 10 project. https://www.owasp.org/index.php/Category:OWASP_Top_Ten_Project#tab=Main. Accessed 14 Aug 2019
5. Insecure Configuration Management. https://www.owasp.org/index.php/Insecure_Configuration_Management. Accessed Dec. 05, 2019
6. NETCRAFT. https://news.netcraft.com/archives/2019/11/27/november-2019-web-server-survey.html. Accessed 05 Dec 2019
7. W3Techs—World Wide Web Technology Surveys. https://w3techs.com/. Accessed 05 Dec 2019
8. Nginx Configuration. https://nginx.org/en/docs/http/ngx_http_core_module.html. Accessed 10 Dec 2019
9. Magic Quadrant for Application Security Testing. https://www.gartner.com/en/documents/3906990/magic-quadrant-for-application-security-testing. Accessed 06 Dec 2019
10. Erdem_Menges: Integrations: Make Sense of SAST and DAST. https://community.microfocus.com/t5/Security-Blog/Integrations-Make-Sense-of-SAST-and-DAST/ba-p/1613919. Accessed 20 Aug 2019
11. 10 Types of Application Security Testing Tools: When and How to Use Them. https://insights.sei.cmu.edu/sei_blog/2018/07/10-types-of-application-security-testing-tools-when-and-how-to-use-them.html?fbclid=IwAR2iSUjZRVBzeYZzwiHOOZxYLsWo1EHzpEB1RNThd_2LbftmUM-gqGsQSCk. Accessed 01 Aug 2019
12. Acunetix. https://www.acunetix.com/vulnerability-scanner/
13. Arachni Web Application Security Scanner Framework. https://www.arachni-scanner.com/
14. B. Eshete, Early detection of security misconfiguration vulnerabilities in web applications, I *Presented at the 2011 Sixth International Conference on Availability, Reliability and Security*. https://doi.org/10.1109/ARES.2011.31.
15. B. Eshete, A. Villafiorita, Confeagle: automated analysis of configuration vulnerabilities in web applications, in *Presented at the 2013 IEEE 7th International Conference on Software Security and Reliability*, https://doi.org/10.1109/SERE.2013.30.

16. A. Wainakh, A.A. Wabbi, B. Alkhatib, Design and develop misconfiguration vulnerabilities scanner for web applications. IRECOS **9**. https://doi.org/10.15866/irecos.v9i10.3840
17. The OWASP Orizon Project. https://orizon.sourceforge.net/. Accessed 22 Aug 2019
18. N. Jovanovic, C. Kruegel, E. Kirda, Pixy: a static analysis tool for detecting Web application vulnerabilities, in *Presented at the 2006 IEEE Symposium on Security and Privacy (S&P'06)*. https://doi.org/10.1109/SP.2006.29
19. S. Kals, E. Kirda, C. Kruegal, N. Jovanovic, SecuBat: a web vulnerability scanner, in *Proceedings of the 15th International Conference on World Wide Web*. https://doi.org/10.1145/1135777.1135817
20. Z. Đurić, WAPTT—Web application penetration testing tool. AECE. https://doi.org/10.4316/AECE.2014.01015
21. A. Ciampa, C.A. Visaggio, M. Di Penta, A heuristic-based approach for detecting SQL-injection vulnerabilities in web applications, in *Proceedings of the 2010 ICSE Workshop on Software Engineering for Secure Systems*. https://doi.org/10.1145/1809100.1809107

Real-Time Access Control System Method Using Face Recognition

Mohammed Abdulhakim Al-Absi, Gabit Tolendiyev, Hoon Jae Lee, and Ahmed Abdulhakim Al-Absi

Abstract Face recognition has been widely studied and studied for many years, but most PC-based face recognition systems have very limited portability and mobility. Face recognition is a process of dynamically capturing facial features through a camera connected to a computer and simultaneously comparing the captured facial features with the facial features previously entered into the personnel library. Face recognition-based person authentication system has been popular among other biometrics recently. This technology can be applied to important departments such as public security, banking, and customs to provide convenient and efficient detection methods. In this paper, we discuss a method of access control system using face recognition technology for entrance limited places. Some face recognition technologies that have shown state-of-the-art performance at their time are discussed. Here we present a method for access control system using facial recognition technology.

Keywords Face recognition · Face verification · Access control · Attendee management

M. A. Al-Absi · G. Tolendiyev
Department of Computer Engineering, Dongseo University, 47 Jurye-ro, Sasang-gu, Busan 47011, Republic of Korea
e-mail: Mohammed.a.absi@gmail.com

G. Tolendiyev
e-mail: tkgabit@gmail.com

H. J. Lee
Division of Information and Communication Engineering, Dongseo University, 47 Jurye-ro, Sasang-gu, Busan 47011, Republic of Korea
e-mail: hjlee@dongseo.ac.kr

A. A. Al-Absi (✉)
Department of Smart Computing, Kyungdong University, Gosung, Gangwon-do, Republic of Korea
e-mail: absiahmed@kduniv.ac.kr

P. K. Pattnaik et al. (eds.), *Proceedings of International Conference on Smart Computing and Cyber Security*, Lecture Notes in Networks and Systems 149,
https://doi.org/10.1007/978-981-15-7990-5_9

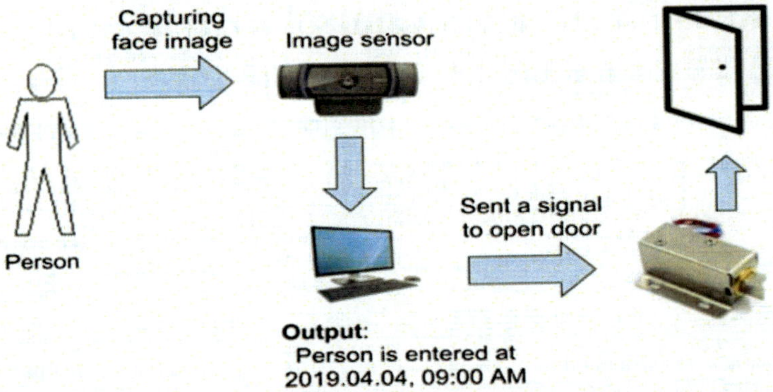

Fig. 1 An illustration of access control system using face recognition

1 Introduction

The research of face recognition technology began in the 1960s. Especially since the 1990s, with the urgent demand of face recognition systems in various industries, a large number of scholars have been engaged in the research of practical face recognition technology.

Recently, facial recognition technology performance has increased thanks to deep convolutional networks. Face recognition is popular than other biometrics such as fingerprint and eye iris recognition. Because it is contactless and non-invasive. We already have experience unlocking phone, verifying payment, and also face recognition access control system being used in the entrance of large-scale event [15] instead of ID card. In near feature, this technique might replace all people who is in charge of verify someone's identity by comparing their actual face appearance with face photography.

Figure 1 depicted an access control system using face recognition. Face recognition identity authentication system is generally authenticated by the client, application system, and face authentication as shown in Fig. 2.

2 Related Works

Face recognition system has been active topic in computer vision and pattern recognition field in last decades. Face recognition is more popular than other biometric systems such as fingerprint, palm vein and eye iris recognition. A big reason of using face recognition is its contactless, non-invasiveness and secureness (Table 1).

Face recognition systems can be classified largely into two types. In case of first type of face recognition, person's intents to be identified by directly looking at the camera and camera are installed in well-conditioned environment for recognition. In

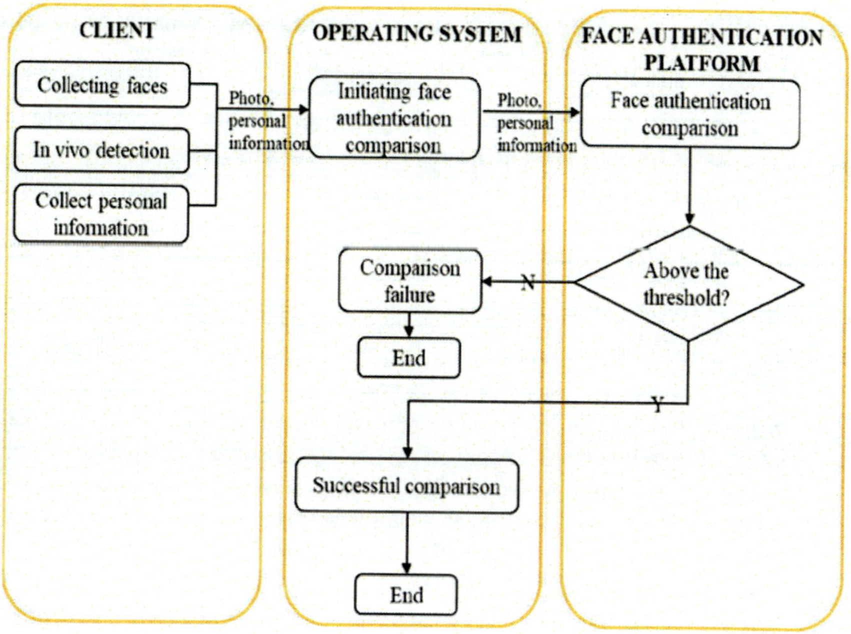

Fig. 2 Face authentication process

Table 1 Comparison of biometric system in term of technical and non-technical weaknesses

Biometric system	Weakness of techniques	Non-technical weakness
Fingerprint	Slow	Contackness
Iris	slow	–
Face recognition	Requires large data to train the model that generalizes well a face image	Contactless, user-friendly

another case, some authorities, e.g., police, security guard intend to try to identify people. In this case, camera might not capture good face image and environment cause difficulties as well. The second case face recognition is more difficult problem to solve than first one.

In face recognition, there are several difficulties:

- Illumination. Less of illumination and direct light effects on image acquiring;
- Pose: persons do not directly look at the camera;
- Occlusion: partial occlusion of face by other objects, i.e., hands;
- Accessories: sunglasses, cap, etc.
- In order to address these problems and achieve better recognition accuracy, many approaches have been proposed:
- Employing large-scale deep CNN containing more than 1.6B parameters.

- Training the model on very huge dataset, i.e., more than 200 M image of 8 M different identities;
- Better loss function then distinguishes two-face images (Contrastive loss [10], triplet loss [11], center loss [12], large margin softmax (L-Softmax) [13], angular softmax (A-Softmax) [14]), and large margin cosine loss [4], and cosine loss [3].

3 Face Recognition System

Face recognition is a technique that identifies a person with his or her facial image by comparing their image stored in database. First of all, we convert cropped face image to 128-dimensional vector. After that measure distance between them. Distance between face images of the same person is less than distance of the face images of the different person. To convert face image to some real number vector, we use convolutional neural networks. This CNN generalizes a face image to vector. To achieve a model that generalizes well a face image to vector, we train the CNN on large data, number of hundreds of million images of million different identities. We can use public available models [1–3] and if necessary, train and build own model on public face datasets [5, 6] and our own data. Popular datasets for testing face verification are Labeled Faces in the Wild (LFW) database, YouTube database (YTD), and MegaFace Challenge.

We developed person authentication system using open-source face recognition model [1]. Figure 2 depicted interface of the system. The system works automatic mode as default without need of any person. However, if it is necessary security person or manager can monitor it. The system also records people enter and exit time, so this can be used as attendee management system as well. One advantage of face recognition to other object identification system is we do not need to train the model for registering a new person. The face recognition model well generalizes a new dataset.

4 Face Verification Method

Euclidean distance calculation similarity is the simplest and most understandable method of all similarity calculation. It takes the items that have been unanimously evaluated as the coordinate axes and then draws the people who participated in the evaluation onto the coordinate system and calculates the linear distance between them. Euclidean distance is the most commonly used distance calculation formula. It measures the absolute distance between points in a multidimensional space. This is a good calculation method when the data is very dense and continuous.

Figure 3 illustrated a method of face verification system. First of all, captured person's face image is cropped, aligned and resized to required size of the CNN, and deep CNN extract face feature, i.e., transforms a face image to 128-dimensional

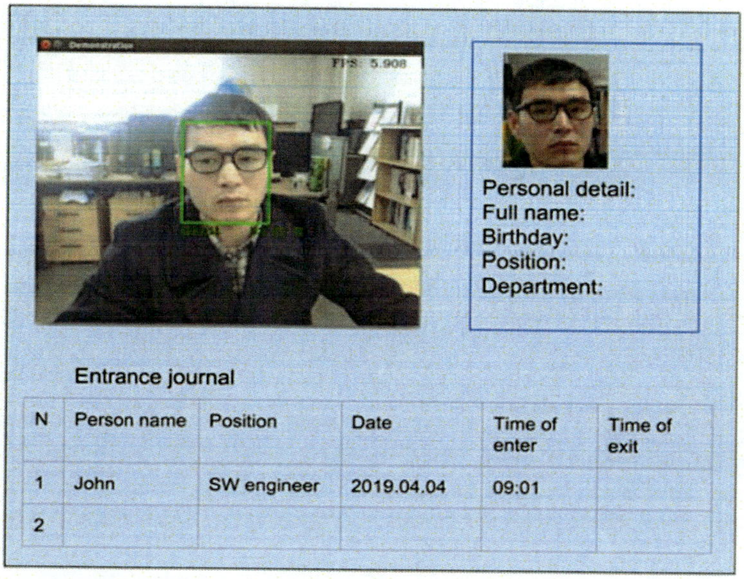

Fig. 3 Interface of the system

vector. After that in order to compare the image with image in database, distance between two vectors of face image is measured. The system makes decision by thresholding distance value, if the distance less than the thresholding value (e.g., 0.3), the system results as the two images of the same person, otherwise different.

When data items *A* and *B* are used as points in the graph, the similarity between the two is r distance (*A*, *B*), which can be calculated by the Euclidean distance. The formula for calculating the Euclidean distance in *n*-dimensional space is measured by Eq. (1).

$$r = \text{sqrt}\left(\sum (x_{i1} - y_{i2})^2\right) \text{ where } i = 1, 2, \ldots n$$

$$r(x, y) = \sqrt{\sum_{i=1}^{n} (x_i - y_i)^2} \tag{1}$$

where x_{i1} represents the *i*-dimensional coordinate of the first point, and y_{i2} represents the *i*-dimensional coordinate of the second point. The *n*-dimensional Euclidean space is a set of points, and each of its points can be expressed as $(x(1), x(2), \ldots x(n))(x(1), x(2), \ldots x(n))$, where $x(i)(i = 1, 2 \ldots n)$ is a real number, called the *i*-th coordinate of *x*, the distance $r(x, y)$ between two points *x* and $(y{=}y(1), y(2), \ldots y(n))$ is defined as the above formula (Fig. 4).

The cosine similarity uses the cosine of the angles of the two vectors in the vector space as the measure of the difference between the two individuals. Compared to

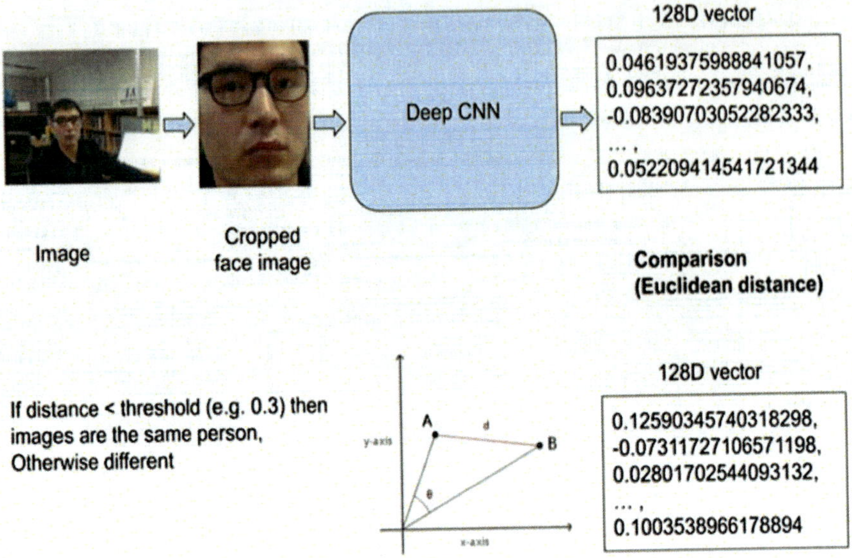

Fig. 4 Face verification method

distance metrics, cosine similarity pays more attention to the difference in direction between two vectors, rather than distance or length. As can be seen from the above figure, the Euclidean distance measures the absolute distance of each point i the space, which is directly related to the position coordinates of each point; and the cosine distance measures the angle between the space vectors, which is more reflected in the direction instead of location. If the position of point *A* is kept unchanged and point *B* is away from the origin of the coordinate axis in the original direction, then the cosine distance remains unchanged at this time(because the angle does not change). The distance between *A* and *B* is points *A* and *B* which is obviously changing, that is the difference between the Euclidean distance and the cosine similarity.

5 Conclusion

Face recognition has been widely studied for many years, but most PC-based face recognition systems have very limited portability and mobility. In special applications such as smart communities, outdoor observation points and personal observation, it is very convenient to use a face recognition system to provide new requirements. Research on face recognition in embedded systems has practical meaning and a wide range of applications. As a point of view, many companies have built-in systems with access control systems based on attendance at this stage, but most are face recognition systems or PC-based structures, an integrated front face image and display recognition results. The face recognition algorithm is implemented in the computer background. As computer technology advances, embedded systems become an integral part of the computer field, compact microprocessor and built-in operating system to improve performance and become the foundation for the integrated face recognition research system.

This paper studied a person authentication method using facial recognition system. Nowadays, there are publicly available face recognition models such as OpenFace, ArcFace model which engineers can use for their project. Face recognition person authentication system is a powerful security among other biometric system such as fingerprint and iris recognition. Face recognition is also being used in our daily life such as payment verification, door unlock, phone unlock. In the feature, all digital door lock devices might include also face recognition feature. A method of person authentication for access control system of entrance limited places is proposed.

References

1. B. Amos, B. Ludwiczuk, M. Satyanarayanan, *OpenFace: A General-purpose Face Recognition Library with Mobile Applications*. Technical report, CMU-CS-16-118, CMU School of Computer Science (2016)
2. F. Schroff, D. Kalenichenko, J. Philbin, *Facenet: A Unified Embedding for Face Recognition and Clustering*. arXiv preprint arXiv:1503.03832 (2015)
3. J. Deng, J. Guo, S. Zafeiriou, *ArcFace: Additive Angular Margin Loss for Deep Face Recognition*. arXiv:1801.07698 (2018)
4. H. Wang, Y. Wang, Z. Zhou, X. Ji, Z. Li, D. Gong, J. Zhou, W. Liu, Cosface: large margin cosine loss for deep face recognition, in *CVPR* (2018)
5. D. Yi, Z. Lei, S. Liao, S. Z. Li, *Learning Face Representation from Scratch*. arXiv preprint ss (2014)
6. H.-W. Ng, S. Winkler, A data-driven approach to cleaning large face datasets, in *Proceedings of the ICIP* (2014). https://vinstage.winklerbros.net/facescrub.html
7. A. Okumura, T. Hoshino, S. Hada, Yugo Nishiyama, M. Tabuchi, Identity verification of ticket holders at large-scale events using face recognition. J. Inf. Process. **25**, 448–458 (2017)
8. P. Li, C. Cadell, *Chine Eyes 'Black Tech' to Boost Security as Parliament Meets*. (March 10, 2018) Retrieved April 4, 2019 from Technology News from Reuters website: https://www.reuters.com/article/us-china-parliament-surveillance/china-eyes-black-tech-to-boost-security-as-parliament-meets-idUSKBN1GM06M?utm_campaign=trueAnthem:+Trending+Content&utm_content=5aa3f9fd04d30121e40e5e73&utm_medium=trueAnthem&utm_source=twitter
9. G. Fleishman, *Face ID on the iPhone X: Everything you Need to Know About Apple's Facial Recognition*, (2017, December 1) Retrieved December 4, 2017 from the Macworld from IDG website: https://www.macworld.com/article/3225406/iphone-ipad/face-id-iphone-x-faq.html
10. S. Chopra, R. Hadsell, Y. LeCun, Learning a similarity metric discriminatively, with application to face verification, in *Conference on Computer Vision and Pattern Recognition (CVPR)* (2005)
11. E. Hoffer, N. Ailon, Deep metric learning using triplet network, in *International Workshop on Similarity-Based Pattern Recognition* (2015)
12. Y. Wen, K. Zhang, Z. Li, Y. Qiao, A discriminative feature learning approach for deep face recognition, in *European Conference on Computer Vision (ECCV)* (2016), pp. 499–515
13. W. Liu, Y. Wen, Z. Yu, M. Yang, Large-margin softmax loss for convolutional neural networks, in *International Conference on Machine Learning (ICML)* (2016)

14. W. Liu, Y. Wen, Z. Yu, M. Li, B. Raj, L. Song, SphereFace: deep hypersphere embedding for face recognition, in *Conference on Computer Vision and Pattern Recognition (CVPR)* (2017)
15. Global Times, *AI, Robots Help Provide Security for SCO Summit in Qingdao.* Retrieved April 4ss, 2019 from Global Times website: https://www.globaltimes.cn/content/1106310.shtmla

Towards a Sentiment Analyser for Low-resource Languages

Dian Indriani, Arbi Haza Nasution, Winda Monika, and Salhazan Nasution

Abstract Twitter is one of the top influenced social media which has a million number of active users. It is commonly used for microblogging that allows users to share messages, ideas, thoughts and many more. Thus, millions of interactions such as short messages or tweets are flowing around among the twitter users discussing various topics that have been happening worldwide. This research aims to analyse a sentiment of the users towards a particular trending topic that has been actively and massively discussed at that time. We chose a hashtag #kpujangancurang that was the trending topic during the Indonesia presidential election in 2019. We use the hashtag to obtain a set of data from Twitter to analyse and investigate further the positive or the negative sentiment of the users from their tweets. This research utilizes RapidMiner tool to generate the Twitter data and comparing Naive Bayes, k-nearest neighbor, decision tree and multi-layer perceptron classification methods to classify the sentiment of the Twitter data. There are overall 200 labeled data in this experiment. Overall, Naive Bayes and multi-layer perceptron classification outperformed the other two methods on 11 experiments with different size of training–testing data split. The two classifiers are potential to be used in creating sentiment analyser for low-resource languages with small corpus.

Keywords Twitter · Sentiment analysis · Low-resource languages · Naive Bayes · K-nearest neighbor · Decision tree · Multi-layer perceptron

D. Indriani · A. H. Nasution (✉)
Informatics Engineering, Universitas Islam Riau, Riau, Indonesia
e-mail: arbi@eng.uir.ac.id

D. Indriani
e-mail: diianindrianii23@student.uir.ac.id

W. Monika
Library Science, Universitas Lancang Kuning, Riau, Indonesia
e-mail: windamonika@unilak.ac.id

S. Nasution
Informatics Engineering, Universitas Riau, Riau, Indonesia
e-mail: salhazan@lecturer.unri.ac.id

P. K. Pattnaik et al. (eds.), *Proceedings of International Conference on Smart Computing and Cyber Security*, Lecture Notes in Networks and Systems 149,
https://doi.org/10.1007/978-981-15-7990-5_10

1 Introduction

A rich sentiment analysis corpus is crucial in creating a good sentiment analyser. Unfortunately, low-resource languages like Indonesian lack such resources. Some prior studies focused on enriching low-resource languages [6–13]. The rapid growth of online textual data creates an urgent need for powerful text mining techniques [1]. Sentiment analysis or opinion mining is a part of text mining. Sentiment analysis basically is a computational research that analyses the textual expression from opinion, sentiment and emotion of the social media users.

[4] It extracts attributes and components of the documented object. Through the sentiment analysis of the text, information such as the public's emotional status, views on some social phenomena and preferences for a product can be obtained [20]. Hence, the perspective of the users either positive or negative could be revealed.

During the Indonesia 2019 presidential election, the competition was quite fierce where there were only two candidates fighting in the battle. Most of supporters from these two candidates were actively campaigning their candidates on social media, and Twitter was the highly used social media chosen by them. Due to huge enthusiasm of those two supporters, most of the time fierce debate among them could not be avoided. One of the trending topics emerged was during the recapitulation of the votes. Twitter users reacted to the several findings showed that the calculation of the votes led to deception. Foremost, supporters from one party, from Prabowo Subianto volunteers, found that many evidences of the wrong data were inputted to the system. Thus, the real count results were irrelevant with the information displayed on the system. This finding made the situation in Indonesia heating up. Supporters from Prabowo Subianto was upset and condemned the General Election Commission as the legal institution to take full responsibility of this matter. To express their disappointment, most of the Twitter users created hashtag #kpujangancurang or "The General Election Commission should not be unfair". However, this issue was objected by the opponent supporters. They argued that this issue was merely caused by human error. The same hashtag actually was being used by the both parties, so that no one knows the exact sentiment of the tweets. Therefore, a sentiment analyser that could analyse the sentiment of the tweets is crucial.

In sentiment analysis, the available corpus in Indonesian language is scarce. The existing machine learning tool such as RapidMiner has two sentiment analysers which are Aylien and Rosette and do not cover Indonesian language. We run an experiment by using the #kpujangancurang hashtag to obtain corpus using RapidMiner to extract the tweets and then analyse the sentiment of users by using four machine learning methods which are Naive Bayes, k-nearest neighbor, decision tree and multi-layer perceptron classification. The objective of this research is to find out which classifier is more suitable to be used in creating sentiment analyser for low-resource languages with small corpus.

2 Literature Study

Several researches have been done on sentiment analysis. A study attempted to analyse the online sentiment changes of social media users using both the textual and visual contents by analysing sentiment of Twitter text and image [19]. Another related study performed linguistic analysis of the collected corpus and explain discovered phenomena to build a sentiment classifier, that is able to determine positive, negative and neutral sentiments for a document [14].

Furthermore, several studies have been done using machine learning method on sentiment analysis, for instance a study showed that a similar research on a Twitter sentiment analysis by applying Naive Bayes classifier method to investigate the sentiment analysis of the Twitter users on the traffic jam in Bandung [18]. Another study focused on data classification using k-NN and Naive Bayes where the corpus was downloaded from TREC Legal Track with a total of more than 3000 text documents and over 20 types of classifications [17]. A study utilized maximum entropy part of speech tagging and support vector machine to analyse the public sentiment. The study used dataset in Indonesian language and implemented machine learning approached due to its efficiency for integrating a large-scale feature into a model. This kind of approach has been successfully implemented in various tasks such as natural language processing [16]. A study proposed a semi-automatic, complementary approach in which rule-based classification, supervised learning and machine learning are combined into a new method to achieve a good level of effectiveness [15]. Another study shows opinion mining for hotel rating through reviews using decision tree classifier. The advantage of using the algorithm is that the rule set can be easily generated and by analysing each level of the tree, so a particular service quality can be improved [3]. Deep learning methods also have been widely used in sentiment analysis tasks [5, 21]. However, these studies show different accuracy from each machine learning method used depending on the size of the corpus.

RapidMiner is an open-source software. RapidMiner is one of the solutions for doing analysis on data mining, text mining and prediction analysis. RapidMiner uses various descriptive techniques and predictions in giving a new insight to the users so that allows and helps users to make a better decision. RapidMiner is a standalone software and enables to be integrated with its own products. RapidMiner provides graphic user interface (GUI) for designing an analytical pipeline. GUI will generate Extensible Markup Language (XML) that defines the analytical process of the users need to be applied on the data. This file is later on read by RapidMiner to be automatically analysed. We use RapidMiner due to several reasons: it eases in getting the dataset, it can be connected to Twitter, it enables to search the topic as query so that the intended topic will emerge and can be saved in excel file; furthermore, it allows extracting plentiful data. A study examined an anomaly detection extension for RapidMiner in order to assist non-experts with applying eight different k-nearest-neighbor- and clustering-based algorithms on their data [2]. However, in this study, we only use RapidMiner to extract data from Twitter.

3 Research Methodology

In this study, we use a dataset that was obtained from the tweets' document. We utilized RapidMiner to obtain the tweets from the hashtag #kpujangancurang. To investigate further about the hashtag #kpujangancurang, we compare Naive Bayes, k-nearest neighbor, decision tree and multi-layer perceptron classification methods to classify the sentiment of the Twitter data. There are two steps of the document classification: the first one is training the document that has been categorized. And the second one is training the uncategorized document. The four methods classify the distribution of the positive and negative sentiments. There are overall 200 labeled data in this experiment. To evaluate the performance of the sentiment analyser, we use accuracy as the evaluation measure.

Sentiment analysis overview is described in details which is depicted in the Fig. 1.

- Data Crawling: It is a process of aggregating data from Twitter using RapidMiner as a tool. The aggregated data from hashtag #kpujangancurang is used as training dataset and testing dataset.
- Preprocessing: It is a process of cleaning the data by deleting common words by referring to stopwords.
- Classification: Naive Bayes method is applied to classify the sentiment into positive and negative sentiments. The rest of methods will be used in the same manner.
- Evaluation: The classification result from classifiers is evaluated with the manual labeling classification. The accuracy of the classification determines whether a

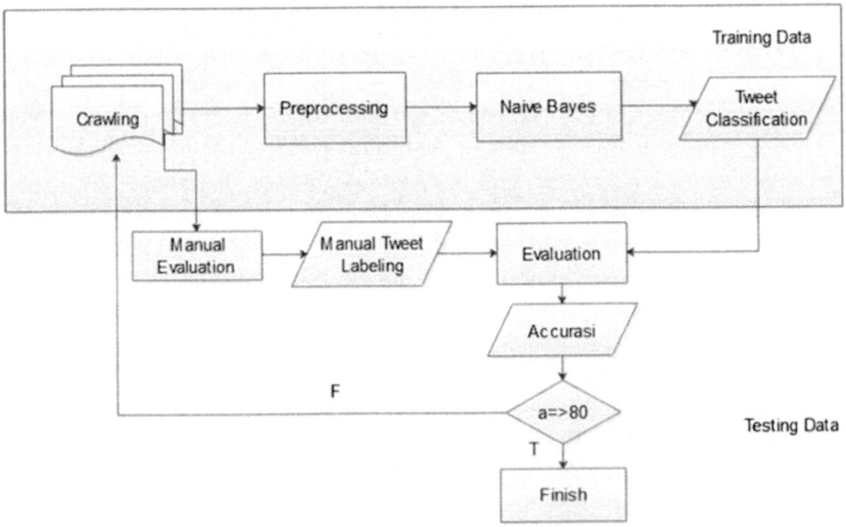

Fig. 1 Example of sentiment analysis workflow using Naive Bayes method

new training dataset needs to be added or not to reach the accuracy threshold of 80%.

Dataset. How do we get the dataset is depicted in Fig. 2. The dataset that we analyse is in Indonesian language. Firstly, the tweet was queried by using the hashtag #kpujangancurang.

Then, the queried data is crawled by using RapidMiner. The result from the query is divided into two parts: training data and testing data. Testing data is classified by using the classifiers, and then, the result was marked with negative and positive sentiment label, whereas the training data is classified manually and the result was marked the same way as testing data is treated. Training data will be used during the evaluation to determine the accuracy of the result. Table 1 shows example of evaluation of the predicted sentiment by the classifiers.

Preprocessing. This process is an important step for the next step which disposes the non-useful attribute that can be noise for the classification process. Data that is imported in this process is a raw data; thus, the result of this process is a high-quality document expected to ease the classification process. Preprocessing process is depicted in Fig. 3.

This step is started with punctuation removal, case folding, tokenizing, and finally stopword removal which is intended to remove words that are not relevant with the topic. If in the tweet document exists irrelevant words, then these words will be removed. An example of each stage of the preprocessing process is listed in Table 2. The detailed preprocessing stage is as follows:

- Removing punctuation. This stage is the initial process in order to get pure text containing words only so that further processing becomes easier.

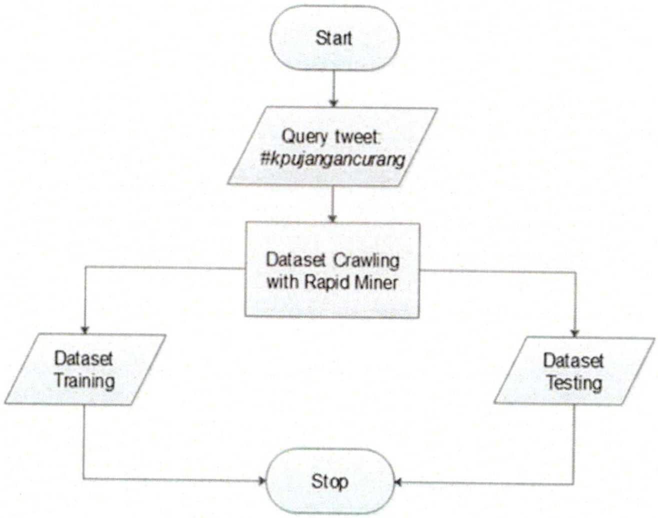

Fig. 2 Dataset flow

Table 1 Example of evaluation of the predicted sentiment

Testing data	Predicted sentiment	Manually labeled sentiment	Accuracy
kalau terus melanggar, hukuman- nya segera diterapkan	Positive	Positive	Accurate
kalau bersih kenapa takut audit forensic	Negative	Negative	Accurate
harus banyak belajar ke @BKNgoid dalam hal penyelenggaraan akbar	Positive	Positive	Accurate
Kebenaran meninggikan derajat bangsa tetapi dosa adalah noda bangsa	Negative	Positive	Inaccurate

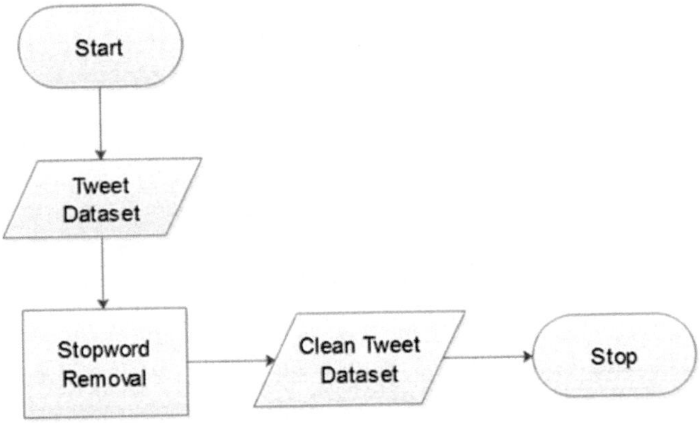

Fig. 3 Preprocessing flow

Table 2 Preprocessing process example

Stage	Before	After
1	Benar juga, kpu yang membuat rakyat resah. Aduh kejamnya kecu-. rangan	Benar juga kpu yang membuat rakyat resah Aduh kejamnya kecu- rangan
2	Benar juga kpu yang membuat rakyat resah Aduh kejamnya kecu- rangan	benar juga kpu yang membuat rakyat resah aduh kejamnya kecu- rangan
3	benar juga kpu yang membuat rakyat resah aduh kejamnya kecu- rangan	-benar- -juga- -kpu- -yang- -membuat- -rakyat- -resah- -aduh- kejamnya- -kecurangan-
4	-benar- -juga- -kpu- -yang- - - membuat- -rakyat- -resah- -aduh- -kejamnya- -kecurangan-	-benar- -kpu- -membuat- -rakyat- resah- -kejamnya- -kecurangan-

- Case Folding. This stage is the process of changing uppercase letters to lowercase letters.
- Tokenizing. In this stage, each word will be separated based on the specified space.
- Filtering. This stage is the removal of unimportant words based on Indonesian stopwords.

Term Frequency-Inverse Document Frequency (TF-IDF). After doing the preprocessing, the next step is to weight the words using the TF-IDF calculation. TF-IDF is a way of giving the weight of a word (term) to words. For single words, each sentence is considered as a document. The following is an example of TF-IDF calculation. The example of documents that want to be weighted is shown in Table 3, and the sample TF-IDF result of Document A is shown in Table 4.

Classifier. The last step is classifying the weighted data with Naive Bayes, k-nearest neighbor, decision tree and multi-layer perceptron classification methods. To evaluate which classifiers are best for scarce corpus, we experimented by changing the size of the training–testing data split from 0.25–0.75 to 0.75–0.25. The evaluation is done by measuring the accuracy of the classifiers for each scenario as shown in Fig. 4.

Table 3 Example of documents

Tweet document	Text
Document A	Jangan ancam rakyat, rakyat indonesia pintar
Document B	Rakyat tidak pernah gagal bernegara, pemerintah yang gagal bernegara
Document C	Suara rakyat dicuri, bagaimana uang rakyat

Table 4 TF-IDF score of document A

Word	TF	IDF	Weight
ancam	1	0.477	0.477
bernegara	0	0.176	0
gagal	0	0.176	0
jangan	1	0.477	0.477
rakyat	0.4	−0.2218	−0.0887
indonesia	1	0.477	0.477
pintar	1	0.477	0.477
tidak	0	0.477	0
pernah	0	0.477	0
pemerintah	0	0.477	0
dicuri	0	0.477	0
bagaimana	0	0.477	0
uang	0	0.477	0

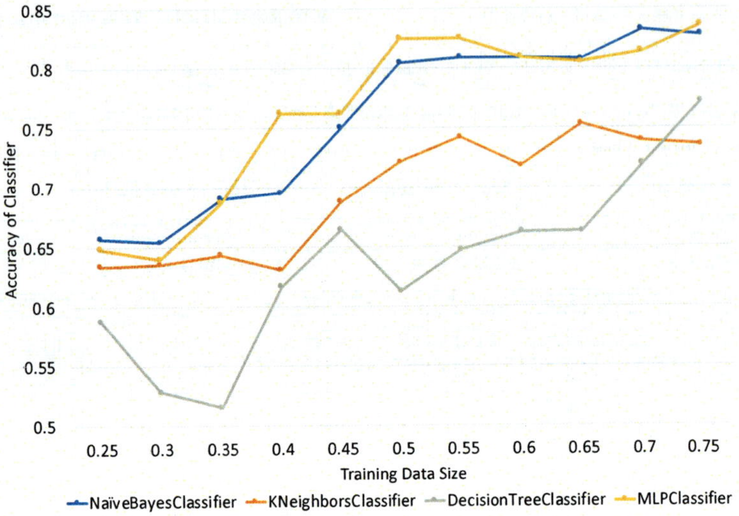

Fig. 4 Accuracy comparison of classifiers

4 Result

We obtained 200 Twitter data using RapidMiner. From the 200 Twitter data, we conducted 11 experiments with different size of training–testing data split. Every classifier shows a trend of increased accuracy on larger size of training data. However, Naive Bayes and multi-layer perceptron classifier outperformed the other two methods in overall experiment as shown in Fig. 4. Decision tree classifier shows a very low performance on small data, while k-nearest neighbor classifier shows accuracy below 0.76 on all combination size of training–testing data split. Both Naive Bayes and multi-layer perceptron classifier have the highest accuracy on all combination size of training–testing data split and show consistent increased of accuracy as the training data size is increased.

5 Conclusion

We have built a sentiment analyser to identify users' sentiment from Twitter hashtag #kpujangancurang toward the General Election Commission. We use the hashtag to obtain a set of data from Twitter to analyse and investigate further the positive and the negative sentiment of the users from their tweets. This research utilizes RapidMiner tool to generate the Twitter data and comparing Naive Bayes, k-nearest neighbor, decision tree and multi-layer perceptron classification methods to classify the sentiment of the Twitter data. There are overall 200 labeled data in this experiment. Overall, Naive Bayes and multi-layer perceptron classifier outperformed the other two methods on 11 experiments with different size of training–testing data split. The two classifiers are potential to be used in creating sentiment analyser for low-resource languages with small corpus. In our future work, we will compare the accuracy of both Naive Bayes and multi-layer perceptron classifier on bigger size of corpus.

Acknowledgements This research is funded by Universitas Islam Riau.

References

1. C.C. Aggarwal, C. Zhai, *Mining Text Data*. Springer Science & Business Media (2012)
2. M. Amer, M. Goldstein, Nearest-neighbor and clustering based anomaly detection algorithms for RapidMiner, in *Proceedings of the 3rd RapidMiner Community Meeting and Conference (RCOMM 2012)* (2012), pp. 1–12
3. S. Gupta, S. Jain, S. Gupta, A. Chauhan et al., Opinion mining for hotel rating through reviews using decision tree classification method. Int. J. Adv. Res. Comput. Sci. **9**(2), 180 (2018)
4. B. Liu, Sentiment analysis and opinion mining. Synthesis Lect. Human Language Technol. **5**(1), 1–167 (2012)
5. S. Mukherjee, A. Adhikari, M. Roy, Malignant melanoma detection using multi layer perceptron with optimized network parameter selection by pso, in *Contemporary Advances in Innovative and Applicable Information Technology* (Springer, Heidelberg, 2019), pp. 101–109
6. A.H. Nasution, Pivot-based hybrid machine translation to support multilingual communication for closely related languages. World Trans. Eng. Technol. Educ. **16**(2), 12–17 (2018)
7. A.H. Nasution, Y. Murakami, T. Ishida, Constraint-based bilingual lexicon induction for closely related languages, in *Proceedings of the Tenth International Conference on Language Resources and Evaluation (LREC 2016)*, Paris, France (May 2016), pp. 3291–3298
8. A.H. Nasution, Y. Murakami, T. Ishida, A generalized constraint approach to bilingual dictionary induction for low-resource language families. ACM Trans. Asian Low-Resour. Lang. Inf. Process. **17**(2), 9:1–9:29 (Nov 2017). https://doi.org/10.1145/3138815
9. A.H. Nasution, Y. Murakami, T. Ishida, Plan optimization for creating bilingual dictionaries of low-resource languages, in *2017 International Conference on Culture and Computing (Culture and Computing)* (Sept 2017), pp. 35–41. https://doi.org/10.1109/Culture.and.Computing.2017.21
10. A.H. Nasution, Y. Murakami, T. Ishida, Similarity cluster of Indonesian ethnic languages, in *Proceedings of the First International Conference on Science Engineering and Technology (ICoSET 2017)*, Pekanbaru, Indonesia (November 2017), pp. 12–27
11. A.H. Nasution, Y. Murakami, T. Ishida, Designing a collaborative process to create bilingual dictionaries of Indonesian ethnic languages, in *Proceedings of the Eleventh International Conference on Language Resources and Evaluation (LREC 2018)*. European Language Resources Association (ELRA), Paris, France (May 2018), pp. 3397–3404
12. A.H. Nasution, Y. Murakami, T. Ishida, Generating similarity cluster of Indonesian languages with semi-supervised clustering. Int. J. Electrical Comput. Eng. (IJECE) **9**(1), 1–8 (2019)
13. A.H. Nasution, N. Syafitri, P.R. Setiawan, D. Suryani, Pivot-based hybrid machine translation to support multilingual communication, in *2017 International Conference on Culture and Computing (Culture and Computing)* (Sept 2017), pp. 147–148. https://doi.org/10.1109/Culture.and.Computing.2017.22
14. A. Pak, P. Paroubek, Twitter as a corpus for sentiment analysis and opinion mining. LREc **10**, 1320–1326 (2010)
15. R. Prabowo, M. Thelwall, Sentiment analysis: a combined approach. J. Informetrics **3**(2), 143–157 (2009)

16. N.D. Putranti, E. Winarko, Analisis sentimen twitter untuk teks berbahasa Indonesia dengan maximum entropy dan support vector machine. IJCCS (Indonesian J. Comput. Cybern. Syst.) **8**(1), 91–100 (2014)
17. R. Setiawan, Performance comparison and op-text document classification using k-NN and classification techniques. Proc. Comput. Sci. **116**, 107–112 (2017). https://doi.org/10.1016/j.procs.2017.10.017
18. S.F. Rodiyansyah, E. Winarko, Klasifikasi posting twitter kemacetan lalu lin-tas kota bandung menggunakan naive bayesian classification. IJCCS (Indonesian J. Comput. Cybern. Syst.) **6**(1) (2012)
19. Q. You, Sentiment and emotion analysis for social multimedia: methodologies and applications, in *Proceedings of the 2016 ACM Multimedia Conference (MM'16)* (2016), pp. 1445–1449. https://doi.org/10.1145/2964284.2971475, https://dl.acm.org/citation.cfm?doid=2964284.2971475
20. H. Yuan, Y. Wang, X. Feng, S. Sun, *Sentiment Analysis Based on Weighted Word2vec and Att-LSTM*, pp. 420–424 (2019). https://doi.org/10.1145/3297156.3297228
21. L. Zhang, S. Wang, B. Liu, Deep learning for sentiment analysis: a survey. Wiley Interdisc. Rev. Data Mining Knowl. Discovery **8**(4), e1253 (2018)

DGA Method Based on Fuzzy for Determination of Transformer Oil Quality

Obhi Thiessaputra, Muhamad Haddin, and Sri Arttini Dwi Prasetyowati

Abstract A common problem in power transformers is the emergence of operational failures, thermal and electrical failures. Thermal and electrical failures generally produce dangerous gases, commonly known as fault gases. Power transformers use oil as an insulator as a cooling system as well as to dissolve harmful gases that do not circulate freely. The solution for this problem is to keep the quality of the oil. This paper presents the DGA method based on fuzzy for determination of transformer oil. DGA becomes an indicator to determine a quality of transformer oil based on fuzzy. The DGA contains three analytical method: total dissolved combustible gas (TDCG), Roger's ratio and Duval triangles used as benchmarks for analyzing test results. Fuzzy is used to determine the output composition from three oil conditions. The analysis result shows that the proposed method is capable of determining the quality of transformer oil.

Keywords Transformer oil quality · DGA · Fuzzy logic

1 Introduction

A common problem in transformer is the emergence of operational failure, both thermal failures and electrical failures. Thermal failure occurs due to the presence of electromotive forced which causes high temperature in the iron core [1], and this thermal failure is directly proportional to the high load processed by the transformer. Due to the high electrical load, the transformer oil has to work extra as a cooling

O. Thiessaputra · M. Haddin · S. A. D. Prasetyowati (✉)
Electrical Engineering, Universitas Islam Sultan Agung, Jl. Raya Kaligawe Km. 4, 5011 Semarang, Indonesia
e-mail: arttini@unissula.ac.id

O. Thiessaputra
e-mail: thiessaputra@std.unissula.ac.id

M. Haddin
e-mail: Haddin@unissula.ac.id

P. K. Pattnaik et al. (eds.), *Proceedings of International Conference on Smart Computing and Cyber Security*, Lecture Notes in Networks and Systems 149,
https://doi.org/10.1007/978-981-15-7990-5_11

system and also as an isolation system. To overcome this problem, several tests must be conducted related to the optimization of transformer oil, among others, testing the breakdown voltage, testing the viscosity and water content, testing acid levels and dissolved gas analysis (DGA) [2]. Some previous researches of determination of oil quality of transformers are researched on failure indicators with DGA [3], DGA based on thermal camera image data [4], failure analysis DGA based on j48 algorithm mining [5] and main parameter test for oil condition analysis [6]. This current research focused on the determination of transformer oil quality using DGA method based on fuzzy. The purpose of DGA test is to find out an indication of failure that occurs in oil samples. So that it can help extend the life of the transformer. As the development of engineering software, artificial intelligence (AI) is also used as a tool in decision making, and fuzzy logic is one of the tools that can help in decision making.

2 Methods

A transformer is a device that functions to move alternating current power from one circuit to another through a magnetic field and based on electromagnetic principles.

The main function of the power transformer is to decrease and increase the electric power. Power transformer has some auxiliary tools such as conservator tank, heat exchanger and fans. These auxiliary tools are used as a cooling system in power transformer, which uses oil as a medium to release the heat. To avoid a thermal fault in transformer, oil condition must be tested by some methods such as breakdown voltage, acid level, viscosity and dissolved gas analysis (DGA) [7].

2.1 *Dissolved Gas Analysis (DGA)*

DGA can literally be interpreted as an analysis of the condition of the transformer which is based on the amount of gas dissolved in the transformer oil. DGA in the industrial world is also known as a blood test on transformer. Testing of dissolved substances (usually gas) in transformer oil (transformer oil analogous to human blood) will provide information related to the health and overall quality of work of the transformer.

From various cases of failures that occur in the transformer and detected through the DGA test, the failures in the transformer can be classified into several classes such as partial discharge (PD), discharge of low energy (D1), discharge of high energy (D2), thermal fault $\leq$300 °C (T1), thermal fault, $300 \leq T \geq 700$ °C (T2) and thermal fault >700 °C (T3).

There are several methods to conduct the analysis as stated in IEE std. c57-104.1991 and IEC 60599, which are total dissolved combustible gases (TDCG), Duval triangle and Roger's ratio.

1. **Total Dissolved Combustible Gases (TDCG)**
   In the use of TDCG, the total amount analysis of combustible dissolved gas will indirectly show the condition of the transformer as shown in Table 1.
2. **Roger's Ratio Method (RRM)**
   In the analysis process, the dissolved gas formula equation is needed. The following is the ratio table used to determine some of the results from DGA manual method. If the results of the analysis are not in accordance with Table 3 thus, it is considered, the diagnosis is not fulfilled as shown in Table 2.
   Table 3 is the failure indicator for Roger's ratio and going to work after the calculation has been done. The failure indicator has codes to match with the result of the calculation.
3. **Duval Triangle Method (DTM)**
   DTM is one of the methods that can visualize clearly the location of the coordinates of a transformer oil failure indication by utilizing the dissolved content as shown in Fig. 1.

Table 1 IEEE standard (TDCG) [8, 9]

Status	Description
Condition 1	Normal
Condition 2	TDCG levels start high
Condition 3	Beware of isolation decomposition
Condition 4	Insulation damage

Table 2 Roger's ratio code

Code ratio	C_2H_2/C_2H_4	CH_4/H_2	C_2H_4/C_2H_6
<0.1	0	1	0
0.1–1	1	0	0
1–3	1	2	1
>3	2	2	2

Table 3 Failure codes based on Roger's ratio table

Failure code			
No *Fault*	0	0	0
PD	1	1	0
D1	1–2	0	1–2
D2	1	0	2
T1	0	2	0
T2	0	2	1
T3	0	2	2

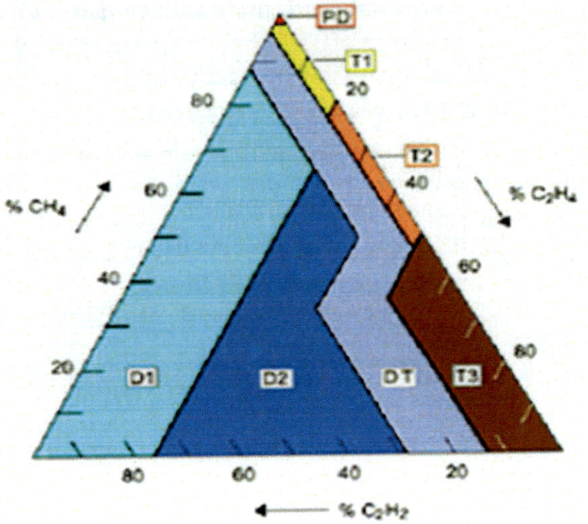

Fig. 1 Duval's triangle

To get the coordinate point on the Duval triangle map, the calculation of the dissolved gas content is then drawn straightly from three directions so that it will be centered on one coordinate point.
Triangle coordinate

$$\begin{aligned} \%CH_4 &= CH_4/(CH_4 + C_2H_4 + C_2H_2) * 100\% \\ \%C_2H_4 &= C_2H_4/(CH_4 + C_2H_4 + C_2H_2) * 100\% \\ \%C_2H_2 &= C_2H_2/(CH_4 + C_2H_4 + C_2H_2) * 100\% \end{aligned} \quad (1)$$

2.2 *Fuzzy Logic*

To find out the quality of the four oil samples, the researcher used fuzzy as a supporting medium in making decision. Fuzzy logic is a tool that is used to determine a complicated system which is difficult to be modeled or there is an abundance of ambiguity on the system [10]. To get one oil quality, there are three inputs needed from the DGA method, such as Roger's ratio, Duval triangle and TDCG. In the determination stage, there are four stages: fuzzification, implication, composition and defuzzification. The scheme of the oil quality function is shown in Fig. 2.

3 Results and Analysis

It contains all results and analysis from each method includes fuzzy determination.

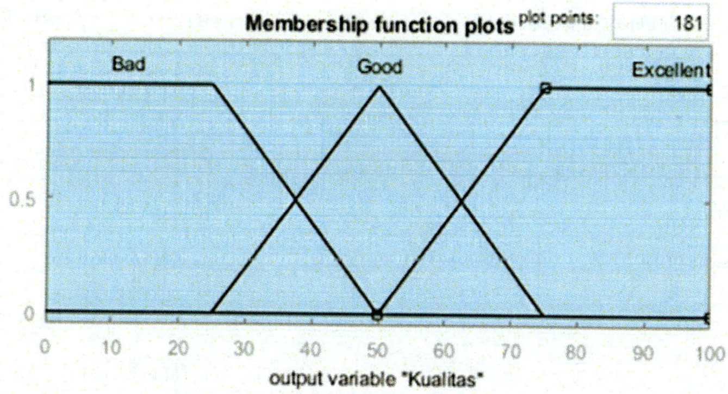

Fig. 2 Quality membership function

3.1 DGA Result

Dissolved gas analysis (DGA) was conducted to detect the quality of the content of some specific types of gas from an oil sample. The parameters that must be considered from the data are the concentration values of various types of fault gases are (hydrogen, methane, ethane, ethylene, acetylene, carbon monoxide and carbon dioxide) the amount of water content (moisture), TDCG values. In addition to the dissolved gas content, oil temperature data was also needed, and thus it was the temperature data in the hotspot as shown in Table 4.

A. **Analysis of Roger's Ratio Method (RRM)**
Analysis of sample data from DGA test results used Roger's ratio approach. The results of the calculation from Roger's ratio method are shown in Table 5.

Table 4 Oil temperature in transformer hotspot

No.	Oil samples	Temperature (°C)
1	GI Srondol oil sample	232
2	GI Srondol 2 oil sample	739
3	GIS Kalisari oil sample	1071
4	GIS Kalisari 2 oil sample	826

Table 5 Roger's ratio result

Ratio	Sample 1	Sample 2	Sample 3	Sample 4
Ratio 1	0	0	0	0
Ratio 2	2	0	0	0
Ratio 3	0	2	2	2
Fault	T1	Unknown	Unknown	T3

Table 6 Duval's triangle method (DTM) result

Oil sample	The percentage of dissolved gases			Fault
	CH_4 (%)	C_2H_4 (%)	C_2H_2 (%)	
Sample 1	92.7	7.3	0	T1
Sample 2	4.8	94	1.2	T3
Sample 3	28.3	71.7	0	T3
Sample 4	41.6	58.4	0	T3

B. **Analysis of Duval's Triangle Method (DTM)**
The results from DGA used the DTM to complete the failure diagnosis of the samples. This method used a specific Duval's calculator; therefore, this analysis can be called a digital analysis. The fault indicator for Duval's triangle is a kind of same as Roger's ratio which are T1, T2, T3, etc. The results for DTM are shown in Table 6.

C. **Determination of transformer oil condition based on TDCG**
Total dissolved combustible gases for this method have a limitation as shown in Table 7, and the gases are H_2, CH_4, C_2H_2, C_2H_4, C_2H_6 and CO.
Based on the limitation, dissolved gas concentration in Table 10 sample 1 is in CONDITION 1. Description on oil condition referring to Table 2 is normal.
Based on the limitation, dissolved gas concentration in Table 10 sample 2 is in CONDITION 2. Description on oil condition referring to Table 2 is that the TDCG level starts to be high.
Based on the limitation, dissolved gas concentration in Table 10 sample 3 is in CONDITION 2. Description on oil condition referring to Table 2 is that the TDCG level starts to be high.
Based on the limitation dissolved gas concentration in Table 10 sample 4 is in CONDITION 2. Description on oil condition referring to Table 2 is that the TDCG level starts to be high.

Table 7 Dissolved gas concentration limit

Concentration limit for dissolved gas							
Status	H_2	CH_4	C_2H_2	C_2H_4	C_2H_6	CO	TDCG
Condition 1	100	120	35	50	65	350	<720
Condition 2	101–700	121–400	36–50	51–100	65–100	351–570	721–1920
Condition 3	701–1800	401–1000	51–81	101–200	101–150	571–1400	1921–4630
Condition 4	>1800	>1000	>80	>200	>150	>1400	>4630

Table 8 Fuzzy output composition

No.	Quality	Range
1.	*Excellent*	50–100
2.	*Good*	25–75
3.	*Bad*	0–50

3.2 Fuzzy Determination

From the composition of fuzzy output, the range of points from the merging of the three DGA methods can be seen. Therefore, the intersection point and the degree of membership can be determined using the centroid approach with the implication of minimum. The composition of fuzzy output for the oil quality is shown in Table 8.

3.3 Fuzzification Sample

The final step is fuzzification, which is the set of fuzzy becomes real numbers. After getting the results of the numbers on the application of fuzzy, the quality of oil is obtained with the centroid approach. Thus, here are the following results:

- Sample 1 shows a value of 62 with the implication of min and the degree of membership is in excellent quality based on the centroid approach on fuzzy Mamdani
- Sample 2 shows a value of 24.5 with the implication of min and the degree of membership is in bad quality based on the centroid approach on fuzzy Mamdani
- Sample 3 shows a value of 20.2 with the implication of min and the degree of membership is in bad quality based on the centroid approach on fuzzy Mamdani
- Sample 4 shows a value of 21.9 with the implication of min and the degree of membership is in bad quality based on the centroid approach on fuzzy Mamdani.

PLN is a state-owned electricity service company. Based on validation data from PLN, there are similarities in the results of the determination that has been made using DGA method based on fuzzy logic as shown in Table 9.

Table 9 Quality results validation

Sample	Oil quality based on fuzzy	PLN P3B JB validity analysis	
		Classification	Category (1/6/9)
1	*Excellent*	1	9—Normal
2	*Bad*	2c	6—Slight Abnormality
3	*Bad*	4	1—Obvious Abnormality
4	*Bad*	2c	6—Slight Abnormality

4 Conclusion

Based on the results of determining oil quality analysis in four units transformer using the DGA method based on fuzzy, conclusions can be drawn as follows:

1. Based on the result and validation data, it shows that DGA method based on fuzzy is capable of determining the quality of transformer oil.
2. Based on the analysis of dissolved gas, the results of the methods have different accuracy to indicate a failure of the transformer. Therefore, it is not enough to only refer to one method to determine the indication of failure caused by transformer oil.
3. Based on DGA test, oil samples from the four units transformer indicate overheat failure.

References

1. S. Salvi, A.P. Paranjape, Study of transformer oil purification (2017)
2. L.M. Adesina, K. Saadu, G.A. Ajenikoko, Transformer oil regeneration as a Panacea for electric power utility company's equipment optimization (2019)
3. A.R. Muhammad Faishal, Analisis Indikasi Kegagalan Transformator dengan Metode Dissolved Gas Analysis (2011)
4. S.A. Aziz, V.L.B. Putri, A. Priyadi, Analisis Kualitas Minyak Transformator Daya 25kva berdasarkan Data Citra Kamera Termal dan Data Hasil Uji *Gas Chromatograph* (2012)
5. A. Pramono, D. Nugroho, M. Haddin, Analisis Minyak Transformator Daya Berdasarkan DGA menggunakan Data *Minning* dengan Algoritma J48 (2016)
6. G. Ilham, M. Setiawan, I. Garniwa, Analisis kondisi minyak Trafo berdasarkan uji parameter utama (2013)
7. A. Nurrahman, R. Maulana S. H., S. Sulaiman, "Transformator Daya," 2015
8. IEEE, *Guide for the Interpretation of Gases Generated in Oil-Immersed Transformers*, IEEE Std C57.104-2008 (2008)
9. IEEE, *Guide for Acceptance and Maintenance of Insulating Oil in Equipment*. IEEE Std C57.106-2006 (2006)
10. Fahmizal, *Fuzzy Logic And Fuzzy Logic Sun Tracking Control*. [Online]. Available: https://fahmizaleeits.wordpress.com/2011/02/25/paper-Fuzzy-logic-and-Fuzzy-logic-sun-tracking-control/. Accessed 20 Aug 2019

Deep Learning-Based Apple Defect Detection with Residual SqueezeNet

M. D. Nur Alam, Ihsan Ullah, and Ahmed Abdulhakim Al-Absi

Abstract Apple defect detection using hyperspectral image has become an interesting research focus since the last decade. It is important for wide range of applications, such as agricultural sector, food processing, and automatic fruits grading system. However, this task is a challenging one due to calyx and stem, their different types and orientation, and similar in their visual appearance. The proposed method based on a deep learning approach using SqueezeNet architecture. However, the apple images are extracted and fed into a deep network for training and testing. The proposed SqueezeNet architecture utilizes convolution neural network to regress a bypass connection between the fire modules across the images. It has been evaluated our own created dataset. The excremental result shows that our proposed methods are efficient and effective. The general detection rate was 92.23%.

Keywords Apple detection · Defect detection · Deep learning · Residual SqueezeNet

1 Introduction

Apples are one of the most common fruits in the world. The cultivation rate of apple with respect to all the other fruits is 21.3% in Korea, which is the highest fruit cultivation ratio of Korea. There are many works related to the apple cultivation. For a long time, the apple industry has been doing a time-consuming manual fruit processing to provide high-quality apples to consumers. But, in these days, automatic apple processing began to be used to decrease the production costs and to compete

M. D. Nur Alam (✉) · A. A. Al-Absi
Department of Smart Computing, Kyungdong University, Global Campus (Gosung) 46 4-gil, Bongpo, Gosung, Gangwondo 24764, South Korea
e-mail: na@kduniv.ac.kr

I. Ullah
Department of Robotics Engineering, Daegu Gyeonbuk Institute of Science and Technology, Daegu 42988, South Korea

P. K. Pattnaik et al. (eds.), *Proceedings of International Conference on Smart Computing and Cyber Security*, Lecture Notes in Networks and Systems 149,
https://doi.org/10.1007/978-981-15-7990-5_12

with countries that have lower production costs. There is continuing a worldwide competition to develop automatic apple classification systems that are more capable, efficient, and accurate. This is because manual fruit grading has drawbacks such as high price, inconsistency, subjectivity, and fatigue.

Hyperspectral imaging systems (HSI) were originally developed to use in the areas of astronomy, solar physics, and earth remote sensing. It takes advantage of the large spatial sampling capability among the different spectra in a neighborhood, but the disadvantage is its complexity and large size of the data. As those disadvantages were solved by the popularizing of high-speed computers with large storage and image processing software, HSI techniques were implemented in many agricultural applications software such as the classification of defects or quality control of apples, eggplants, pears, cucumbers, and tomatoes [1, 2].

Image is an important media of information source for human communication and understanding. To achieve better results, the artificial intelligence-based application is introduced in image processing [3]. In the convolutional neural network, semantic segmentation is important in high-level computer vision applications. Machine learning algorithms are more capable of handling the inhomogeneous intensity of images [4]. The fruit surface defects are classified using radial basis probabilistic neural network. The attributes of the defected area are extracted by manipulating a gray level co-occurrence matrix [5]. It has been observed that the convolutional neural network (CNN) detects mangosteen defects with an accuracy of 97% [6]. The fruit quality depends on the defects type, fruit size, and its skin color. The fruits maturity and grading into their relevant category are attained by developing a pattern recognition system. The system involves two stages. First, training and then recognition of pattern by using the backpropagation diagnosis model [7]. A multi-level thresholding algorithm improves the segmentation response and reduces the processing time. Then, region of interest is analyzed for identification of defects in fruit surface [8]. A unified convolutional neural network centered on a matrix-based convolution neural network is implemented to overcome this challenge [9]. CNN requires a large number of datasets to facilitate feature extraction and classification. This problem is overcome by employing transfer learning that involves a pre-trained CNN. The datasets are sufficient for classification later optimized explicitly to provide a consistent result [10]. The external properties such as shape, size, and color aid for quality inspection of fruits. The local feature extraction process describes object recognition for rapid quality inspection [11]. A deep convolutional neural network improves image classification accuracy. The classification framework is independent and minimizes the loss between class similarity and within-class variance [12]. Deep learning classification is very challenging for high precision diagnosis.

A lot of researches has been carried out on apple detection and fruits grading in hyperspectral images over the years. These works can be deep learning-based methods. In this section, we briefly introduce the latest works carried out in our proposed method. We solve the problem of apple detection and counting as a supervised learning problem. We solve the problem by utilizing the convolutional neural network (CNN). Thus, in this study, a smaller CNN architecture named SqueezeNet is

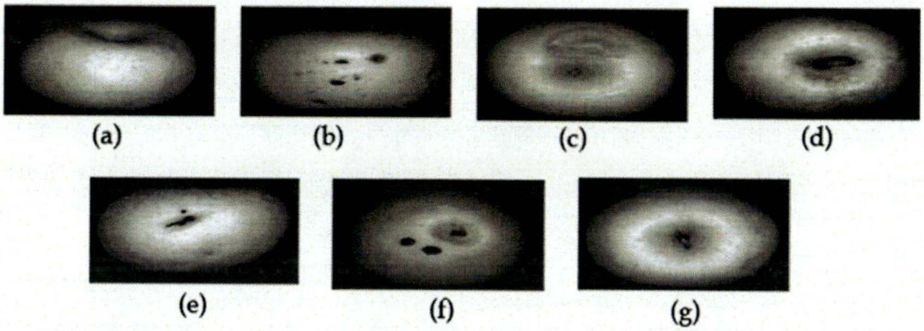

Fig. 1 Dataset with different apple skin types: normal **a** scab **b** bruise **c** crack **d** cut **e** defect with stem **f** and calyx **g** Images are from the 729.3 nm wavelength

adopted, which requires much fewer parameters, using a bypass connection-inspired version of SqueezeNet, and improvement in recognition accuracy.

The rest of the paper is organized as follows: Sect. 2 excremental information. Section 3 describes the proposed system. Section 4 introduces datasets, evaluation procedures, and experimental results. This paper is concluded in Sect. 5.

2 Experimental Information

2.1 Apple Collection

We used red delicious apples were used in this study. The image acquisition and samples were provided by the Agro-Food Quality & Safety Sensing Lab from the Rural Development Administration (RDA) in the Republic of Korea (ROK). A total of 550 samples were separated into normal and defect apples using a visual inspection. The apple samples comprise groups with normal skin and six common skin-defect conditions. Fig. 1 shows a sample with the common defects scab, bruise, crack, cut, defect with stem, and calyx. A total of 550 red delicious apples, 90 normal-skin samples and 460 defect apple samples were used to develop the algorithm. Figure 2 shows the conceptual view of a hyperspectral imaging system [13].

3 The Proposed Method

3.1 Residual SqueezeNet Architecture

Firstly, we created the patches using sliding window technique. With this technique, each image is decomposed into small patches of size 227 × 227. After creating

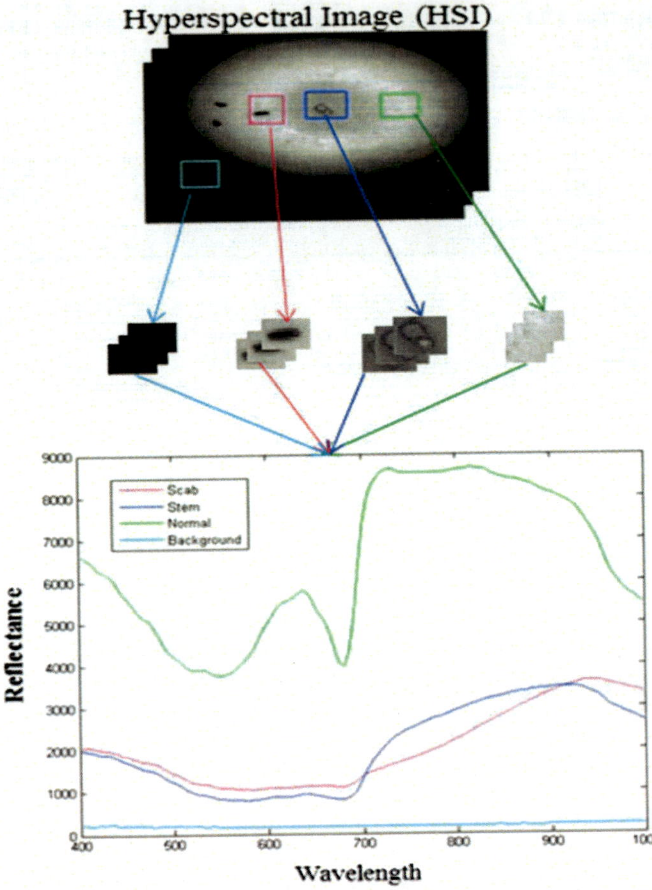

Fig. 2 Conceptual view of a hyperspectral imaging system of apple different kinds of skin condition with spectral reflectance (403–998 nm)

patches, we feed these patches into a deep neural network that recognizes apple defects classes, including backgrounds. A smaller CNN called SqueezeNet [14] is used in this study. This can achieve performance comparable to other CNN frames like AlexNet while having less than real-time parameters.

The basic SqueezeNet begins with a convolution layer, followed by eight fire modules, and ends with an extra convolution layer, shown in Fig. 3a. The number of filters per fire module is gradually increased from the start to the end of the network. The two-stage max-pooling occurs after layers conv1, fire4, fire8, and conv10. ReLU is adopted as the activation function and 0.5 dropout after the fire9 module is used.

To boost recognition performance, the updated SqueezeNet is built by inserting a few simple bypass connections to the SqueezeNet between certain fire modules. The bypass connections around fire modules 3, 5, 7, and 9 are added to our simple bypass architecture, allowing these modules to learn a remaining feature between input and

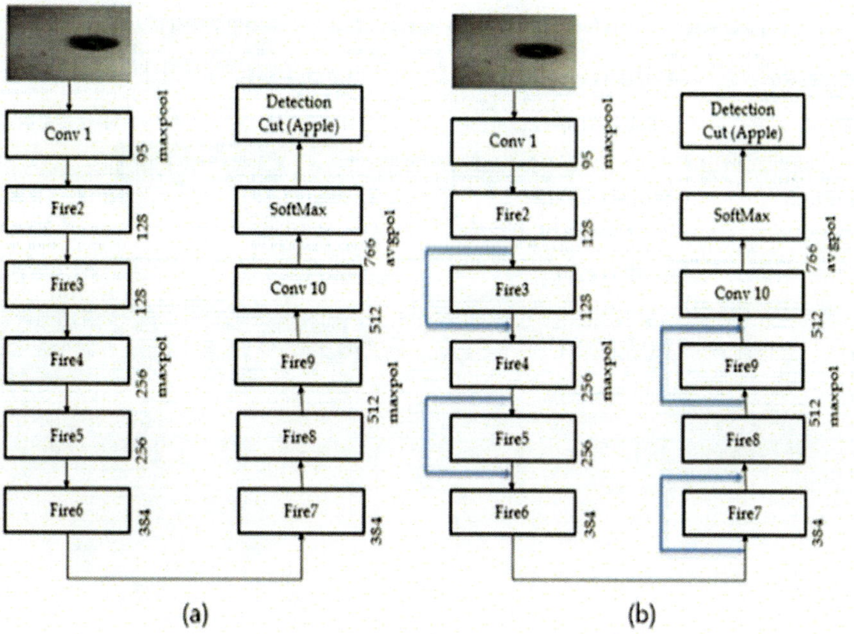

Fig. 3 Illustration of SqueezeNet-based MMR system. **a** SqueezeNet architecture. **b** Our proposed residual SqueezeNet architecture

output as shown in Fig. 3b. Just like in ResNet, we set the Fire4 input for the Fire2 + output of Fire3 to create the bypass link around Fire3, where the + operator is an element-wise addition as shown in Fig. 4. This improves the regularization of these fire modules' parameters and will increase the precision and/or the trainability of the entire model according to the ResNet. Throughout training, the weight parameters are initialized randomly with a Gaussian distribution. Initially, the learning rate is set as 0.01 and decreases with step size 10. Furthermore, the RMSprop is the optimizer.

3.2 The Hardware and Software

In our setup, we used a GeForce GTX 1080 to experience accelerated training in deep learning frameworks with Cuda and cuDNN. The CPU used is the i7-4790 Intel(R) core with eight cores and 3.60 GHz.

Nvidia DIGITS are used as a second framework on top of Caffe or, more specifically, a separate caffe fork by Nvidia of the official repository. For training, DIGITS have on-the-fly data augmentation capabilities for random crops of fixed size, for example, by random taking 227–227 regions of a 256–256 image and by random horizontal flipping. In addition, DIGITS visualize the decrease in the learning rate. On a continuously modified dynamic graph, visualization of the loss of preparation,

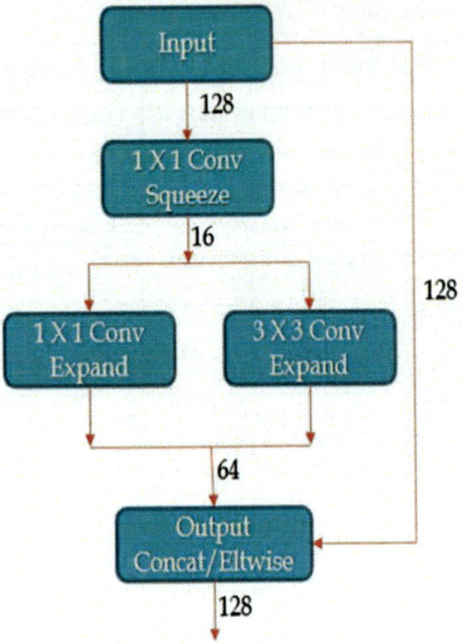

Fig. 4 Proposed residual SqueezeNet architecture with simple bypass connections

performance failure, and testing accuracy make it easier to see whether the preparation went well or not. A downside is that because DIGITS operate on a Caffe branch, new developments in the main Caffe repository can only be merged with it if DIGITS developers decide to do so.

4 Experimental Result

As shown in Table 1, we trained the SqueezeNet model with training samples of 9000, containing eight classes for 30 epochs, and validated the model with 1800 samples, containing all eight classes. The SqueezeNet model obtained a grade-1 accuracy of 83.95%. Upon adding bypass connections to fire modules 3, 5, 7, and 9, these modules learn the residual function between input and output. Interestingly, the simple bypass made it possible to increase accuracy compared to the bypass-free architecture. Adding simple bypass connections resulted in an improvement of 2.1 percentage points in accuracy without increasing the size of the model.

Table 1 Average accuracy of patch-wise input and entire image

Inputs images	SqueezeNet	Residual SqueezeNet
Entire apple image	83.96	86.72
Patch	**91.97**	**92.23**

Table 2 Accuracy of two folds with patches and entire image

Inputs images	SqueezeNet	Residual SqueezeNet
Entire apple image -Fold1	84.23	86.33
Entire apple image -Fold2	83.68	87.10
Patch -Fold1	91.31	92.10
Patch -Fold1	92.62	92.35

To learn a robust classifier, we extract small image patches, including the apple portions, and then learn the informative spatial features for the classification network. The proposed method enables accurate classification of the classes of apple detection. Using patches instead of entire image achieves better accuracy of 91.97% using SqueezeNet. While using bypass connection achieves even more better accuracy with improvement 0.26 percentage points compared to patch-wise SqueezeNet. Table 2 shows the results of each fold of SqueezeNet and residual SqueezeNet by utilizing entire image and patches.

In this section, we introduce the dataset used for training the proposed system and the results of the proposed system in comparison with the sate-of-the-art-methods.

5 Conclusion

We proposed and investigated a highly robust and real-time automated defect detection approach in apple based on memory-efficient CNN architectures. To this end, we showed the effectiveness of SqueezeNet deep networks used to tackle challenges in apple detection. The SqueezeNet architectures allow real-time use due to compact model sizes without a significant change in recognition accuracy of 92.23%. Experimental results demonstrated the superiority of our proposed apple detection method. We intend to further optimize residual SqueezeNet architecture to achieve better accuracy for future research.

References

1. G. ElMasry, N. Wang, C. Vigneault, J. Qiao, A. ElSayed, Early detection of apple bruises on different background colors using hyperspectral imaging. LWT—Food Sci. Technol. **41**(2), 337–345 (2008)
2. P. Baranowski, W. Mazurek, J. Wozniak, U. Majewska, Detection of early bruises in apples using hyperspectral data and thermal imaging. J. Food Eng. **110**(3), 345–355 (2012)
3. X. Zhang, W. Dahu, Application of artificial intelligence algorithms in image processing. J. Vis. Commun. Image Represent. **61**, 42–49 (2019)
4. A. Pratondo et al., Integrating machine learning with region-based active contour models in medical image segmentation. J. Vis. Commun. Image Represent. **43**, 1–9 (2017)

5. Image Metrology, http://www.imagemet.com/WebHelp6/Default.htm#PnPParameters/Measure_Shape_Parameters.htm. Accessed 7 July 2019
6. G. Capizzi, G. Lo Sciuto, C. Napoli, E. Tramontana, M. Woźniak, Automatic classification of fruit defects based on co-occurrence matrix and neural networks, in 2015 *Federated Conference on Computer Science and Information Systems (FedCSIS)*, Lodz, 2015, pp. 861–867
7. L.M. Azizah, S.F. Umayah, S. Riyadi, C. Damarjati, N.A. Utama, Deep learning implementation using convolutional neural network in mangosteen surface defect detection, in *2017 7th IEEE International Conference on Control System, Computing and Engineering (ICCSCE)*, Penang, 2017, pp. 242–246
8. Z. Effendi, R. Ramli, J.A. Ghani, M.N.A. Rahman, Pattern recognition system of Jatropha curcas fruits using back propagation, in *2009 IEEE International Conference on Signal and Image Processing Applications*, Kuala Lumpur, 2009, pp. 58–62
9. O.C. Morene, D.M.M. Gila, D.A. Puerto, J.G. García, J.G. Ortega, Automatic determination of peroxides and acidity of olive oil using machine vision in olive fruits before milling process, in *2015 IEEE International Conference on Imaging Systems and Techniques (IST)*, Macau, 2015, pp. 1–6
10. Z. Lin et al., A unified matrix-based convolutional neural network for fine-grained image classification of wheat leaf diseases. IEEE Access **7**, 11570–11590 (2019)
11. S. Akcay, M.E. Kundegorski, C.G. Willcocks, T.P. Breckon, Using deep convolutional neural network architectures for object classification and detection within X-Ray baggage security imagery. IEEE Trans. Inf. Forensics Secur. **13**(9), 2203–2215 (2018)
12. Yogesh, A.K. Dubey, Fruit defect detection based on speeded up robust feature technique, in *2016 5th International Conference on Reliability, Infocom Technologies and Optimization (Trends and Future Directions) (ICRITO)*, Noida, 2016, pp. 590–594
13. M.D. Nur Alam, I. Pineda, J.G. Lim, O. Gwun, Apple defects detection using principal component features of multispectral reflectance imaging. Sci. Adv. Mater. **10**, 1051–1062 (2018)
14. G.E. Hinton, R.R. Salakhutdinov, Reducing the dimensionality of data with neural networks. Science **313**, 504–507 (2006)

Smart Parking Management System in Shopping Malls

S. Aravinthkumar, Shreya Makkar, and Ahmed Abdulhakim Al-Absi

Abstract The smart parking management system makes parking management ease in shopping mall. This system will make availability of parking during weekends/rushing hours in shopping malls. This system includes two steps. One is pre-reservation of parking for vehicle owners, and the other is parking management system for vendor. Smart parking system will provide current parking availability status for vehicle owners. If available, they can reserve the parking after making online payment or else they can find another parking area. The application used by parking management will manage all parking areas and will be used for making reservation on entry of vehicle. This will be very helpful in saving the time and a lot of efforts for finding a space to park the vehicle, especially during the weekends. User can connect to the smart parking system from anywhere with their smart phones in any browser. And this information is also available to the parking operators to determine the free parking areas, and statistics can be measured at different times in a day on each parking space.

Keywords Smart parking system · Pre-reservation · Vendor management

S. Aravinthkumar (✉)
SRM University, Sonepat, India
e-mail: tsaravintkumar@gmail.com

S. Makkar (✉)
VIT University, Vellore, India
e-mail: makkar.shreya95@gmail.com

A. A. Al-Absi (✉)
Dean of International Faculty (Academic), Director of Global Academic Collaboration Centers,
Kyungdong University, Global Campus, Yangju, South Korea
e-mail: absiahmed@kduniv.ac.kr

P. K. Pattnaik et al. (eds.), *Proceedings of International Conference on Smart Computing and Cyber Security*, Lecture Notes in Networks and Systems 149,
https://doi.org/10.1007/978-981-15-7990-5_13

1 Introduction

Parking guidance system is systems that obtain information about available parking spaces in malls, process it and then present it to drivers by means of variables message signs [1]. This system can be implemented in two ways, which guide drivers in malls to the parking facility with empty parking lots. The latter guidance system addresses driver's need for information about the position and number of the parking lots that are actually available within a parking structure [2–4]. This system reduces time and fuel while searching for empty parking lots and helps the car park to operate efficiently. Car park inside a building has become popular because it is user friendly and prevents vehicle from exposing to the sunlight, especially in the place where parking lots are in high demand. This type of car park poses a parking guidance system that primarily based on the use of message signs to give drivers information regarding parking availability inside the car park [5]. The availability of parking lot inside car park normally is obtained from the sensors that count the number of cars entering and exiting or, in other cases, by comparing the ticket issued at machines. This information is sent to a computer that processes it, and availability is generally expressed in terms of "full" or "empty." In some cases, the actual numbers of parking lots are given. Every car park owner has their different parking guidance systems [6, 7]. The wider use of high technology like sensor and wireless communication in some parking guidance systems has successfully attracted and met the requirement of the customers. Car parks and parking guidance system are listed below as follows.

1.1 Case 1: The Baltimore/Washington Airport Car Park

Smart Park, Baltimore/Washington Airport parking system is one of the examples of parking system to guide car drivers within parking facilities to empty parking lots [8, 9]. It is an automated parking guidance system intended to make finding a parking lit quick and hassle free at the Baltimore/Washington Airport. This system is currently available in all parking lots in the new Daily a Garage and on Level 2 of Hourly Garage. At car park entrance, ultrasonic sensors mounted over each parking lot are used to monitor the vacancy status of each parking lot and a display board is used to display the number of parking lots available in car park [10, 11].

1.2 Case 2: Plaza Singapore Car Park System

Plaza Singapore has introduced ST electronics' user-friendly parking guidance system. With the multiple space display boards positioned at the strategic locations, drivers can get real-time information guiding them to the areas where parking lots are available. With this system, even when car park is full, car drivers need not wait at the car park entrance and be allowed to enter the car park anytime. This is because whenever a car park lots is vacated, a sensor will automatically signal the car parking management system allow another vehicle to enter [12, 13]. This will prevent a long queue of cars from forming at the entrance point. Figure below shows at the entrance to the car park a main parking lot display board indicates the availability of parking lots for the various levels within the car park. This display board allows car drivers to know the total parking lot at each level of car park [14, 15].

1.3 The Comparison of Parking Guidance System in Car Park Case Studies

The comparison of parking guidance system in car park case studies is shown in table below. This table clearly shows some foreign car parks already improved the parking guidance system inside car park by applying sensor to detect the availability of parking lot inside car park and displayed the availability of parking lots inside car park in message display board to help car drivers to easily locate the empty parking lot. Therefore, these foreign car parks are more user friendly than some local car parks. So from the above examples, the management can be much easier in our environment. People can save lots of efforts, time, fuel, and things will go in systematic manner [16–18].

The existing system, in which the things are happening manually, is time consuming. It is inevitable for the people to update with the growing technology. And generally, people are facing problems on parking vehicles in parking slots in a city. In this study, we design a smart parking system (SPS) which enables the user to find the nearest parking area and gives availability of parking slots in that respective parking area. New approaches using smart parking systems look to provide a more balanced view of parking that better manages the relationship between supply and demand. Smart parking is also made viable by innovation in areas such as smart-phone apps for customer services, mobile payments. At the heart of the smart parking concept is the ability to access, collect, analyze, disseminate and act on information on parking usage [19–21]. Increasingly, this information is provided in real time from intelligent devices that enable both parking managers and drivers to optimize the use of parking capacity. Drawbacks of existing system are in case of no access or limited access to the internet the site will not be accessible. It may be a bit confusing for unfamiliar users. It is not recommended for high peak hour volume facilities. It is inaccessible in case of the site is down.

In this era where everything is getting digitize, this can be a new trend. Smart parking system (SPS) enables the user to find the nearest parking area and gives availability of parking slots in that respective parking area [22]. It mainly focusses on reducing the time in finding the parking lots, and also, it avoids the unnecessary traveling through filled parking lots in a parking area. Thus, it reduces the fuel consumption which in turn reduces carbon footprints in an atmosphere. Most of the existing car parking systems are manually managed and a little inefficient. In urban areas, where the number of vehicles is higher as compared to the availability of parking spaces, generally parking locations are full. Hence, online booking parking system is a proposed method where users can pre-book their parking places [23].

Car parking in the malls is very difficult to manage, and this issue occurs especially during the weekends. Car owners have to wait for long in queues to get their vehicle parked in the malls [24]. So, it is a better initiative for the car drivers to pre-book a parking area in a particular time slot to avoid the mess of standing in the long queues and waiting for long time. SPS is very effective in changing the trend.

The objective of smart parking website is booking the parking slot online the payment can be done easily, slot can be available for the required time, monitoring can be done by the parking operator to check the available slots, management task is simple for tool administrators, manual efforts and time will be reduced by automated tasks by using this tool, and ease of access for the users in GUI mode as the website will be very user friendly.

2 Materials and Methods

In this section, we give the materials and methods used to smart parking system. Architecture of the proposed method is illustrated in Fig. 1.

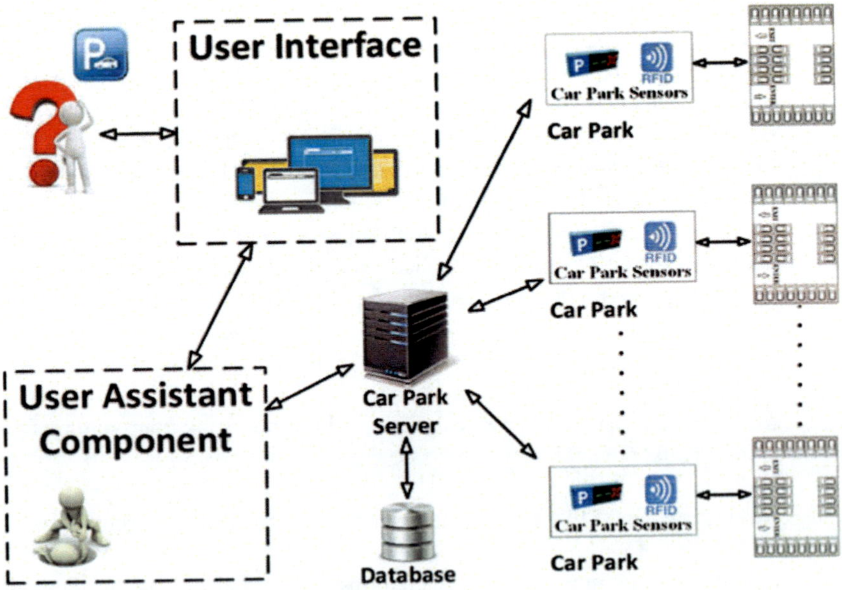

Fig. 1 Architectural diagram of proposed system

The proposed application contains two user interface, one for the vehicle owner for pre-booking parking area and another for parking management which is used by allocating parking area for vehicles. It will help in management of the long queues of the cars which are in search of the parking space. This will be helpful in the reduction of the manual efforts, as payment can be done online. The vehicle owner has to just show the booking details at the entry point. For vehicles which have not reserved their parking space will be allotted parking area on entry based on the availability. At the exit gate, the vehicle owner will be charged according to number of hours parked.

However, a smart car park system should provide more convenience and automation to both the business and customers. It should also satisfy the following requirements:

- The system should provide plenty of informative instructions or guidelines to help drivers to find an available parking lot.
- The system should provide powerful functions to facilitate administrators and managers to manage a car park.
- Reduction of manual efforts: The payment can be done online for the confirmation of the allocated area; and before the particular time, the booking can be canceled and some amount of money can be refunded back
- Decreased management costs—More automation and less manual activity save on labor cost and resource exhaustion.
- Reduced pollution—Searching for parking burns around one million barrels of oil a day. An optimal parking solution will significantly decrease driving time, thus lowering the amount of daily vehicle emissions and ultimately reducing the global environmental footprint.

2.1 *Advantages of Proposed System*

- There is a greater sense of security due to the fact that patrons do not actually walk to and from their own space.
- It is highly feasible for extremely small sites that are unable to accommodate a conventional parking structure.
- There is high parking efficiency (i.e., sf/space and cf/space).
- There is no need for driving while looking for an available space.
- Emissions are greatly brought down and reduced.
- The patrons wait for their car in a highly controlled environment.
- There are less chances for vehicle vandalism.
- There is a minimal staff requirement if it is used by known parkers.
- It is possible that the retrieval time is lower than the combined driving/parking/walking time in conventional ramped parking structures.
- There is an easier facade integration since there are no ramping floors or openings in exterior walls.

2.2 *Modules*

Smart parking system is divided into three modules. They are user application module, entry point management module, exit point management module. Workflow diagram for smart parking system is shown in Fig. 3.

2.2.1 User Application Module

This module contains user data and user interface which will be used by vehicle owner to book parking slot with timing slot available. User will login with its valid credentials and book parking slot as per availability of parking area. User will have to make online payment to confirm parking. After successful payment, user will be provided a unique booking ID. User can also see his booking details in view booking section. User can also change his credentials. This booking details need to be displayed on entry point.

2.2.2 Entry Point Management Module

This module contains managing parking area and booking parking area at entry point of parking area. If vehicle has made entry without pre-booking, then parking area will be assigned to that vehicle based on car number and entry will be granted. If vehicle owner has made pre-booking, then only display of booking details at entry point will be grant entry to that vehicle.

2.2.3 Exit Point Management Module

This module contains calculating parking charges for vehicle based on duration and pre-booking details. While exiting the parking area, parking charges will be charged according to the duration and type of booking (pre-book/on arrival booking). On successful payment of parking charge, the vehicle will be granted exit from the parking area.

3 Results and Discussion

3.1 *Admin Panel*

See Figs. 2, 3, 4, 5, 6 and 7.

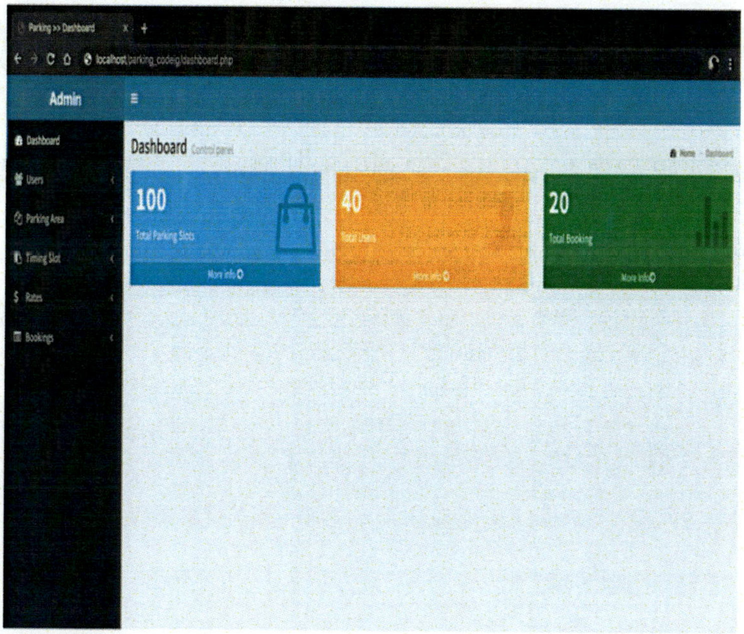

Fig. 2 Dashboard page

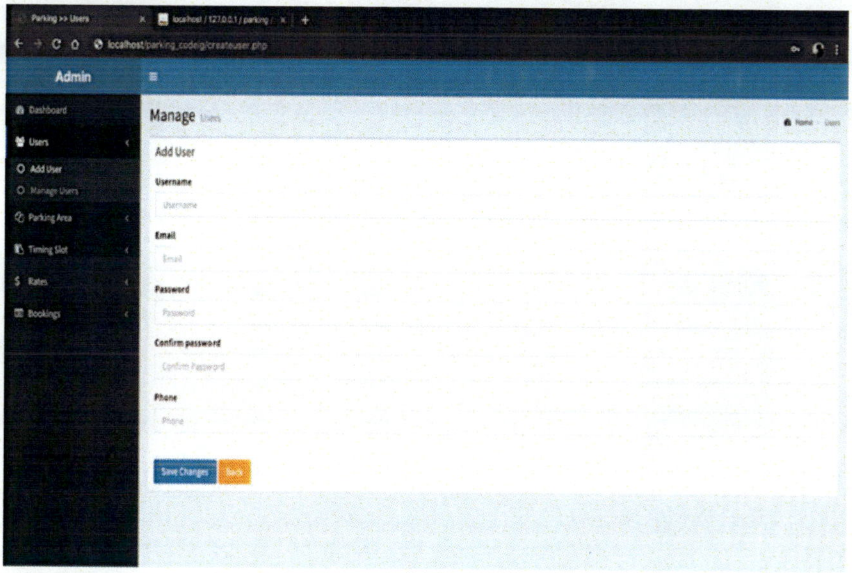

Fig. 3 Create user

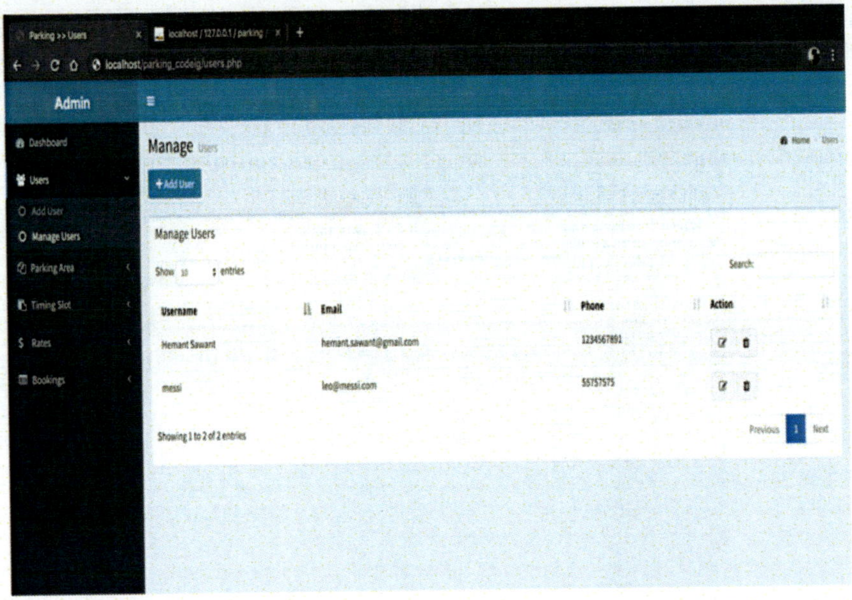

Fig. 4 All user list

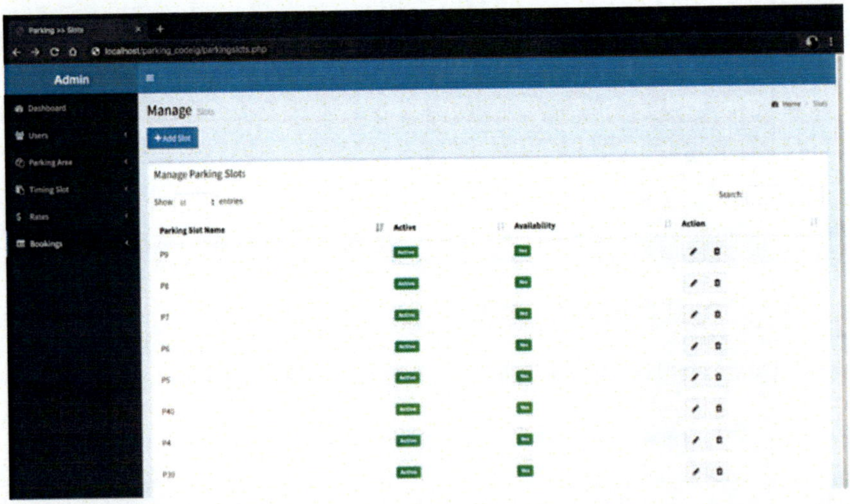

Fig. 5 Parking slot List

3.2 *Front-End*

See Fig. 8 and 9.

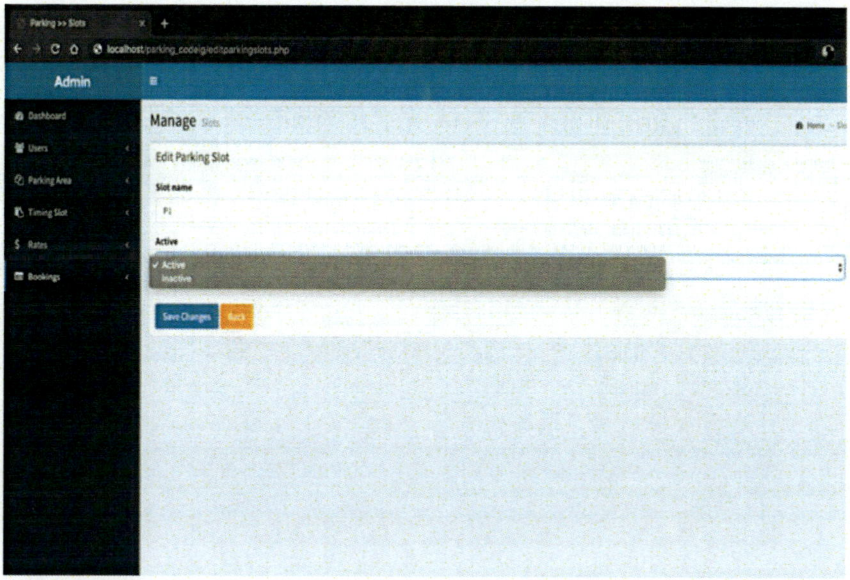

Fig. 6 Edit parking slot

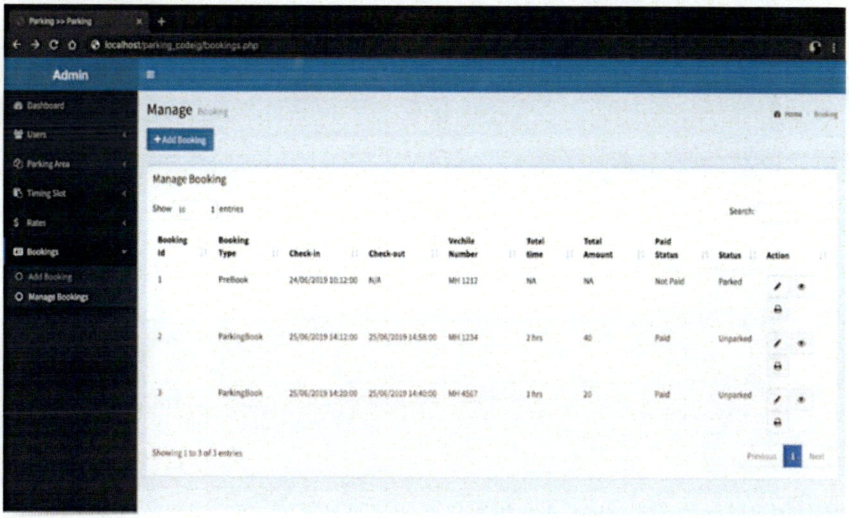

Fig. 7 Book parking list

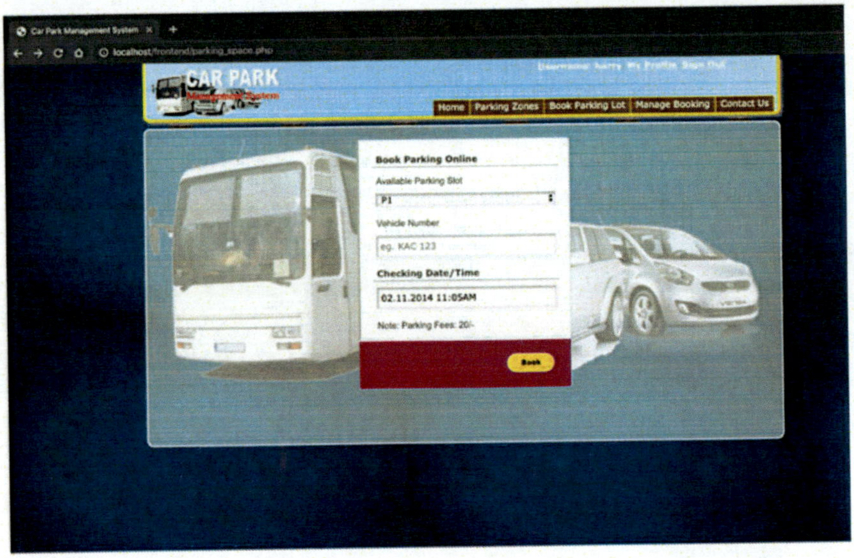

Fig. 8 Book parking

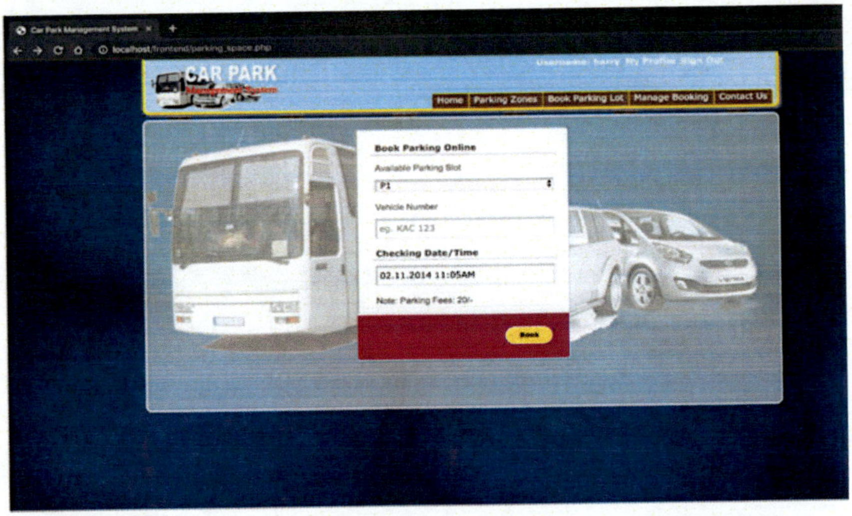

Fig. 9 Contact us

4 Conclusion

Smart car parking in shopping malls were implemented successfully. This system is used for availability of parking during weekends/rushing hours in shopping malls. This system included two steps. One is pre-reservation of parking for vehicle owners, and the other is parking management system for vendor. Smart parking system provided current parking availability status for vehicle owners. If available, they can reserved the parking after making online payment or else they can find another parking area. The application used by parking management managed all parking areas and was used for making reservation on entry of vehicle. This system was very helpful in saving the time and a lot of efforts for finding a space to park the vehicle, especially during the weekends. User can connected to the smart parking system from anywhere with their smart phones in any browser. And, this information is also available to the parking operators to determine the free parking areas, and statistics can be measured at different times in a day on each parking space.

References

1. M. Magrini, D. Moroni, G. Palazzese, G. Pieri, G. Leone, O. Salvetti, Computer vision on embedded sensors for traffic flow monitoring, in *Proceedings of the 18th IEEE International Conference on Intelligent Transportation Systems, ITSC 2015*, pp. 161–166, September 2015
2. G. Amato, F. Carrara, F. Falchi, C. Gennaro, C. Vairo, Car parking occupancy detection using smart camera networks and Deep Learning, in *Proceedings of the 2016 IEEE Symposium on Computers and Communication, ISCC 2016*, pp. 1212–1217, July 2016
3. C.-C. Huang, Y.-S. Dai, S.-J. Wang, A surface-based vacant space detection for an intelligent parking lot, in *Proceedings of the 2012 12th International Conference on ITS Telecommunications, ITST 2012*, pp. 284–288, IEEE, November 2012
4. Z. Shelby, Embedded web services. IEEE Wirel. Commun. Mag. **17**(6), 52–57 (2010)
5. G. Maria, E. Baccaglini, D. Brevi, M. Gavelli, R. Scopigno, A drone-based image processing system for car detection in a smart transport infrastructure, in *Proceedings of the 18th Mediterranean Electrotechnical Conference, MELECON 2016*, pp. 1–5, April 2016
6. C. Gálvez del Postigo, J. Torres, J.M. Menéndez, Vacant parking area estimation through background subtraction and transience map analysis. IET Intel. Transport Syst. **9**(9), 835–841 (2015)
7. A. Azzara, M. Petracca, P. Pagano, The icsi m2m middleware for iot-based intelligent transportation systems, in *Proceedings of the 2015 IEEE 18th International Conference on Intelligent Transportation Systems*, pp. 155–160, 2015
8. E.C. Neto, E.S. Reboucas, J.L. De Moraes, S.L. Gomes, P.P.R. Filho, Development control parking access using techniques digital image processing and applied computational intelligence. IEEE Latin Am. Trans. **13**(1), 272–276 (2015)
9. L. Ruizhi, R. Cristian, B. Peter, O. Shumao, C. Liping, "Crowdsourcing on-street parking space detection (2016). https://arxiv.org/abs/1603.00441
10. V.W.S. Tang, Y. Zheng, J. Cao, An intelligent car park management system based on wireless sensor networks, in *Proceedings of the SPCA 2006: 1st International Symposium on Pervasive Computing and Applications*, pp. 65–70, August 2006
11. S. Funck, N. Möhler, W. Oertel, Determining car-park occupancy from single images, in *Proceedings of the 2004 IEEE Intelligent Vehicles Symposium*, pp. 325–328, June 2004
12. M. Magrini, D. Moroni, G. Pieri, O. Salvetti, Smart cameras for ITS in urban environment, in *Intelligent Transport Systems: Technologies and Applications*, pp. 167–188, 2015
13. Q. Wu, C. Huang, S.-Y. Wang, W.-C. Chiu, T. Chen, Robust parking space detection considering inter-space correlation, in *Proceedings of the IEEE International Conference on Multimedia and Expo, ICME 2007*, pp. 659–662, July 2007
14. T. Lin, H. Rivano, F. Le Mouel, A survey of smart parking solutions. IEEE Trans. Intell. Transp. Syst. **18**(12), 3229–3253 (2017)

15. J. Vera-Gómez, A. Quesada-Arencibia, C. García, R. Suárez Moreno, F. Guerra Hernández, An intelligent parking management system for urban areas. Sensors **16**(6), 931 (2016)
16. K. Mouskos, M. Boile, N.A. Parker, Technical solutions to overcrowded park and ride facilities. New Jersey Department of Transportation Tech. Rep. (2007)
17. D. Gavalas, D. Economou, Development platforms for mobile applications: status and trends. IEEE Softw. **28**(1), 77–86 (2011)
18. M. Wang, C. Perera, P.P. Jayaraman, M. Zhang, P. Strazdins, R. Shyam-sundar, R. Ranjan, City data fusion: Sensor data fusion in the internet of things, in *The Internet of Things: Breakthroughs in Research and Practice* (IGI Global, 2017), pp. 398–422
19. I. Samaras, N. Evangeliou, A. Arvanitopoulos, J. Gialelis, S. Koubias, A. Tzes, Kathodigos-a novel smart parking system based on wireless sensor network. Intell. Transp. Syst. **1**, 140–145 (2013)
20. L. Wenghong, X. Fanghua, L. Fasheng, Design of inner intelligent car parking system, in *International Conference on Information Management Innovation Management and Industrial Engineering*, 2008
21. C. Wenzhi, L. Bai, A smart roadside parking navigation system based on sensor networks for its, in *IET International Conference on Wireless Mobile and Multimedia Networks, IET*, pp. 1–4, 2006
22. S. Hanche, P. Munot, P. Bagal, K. Sonawane, P. Pise, Automated vehicle parking system using rfid, **1**(2) (2013)
23. H. Wang, W. He, A reservation-based smart parking system, in 2011 *IEEE Conference on Computer Communications Workshops (INFOCOM WKSHPS)*. IEEE, pp. 690–695, 2011
24. M.A. Razzaque, M. Milojevic-Jevric, A. Palade, S. Clarke, Middle-ware for internet of things: a survey. IEEE Internet of Things J. **3**(1), 70–95 (2016)

Blockchain-Based Solution for Effective Employee Management

Yuli Nurhasanah, Dita Prameswari, and Olivia Fachrunnisa

Abstract The Fourth Industrial Revolution (IR 4.0) aims to disrupt business in various sectors like manufacturing, services, mining, fashion, etc., with the use of digital technology, one of them blockchain. The use of blockchain in the human resource management function will help collaboration between parties in updating employee skills and knowledge. Information generated from the blockchain process can also be used as a source of policy makers to set competency standards among employees. In addition, HR blockchain will run an automated process to make consensus between the parties involved. The purpose of this article is to develop a blockchain framework for managing employee performance and career progression in a particular organization. The blockchain design will focus on improving and developing the quality of the workforce, especially for sharing information between employees and managers. This will facilitate the process of identifying and analyzing employee skills, knowledge, and attitude data. In addition, data transparency and access opportunities by employees and company managers will exist without being misused. This will make managers easily track records of each employee's performance items, which in turn can be used to develop employee careers.

Keywords Blockchain · Employee management · Employee performance

Y. Nurhasanah · D. Prameswari · O. Fachrunnisa (✉)
Department of Management, Faculty of Economics, Universitas Islam Sultan Agung, Jl. Kaligawe Raya KM. 4, Semarang 50117, Indonesia
e-mail: olivia.fachrunnisa@unissula.ac.id

Y. Nurhasanah
e-mail: yulinurhasanah@std.unissula.ac.id

D. Prameswari
e-mail: prameswari@std.unissula.ac.id

P. K. Pattnaik et al. (eds.), *Proceedings of International Conference on Smart Computing and Cyber Security*, Lecture Notes in Networks and Systems 149,
https://doi.org/10.1007/978-981-15-7990-5_14

1 Introduction

Currently, Industry 4.0 brings innovation and smart manufacturers, in which one of them is digital transformation. Digital transformation also becomes an inevitable necessity, including encouraging disruption in the HR world. Utilization of this technological revolution is one of the prerequisites for improving organizational management. To deal with an era of disruption, organizations need skill up or continue to develop the required skills. There is no denying that business will be affected by the rapid advancement of technology. The Internet, automation, and artificial intelligence are changing the way business operates. This becomes important as the social implications of these changes become more widespread. As a result, the HR team will play an important role to help companies navigate the business into new fields.

The skill gap will become more acute along with the development of technology and the struggle for talent. Considering this is important for companies to immediately understand what advantages they have and what they will need in the future. The HR field is currently undergoing a rapid transformation. From the beginning, it was considered as a support function, and now, the HR team has become an important tool in digital transformation. HR leaders began to apply digital culture in their respective organizations by utilizing tools and applications to support successful transformation. At a certain point, technological innovation influences not only HR technical competence but also stimulates HR "technology knowledge" within the organization, namely the form of basic assumptions and HR perspective on internal processes and relationships between organizational units [7].

At present, HR in the company is the most valuable asset. HR development is an effort made to form quality personnel by having skills, work abilities, and work loyalty to a company or organization. Quality human resources will help the company to further develop and achieve company goals. The HR development of the company is a means for employees to develop their potential and increase employee knowledge in certain fields. This will affect the individual performance productively [5].

In an effort to increase work productivity, a number of things are related to HR development, namely by using a technology called blockchain. Blockchain is a database that serves to store a collection of data that continues to grow or increase. Blockchain is a potential technology behind the bitcoin cryptocurrency because it could be able to solve a problem in achieving end-to-end transparency [4]. Blockchain is often associated with digital currency bitcoin, because it was originally created to monitor bitcoin transactions. Over time, blockchain technology has evolved and has now entered the HR industry because companies have begun to use blockchain to facilitate HR functions [13].

In addition, blockchain could have a major potential on both sides of the employment relationship, that is from the ability in maintaining and controlling access to a comprehensive and trustworthy of blockchain-based record of their education, skills, training and performance in the workplace. Individuals would be able to turn their skills, training, and experience into genuine value if they were provided to access "value passport" in organization. By applying data analytically, companies would be able to match individuals to the roles accurately and effectively. This could be particularly relevant, as skills requirements have changed in the fourth of industrial revolution.

Blockchain's ability in supporting behavioral and cultural changes under way in organizations and the workforce is necessary. The benefits of the individual's highly portable and up to date "value passport" will become all the greater—both to themselves and employers—as the trend toward the gig economy continues and younger people change jobs more frequently or opt for portfolio careers. Also, with the younger generation generally being more relaxed than their predecessors about sharing personal information, blockchain provides them with the opportunity to do this in a more secure and trustworthy way.

In blockchain technology, data created by one server can be replicated and verified by another server. Therefore, the blockchain is often likened to a bank master cash book that contains all customer transaction data. With blockchain, a transaction no longer needs to depend on one server, because the transaction will be applied to the entire network [11]. Because the nature of the network is peer to peer, blockchain users can also avoid various frauds that can occur due to data modification or hacking [10]. Each of these blocks contains hashes from the previous block. In this, systems are interrelated and if there is an effort in changing data in one block, it must change the data in the other block. Each block protected by cryptography is interconnected to form a network [8].

Data is increasingly valuable in business in the midst of technological developments as it is today. With a data-driven approach, data collection in business has an important role as a fuel for various improvisations [2]. One transformation carried out with data is to use analysis to evolve the role of HR (Human Resource) or Human Resources (HR) in the company. It can also have an impact on the pace of company development. In a data-driven approach, data analysis is indeed a central role as the main reference for decisions taken by the HR department. Utilization includes finding out the reasons for employee turnover, choosing who is worthy of holding the next leadership, and some other effectiveness in managing talent.

Ideally, companies or businesses identify employees who need training to optimize their performance and make the company successful. Data analysis enables offering HR insights into initiative programs that can help improve HR, for example, such as rewards and training programs. By storing data across its network, blockchain eliminates the risk that comes with centralized data. Decentralized blockchain can use ad hoc messages that pass and distribute the network. Data stored on blockchain is generally considered to be undamaged. While centralized data is more controlled, manipulation of information and data is common. By decentralizing it, the blockchain makes data transparent for everyone involved [1].

This article aims to discuss how the blockchain works and can help managers to improve the quality of employee performance in the company, so that employees can develop according to their abilities and company desires.

2 Blockchain for Employee Performance

Basically, blockchain is a distributed database system that recording a transactional data or other information which secured by cryptography and governed by a consensus mechanism [14]. Specifically, in processing and verifying transactions on the ledger, blockchains use cryptography so the data encryption and coding in a blockchain could improves transparency, efficiency, and trust in sharing an information [9].

Based on [12] stated that there are four main facets of a blockchain. First, it has been designed to distribute and synchronise across networks, it encourages businesses to share data and it is ideal for multi-organisational business networks, such as supply chains or financial consortia. Second, in blockchains, there are smart contracts which is an agreement made by participants in advance and stored in the blockchain. The used of a smart contract is intended to facilitate digitally, verify or enforce the terms that have negotiated in a contract that allowing for transactions credibly without of third-party interventions as they are automatically. To verify a valid transaction, the third characteristic of the blockchain is built P2P networks, whereby there must be agreeableness between all relevant parties which serves to keep inaccurate or potentially fraudulent transactions out of database. And the last, transactions that have been agreed and recorded cannot be changed as provides provenance of assets, which it means that for any asset is possible to know of its existence, where it has been happened throughout its lifetime.

Blockchain takes the general ledger with transaction details, and distributes it to P2P network instead of placing it on a particular organizations central server, so that it can be managed together and solve high costs of management and gacking problem [4]. Blockchain technology was implemented firstly in 2009 that blockchain as the underlying platform designed in solving the problem "double-spending" for bitcoin that is how to transfer digital value without relying on a trusted third party. However, the attributes that make blockchain technology essential and used for bitcoin can solve a variety of other problems [6].

At companies, blockchain can be used to help managers improve employee performance. Some of them are in the process of identifying and analyzing data; it is to improve the quality of human resources and company performance using a blockchain on employee performance related to their respective fields of work. Based on Wang et al. [15], by adopting blockchain technology, it will improving security. Meanwhile, management system in organizations become more open and it could be effective solution for employee management. These are the following details employee performance data that will be managed on the blockchain.

2.1 Human Resources Department

In the HRD field, there are a number of things that can be done with blockchain:

Quality orientation
The ability to perform some tasks completely, on time, and also having an excellent quality of work and performance, even above established quality standards. With blockchain, HR can use it to record and map the skills and abilities of employees in the company such as knowing who employees have the right competencies for a project.
Problem-solving skills
The ability in analizing a problem, identify the source of the problem and the relationship between various factors of the problem, and then formulate alternative solutions that are relevant and applicable.
Planning skills
The ability to prepare planning of work in a systematic and well-scheduled manner, to allocate resources based on the results of planning, and monitoring to ensure that the work plan can run effectively and efficient.
Teamwork
The ability to coordinate and communicate with various parties involved, formulate common goals and share tasks to achieve the work goals that have been planned, and respect each other's opinions and input to improve team performance.
Self learning capacity
The ability to carry out learning activity processes both independently and in groups shows an adequate interest in continuing to develop self-skills and be proactive to share knowledge among fellow employees.

2.2 Marketing Department

There are a number of things that can be done using the blockchain in the marketing development field and are as follows:

- Data on sales growth compared to the previous year;
- Customer satisfaction score;
- Total new product development;
- Market share and distribution of products/ services;
- Average hours to completely resolve customer complaints;
- ROI (Rate of Investment) of the marketing program.

2.3 *Finance Departement*

In the finance department, things that can be included in the blockchain system include:

- Recording transactions;
- Financial reporting and analysis;
- Funding;
- Transparent business audits.

3 How to Work Blockchain in Improving the Quality of Employee Performance

The use of blockchain technology in the company and supply chain management field. Data and results of employee performance above will be recorded in the blockchain system. Previously, employees would be given their respective servers to get an account in the blockchain to enter employee performance data that had been done, so that each employee could input themselves what had been done in the company, so managers and other employees could see what performance results have been done and can evaluate what are the deficiencies and errors so that the employee concerned can fix it and produce maximum performance.

In using the blockchain itself, all employees have a blockchain server and account so that it will be easy to enter data on the performance results of each, for example, financial statements and sales transactions must always be input in the system in order to reduce employee dishonesty in funding reports. Data in the blockchain cannot be replaced because every server will have a copy, so that it will reduce the level of fraud and fraud at work.

All of information in a digital ledger has been stored in a data block; so everything that has been recorded in the blockchain cannot be changed by a unique cryptographic identifier (or "hash"). Each subsequent data block includes the hash of the previous block to create a chain that connects all the way back to the first data block (hence the "blockchain"). If data in any block in the chain is then unreasonably changed in any version of the ledger node, the hash for it and each subsequent block must change, making the modified ledger easily identified as an invalid version. That invalid version is then rejected by consensus among the nodes [6]. Therefore, every week managers will evaluate employee performance by looking at the data on the blockchain, so that their quality of work would be better and improved.

By using the blockchain, all employees and leaders in the company can find out the results of each performance. All employees can learn from the data that has been stored in the blockchain, so that not only focus on the field of work but also understand other fields of work. That way, employees will indirectly gain knowledge other than their field of work, and in the evaluation will also be done sharing of performance reports contained in the blockchain, each employee can provide criticism and suggestions for the data that has been reported in the blockchain, and it will be a learning and new knowledge for all employees.

Existing and passing data on the blockchain network will be encrypted and stored copies to all blockchain network owners based on an agreement (consensus). The incoming data will later be guarded by a blockchain whose nodes are installed on several parties, so that it cannot be tampered with by anyone, and is protected from various forms of data misuse due to modified data, server down, or hacked accounts [3]. Transaction history that has been locked into a block will make it easier for managers to track and track every item, data and lost and lack of resources.

Based on the existing research said that by building a model of human resource management based on blockchain technology can provide effective information for human decision making especially for employee management. This model is more applicative and valid for HRM. Meanwhile, blockchain is the best solution in problem solving and security, also promote the use efficiency and effect of HRM information [15]. That way, managers or leaders of companies can assess their employees by looking at the results of company performance on the blockchain, can also find out the shortcomings of each field of work and the company can make training and development of employees about the field of work that is still lacking so that employee performance in the company can improve. However, it is also important to get to know the people who will manage the blockchain at the company. Must look for the most quality and understand the culture of the company. Likewise, with those who become the liaison of the company, they should be familiar with the programming language.

4 Conclusion

With the ownership of data for each field of work in the blockchain network, each employee can find out the results of each other's performance. And all employees can share to find out work experience in the company, exchange ideas among fellow HR, and in that way, the company can determine what HR development is in an effort to improve the quality of HR performance. And also the blockchain can be used to change a company's system by referring to the performance results that have been listed in the blockchain network, for example, in a field of work, there are deficiencies and challenges related to performance, then automatically repairing the existing system in it to develop employee potential.

In essence, the blockchain here as a means to provide facilities for employees to share information so that employees in the company have equal knowledge and skills and are not unequal to each other. Moreover, it can increase the profitability and credibility of the company. With blockchain, data is stored securely, cannot be changed, and companies do not need to waste a lot of time. With this blockchain technology, it will be easier to identify if there is a fraud in one of the supply chain processes.

References

1. A. Bahga, V.K. Madisetti, Blockchain platform for industrial Internet of Things. J. Softw. Eng. Appl. **9**(10), 553–546 (2016)
2. M. Casey, P. Wong, Global supply chains are about to get better, thanks to blockchain. Harv. Bus. Rev. **13** (2017)
3. B.J. Cho, S.Y. Park, Blockchain technology: the 3rd information revolution? Asia Pac. J. Innov. Entrep. **11**(1) (2017)
4. R. Cole, M. Stevenson, J. Aitken, Blockchain technology: implications for operations and supply chain management. Supply Chain Manag.: Int. J. (2019). https://doi.org/10.1108/scm-09-2018-0309
5. O. Fachrunnisa, A. Adhiatma, H.K. Tjahjono, Cognitive collective engagement: relating knowledge-based practices and innovation performance. J. Knowl. Econ. (2018). https://doi.org/10.1007/s13132-018-0572-7
6. R.P. George, B.L. Peterson, O. Yaros, D.L., Beam, J.M., Dibbell, R.C. Moore, Blockchain for business. J. Invest. Compliance (2019). https://doi.org/10.1108/joic-01-2019-0001
7. J. Han, C.-M. Park, Case study on adoption of new technology for innovation. Asia Pac J. Innov. Entrep. **11**(2)
8. C. Harris, The History of Bitcoin Crypto Currency News (2018). https://cryptocurrencynews.com/thehistory-of-bitcoin/, diakses tgl 20 Juni 2018
9. P. Misra, 5 Ways blockchain technology will change the way we do business. available at (2018). www.entrepreneurcom/article/309164. Accessed 1 July 2018
10. S. Nakamoto, *Bitcoin: A Peer-to-Peer Electronic Cash System* (2008). https://bitcoin.org/bitcoin.pdf, diakses tgl 20 Juni 2018
11. J. Oh, I. Shong, A case study on business model innovations using Blockchain: focusing on financial institutions. Asia Pac. J. Innov. Entrep. **11**(3), 335–344 (2017). https://doi.org/10.1108/apjie-12-2017-038
12. I. Pattison, 4 Characteristics that set blockchain apart. available at (2017). www.ibm.com/blogs/cloud-computing/2017/04/11/characteristics-blockchain/. accessed 1 July 2018
13. A. Rai, R. Patnayakuni, N. Seth, Firm performance impacts of digitally enabled supply chain integration capabilities. MIS Q. 225–246 (2006)
14. M. Swan, *Blockchain: Blueprint for a New Economy* (O'Reilly Media, Sebastopol, CA, 2015)
15. X. Wang, et al., Human Resource Information Management Model based on Blockchain Technology. Dalian Maritime University China Institute of Electrical and Electronics Engineers (2017)

Implementation of Motorcycle Monitoring Using Bluetooth with an Android-Based Microcontroller Using Arduino

Yudhi Arta, Evizal Abdul Kadir, Ari Hanggara, Des Suryani, and Nesi Syafitri

Abstract Vehicle safety at this time is a priority that must be considered by the motorcycle manufacturer. For this security, it now needs a system that works systematically and uses the latest technological devices. The system offered for current technology is an Arduino microcontroller and is also based on android to design an engine control system ON/OFF on a motorcycle by using android via Bluetooth which can control the engine remotely or close. This equipment data communication protocol is via Bluetooth installed on the device motorcycle using 6 or 12 volts to ignition systems, starters, lights, and other electrical components. The distance generated by this Bluetooth device also needs to be calculated so that there is an estimation of the distance and the length of the data generated both for near and far scale. This equipment has been tested and can be used to control the engine on a motorcycle from a distance or near without having to go through the contact of a motorcycle installed.

Keywords Arduino · Android · Microcontroller · Bluetooth

Y. Arta (✉) · E. A. Kadir · A. Hanggara · D. Suryani · N. Syafitri
Department of Informatics, Universitas Islam Riau, Kaharuddin Nasution Street 117, Pekanbaru, Indonesia
e-mail: yudhiarta@eng.uir.ac.id

E. A. Kadir
e-mail: evizal@eng.uir.ac.id

A. Hanggara
e-mail: arihanggara@eng.uir.ac.id

D. Suryani
e-mail: des.suryani@eng.uir.ac.id

N. Syafitri
e-mail: nesisyafitri@eng.uir.ac.id

P. K. Pattnaik et al. (eds.), *Proceedings of International Conference on Smart Computing and Cyber Security*, Lecture Notes in Networks and Systems 149,
https://doi.org/10.1007/978-981-15-7990-5_15

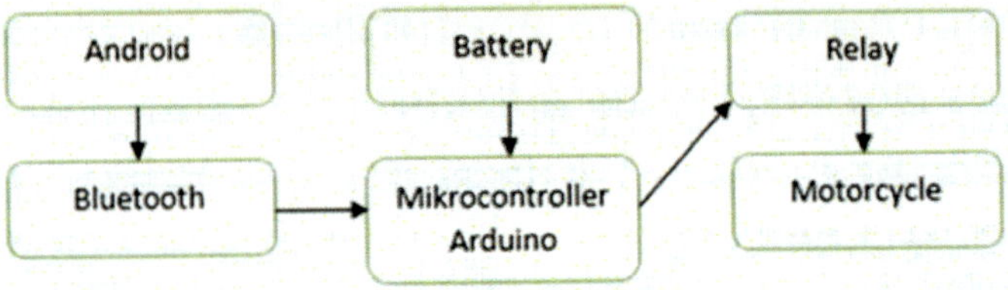

Fig. 1 Block diagram of a motorcycle control system

1 Introduction

At present, motorcycle is typically a possible means of local transportation that is widely used in everyday life. With the development of bicycle technology at this time, the need for a security system on motorbikes is a much-needed requirement for motorcycle owners. Many security systems are offered and used by consumers at this time in the form of non-electric and electric safety, such as handlebar locks, padlock, and the alarm that uses sound as an indicator which is the standard security of an alarm and others. [1].

On the other hand, technological advances are increasingly rapid, so that almost every individual already has a gadget, both in the form of smartphones and tablets [2]. Now smartphones are a special device for everyone because of the importance of their function as a medium of communication and information that can be accessed instantly wherever and whenever. Therefore, smartphones are always carried by everyone [3]. For this reason, the author tries to improve the functionality of a smartphone which is usually only used as a communication tool so that it can be used as a media control device [4].

For many security systems that have been created only use the RFID system. This system makes RFID work as security verification on motorbikes. Then, RFID reads the code and some other numeric so that when an RFID tag responds to an RFID reader it can read the data stored in the tag and send this data to be compared with the data in the database. In the case of information reading from the wrong tag, conditions are recorded in the database [5].

The security system for vehicle users is easy to save in all gadget devices, saving in terms of cost and practicality. This system utilizes the development of smartphones and Arduino technology as a communication tool and single-board microcontroller that has experienced many developments nowadays, such as motorbike controllers combined with microcontroller components and utilizing Arduino and Bluetooth facilities available on android smartphones [6–8].

2 Architecture Design

In this design in Fig. 1, at this stage, the thing that needs to be done is by inputting the control application that is available on the android smartphone device. In the application, standard input (button ON/OFF) is available to be used. Data sent from an android smartphone will be received by a Bluetooth module connected to the Arduino microcontroller system. The serial data is translated by the Arduino microcontroller into parallel data [9]. Parallel data generated by the Arduino microcontroller is forwarded to the relay via the LED indicator which functions to turn off the motorcycle. Then, the relay will forward the data used to turn on or turn off the vehicle [10].

2.1 Motor Starter System Components

The motor starter system has several components including

- The battery is an electrochemical device that is made to supply low-voltage electrical energy (on motorbikes using 6 volts or 12 volts) to ignition systems, starters, lights, and other electrical components. Batteries store electricity in the form of chemical energy, which is released when needed according to the load and system that needs it.
- Ignition key serves as the main switch to connect and disconnect (on/off) the motorbike electrical circuit.
- Relay starter is the main relay of the starter system which serves to reduce the voltage loss that is supplied from the battery to the starter motor.
- The starter switch functions as a motorbike starter switch that works when the key is in the ON position.
- The starter motor is an electric starter that functions to convert the battery's chemical power into a rotary power that is capable of rotating the crankshaft to start the engine (Fig. 2).

The start engine on this motor is made using Arduino Uno as the main part of the program using the Arduino 1.8.4 software. In Fig. 3, tools such as the Bluetooth HC-05 module, Relay, and jumper cables then installed on the electrical part of the motorcycle. The essence of making a system on this motorbike is to make it easier for humans to improve the safety of a motorcycle without using a key so that it only uses an android smartphone. The development of the system in this motorbike requires several stages of analysis that must be passed; at this stage, modeling is done using flowcharts and coupling on Arduino. For the design of the tool, we will use several stages including:

- Designing a tool by assembling a relay to an Arduino using a jumper cable.
- Installing the Bluetooth HC-5 module into Arduino using a jumper cable.
- Programming Arduino on software using Arduino 1.8.5.

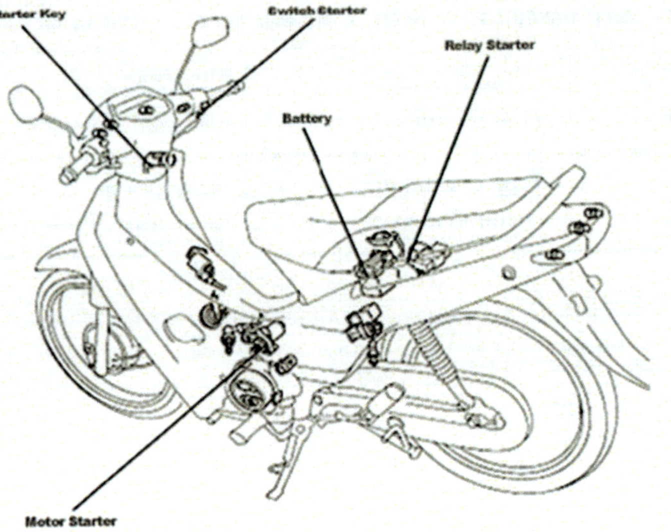

Fig. 2 Design motorcycle control system

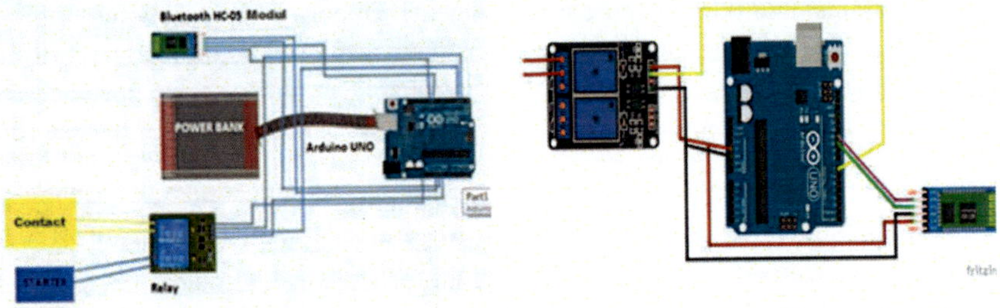

(a) (b)

Fig. 3 Arduino module design **a** top and **b** bottom views

- Installation of electric motors.

2.2 *Design of Blocks Using App Inventor*

The design of the android application design block with the inventor app works so that the application can function properly. The design block of the inventor app is also enabled so that the application can be connected to Arduino and can be used according to the functions of the application, while the blocks made can be seen in Fig. 4.

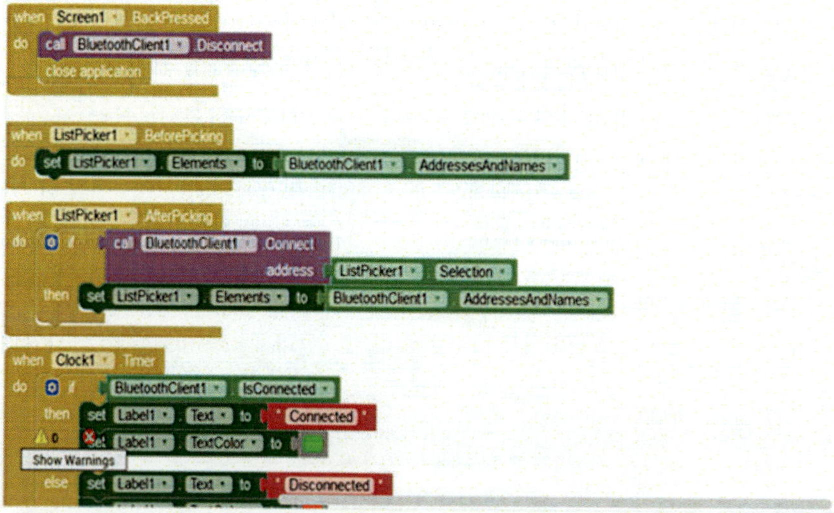

Fig. 4 Designing application design blocks using app inventor

A list of Bluetooth devices connected to android can be taken as a reference address. In Fig. 5, before being used, the 'ListPicker' element is used as an available Bluetooth list container. If the items in the list have been selected, the procedure 'BluetoothClient1' will be called with the selected address variable.

In Fig. 5, with the help of 'clock.Timer,' each time interval is checked when data is received via Bluetooth and then takes it with the function 'Label1.' But, the above functions have several disadvantages, namely

- The data received is not parsed, so any data received is a string of strings.
- If in a one-time interval some data is received, then the data is considered a unit.
- If the process of sending data takes place when a timer interruption occurs, then the data will be truncated (data is not intact).

3 Result and Discussion

This test is done to see whether the system is good or not when it is used. This test uses two motorbikes as a test medium and tested 10 times repeatedly. This is done so that testing can produce more accurate data.

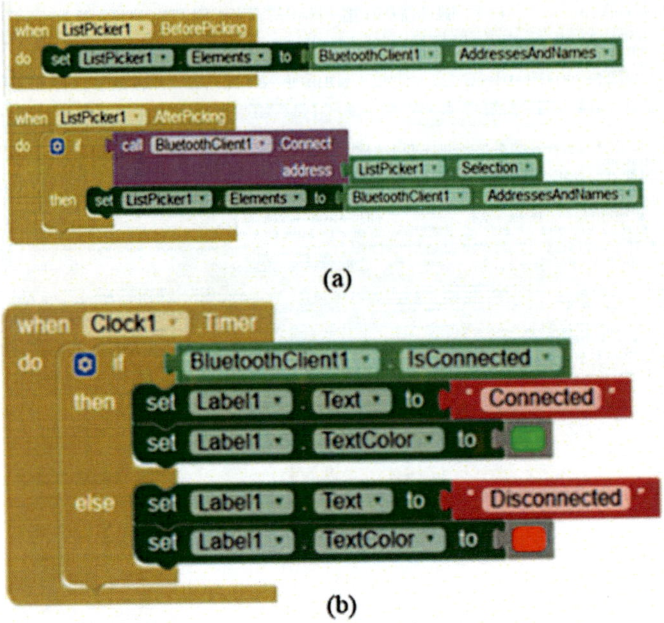

Fig. 5 Application BLOCK CODE **a** Connects Bluetooth to the tool and **b** receiving data from Bluetooth

3.1 Bluetooth Connectivity Distance Testing to Turn on the Engine on a Motorcycle

This test is done to find out how much the maximum distance that can be done to be able to connect between Bluetooth to turn on the motorcycle engine. Testing is done by turning on the circuit switch to turn on the appliance and trying to connect between the Bluetooth module and Bluetooth on the handphone. The results of this test can be seen in Table 1.

Table 1 Testing for Bluetooth connectivity to turn on the engine

Height (METER)	Testing	Connected	Connected	Not connected	Success	
Delay Time						
2	10	Connected	10	–	100%	1 s
5	10	Connected	10	–	100%	1 s
6	10	Connected	10	–	100%	1 s
8	10	Connected	–	10	–	–
9	10	Connected	–	10	–	–

Table 2 Testing for Bluetooth connectivity to turn on the engine

Length (Meter)	Testing	Connected	Connected	Not Connected	Success	Delay Time
2	10	Connected	10	–	100%	1 s
4	10	Connected	10	–	100%	1 s
6	10	Connected	10	–	100%	1 s
7	10	Connected	10	–	100%	1 s
9	10	Connected	10	–	100%	1 s
10	10	Connected	7	3	70%	1 s
12	10	Connected	–	10	–	–
13	10	Connected	–	10	–	–

From this distance test, the results are obtained where the Bluetooth connection to turn on the engine on the motorbike will function properly at a distance of 12 meters, besides that at a distance of 13 meters Bluetooth connectivity cannot be connected properly, but Bluetooth is still possible to connect. The Bluetooth connection will not be able to connect if the distance between the user and the motorcycle is more than 13 meters.

From Table 2, it can be seen how the results of the connection between Bluetooth and controller to turn on a motorcycle engine with a certain distance, the results will be obtained where the Bluetooth connection will be connected properly at a distance of 9 meters. In this test, the barrier used is in the form of a wall with a height of 2.5 meters and has a wall thickness of 11 cm. With this barrier, the ability of the Bluetooth connection to be connected becomes reduced.

From the tests that have been carried out in Tables 1 and 2, then in Table 3, we can see how the connection between Bluetooth and controller with a certain height range with the position of the smartphone vertically aligned with the motorcycle. In

Table 3 Testing for Bluetooth connectivity to turn on the engine

Length (Meter)	Testing	Connected	Connected	Not Connected	Success	Delay Time
2	10	Connected	10	–	100%	1 s
4	10	Connected	10	–	100%	1 s
6	10	Connected	10	–	100%	1 s
7	10	Connected	10	–	100%	1 s
9	10	Connected	10	–	100%	1 s
10	10	Connected	10	–	100%	1 s
12	10	Connected	10	–	100%	1 s
13	10	Connected	6	4	60%	1 s
14	10	Connected	–	10	–	–
15	10	Connected	–	10	–	–

Table 4 Bluetooth connectivity testing to turn on the engine from height

Length (Meter)	Testing	Connected	Connected	Not Connected	Success	Delay Time
2	10	Connected	10	–	100%	1 s
4	10	Connected	10	–	100%	1 s
7	10	Connected	10	–	100%	1 s
9	10	Connected	10	–	100%	1 s
10	10	Connected	10	–	100%	1 s
12	10	Connected	10	–	100%	1 s
13	10	Connected	6	4	60%	1 s
15	10	Connected	–	10	–	–

this test Bluetooth can be connected properly at an altitude of 6 meters and has no connection at a distance of 8 meters.

3.2 *Bluetooth Connectivity Distance Testing to Turn off the Engine on a Motorcycle*

This test is done to find out how much the maximum distance that can be done to be able to connect between Bluetooth to turn off the motorcycle engine. In Table 4, the test is done by turning off the relay on the controller which aims to turn off the motorcycle while it is on.

In Table 4, it can be seen how the connectivity based on different distances to turn off the motorcycle engine can run well. From this distance testing, the results are obtained where the Bluetooth connection to turn off the engine on the motorbike will function properly at a distance of 12 meters without any obstacles.

From Table 5, it can be seen how the results of the connection between Bluetooth and the controller to ensure a motorcycle engine with a certain distance is obtained where the Bluetooth connection will be connected properly at a distance of 9 meters if there is a barrier between the smartphone and controller. In this test, the barrier used is in the form of a wall with a height of 2.5 meters and has a wall thickness

Table 5 Bluetooth connectivity testing to turn on the engine from height

Length (Meter)	Testing	Connected	Connected	Not connected	Success	Delay time
2	10	Connected	10	–	100%	1 s
5	10	Connected	10	–	100%	1 s
6	10	Connected	10	–	100%	1 s
8	10	Connected	–	10	–	–
9	10	Connected	–	10	–	–

Table 6 Bluetooth connectivity testing to turn on the engine from height

Length (Meter)	Testing	Connected	Connected	Not Connected	Success	Delay Time
2	10	Connected	10	–	100%	1 s
4	10	Connected	10	–	100%	1 s
6	10	Connected	10	–	100%	1 s
7	10	Connected	10	–	100%	1 s
9	10	Connected	10	–	100%	1 s
10	10	Connected	7	3	70%	1 s
12	10	Connected	–	10	–	–

of 11 cm. With this barrier, the ability of the Bluetooth connection to be connected becomes reduced.

From the test in Table 6 above, the connection between Bluetooth and controller with a certain height range is obtained where the Bluetooth connection that is enabled to turn off the motorcycle engine will be connected properly at an altitude of 6 m. This test is carried out at altitude and is parallel to the motorbike below; the results of this test can be seen in Table 4.6, but Bluetooth cannot be connected to the controller if the smartphone is at an altitude of more than 7 m.

4 Conclusion

Based on the results of testing that has been done on the security system at this motorbike, system can be controlled properly according to the specified distance. This system works with relay switching schemes which can be activated via a micro-controller using Bluetooth on the smartphone. The results of the tests that have been done have shown that the system has been able to work according to the scheme designed with the maximum distance of communication between smartphones and motorbikes via Bluetooth media which is ±12 m. Therefore, with the use of this system, the level of vehicle safety can be increased.

References

1. A. Septryanti, F. Fitriyanti, Rancang Bangun Aplikasi Kunci Pintu Otomatis Berbasis Mikrokontroler Arduino Menggunakan Smartphone Android. Comput. Eng. Sci. Syst. J. **2**(2), 59–63 (2017)
2. P. Faruki et al., Android security: a survey of issues, malware penetration, and defenses. IEEE Commun. Surv. tutorials **17**(2), 998–1022 (2014)
3. Y. Arta, E.A. Kadir, D. Suryani, KNOPPIX: parallel computer design and results comparison speed analysis used AMDAHL theory, in *Information and Communication Technology (ICoICT), 2016 4th International Conference on*, 2016, pp. 1–5

4. W. Enck, M. Ongtang, P. McDaniel, Understanding android security. IEEE Secur. Priv. 7(1), 50–57 (2009)
5. N. Jinaporn, S. Wisadsud, P. Nakonrat, and A. Suriya, Security system against asset theft by using radio frequency identification technology, in *2008 5th International Conference on Electrical Engineering/Electronics, Computer, Telecommunications and Information Technology*, 2008, vol. 2, pp. 761–764
6. M. Banzi, M. Shiloh, *Getting Started with Arduino: the Open Source Electronics Prototyping Platform* (Maker Media, Inc., 2014)
7. Y.A. Badamasi, The working principle of an Arduino, in *2014 11th International Conference on Electronics, Computer and Computation (ICECCO)*, 2014, pp. 1–4
8. S.F. Barrett, *Arduino Microcontroller Processing for Everyone!* (Morgan and Claypool Publishers, 2010)
9. M. Syahwil, Panduan Mudah Simulasi dan Praktek Mikrokontroler Arduino, Yogyakarta Penerbit Andi (2013)
10. D.A. Prasetyo, Perancangan Dan Pembuatan Alat Pengendali Kendaraan Roda Dua Berbasis Android (2016)

A Comparative Analysis of Data Mining Analysis Tools

Eugene Istratova, Dina Sin, and Konstantin Strokin

Abstract The paper presents the results of a comparative analysis of data mining tools where various methods were considered and categorized based on their functionality. The key comparison criteria were identified, and analysis was carried out. As a result, the most optimal toolkit for the comprehensive implementation of research projects was the data mining suite tool (DMST) method which was the best in terms of providing a complete set of analytical tools for establishing the processes of search, processing, analysis, visualization, storage, and replication of data. This comparative analysis can help in the choice and selection process of data mining tools.

Keywords Data mining · Data mining tools · Scientific research · Data analysis

1 Introduction

Nowadays, the use of information technology in scientific research is not limited to the automation of data collection, but also extended to data analysis. As a result, the collection, processing, and analysis of data in scientific research become no more separated methods as various toolkits combined into one information platform. Such platforms include intelligent systems that not only processing but also storing data [1, 2].

E. Istratova
Department of Automated Control Systems, Novosibirsk State Technical University, K. Marksa Pr. 20, 630073 Novosibirsk, Russia
e-mail: istratova@mail.ru

D. Sin (✉) · K. Strokin
Department of Construction, Technical Oil and Gas Institute, Sakhalin State University, Pogranichnaya Street 2, 693007 Yuzhno-Sakhalinsk, Russia
e-mail: shin.dina16.01@gmail.com

K. Strokin
e-mail: strokin07@rambler.ru

P. K. Pattnaik et al. (eds.), *Proceedings of International Conference on Smart Computing and Cyber Security*, Lecture Notes in Networks and Systems 149,
https://doi.org/10.1007/978-981-15-7990-5_16

One of the relevant areas of application of intelligent systems in practice is the data mining technology, which is a tool for intelligent data search and analysis. The main feature of this technology can be considered the ability to visually visualize the results of scientific research and calculations. At the same time, unlike classical artificial intelligence systems, this technology does not simulate natural intelligence, but only strengthens it with using a combination of data warehouses that are significant in volume, powerful computing servers, and search engines [3].

The use of data mining technology allows revealing the fullness and diversity of interactions in the data structure, determining the patterns characteristic of individual data samples, and associating them with the values of the entire population [4, 5]. This is due to the fact that data mining methodology provides the opportunity to solve not only direct but also inverse problems, develop multivariate experimental models, and predict the results of experiments that were just planned [6]. That is why the application of data mining methods is a fundamentally new approach to conducting both applied and fundamental scientific research.

2 Literature

In the literature, there are confirmations of the application of data mining methods in the development of intelligent systems [2], for quantitative analysis of data extracted from digital information resources [7], in structuring metadata in the history of science [8]. In [9], an algorithm for solving a specific practical problem using data mining tools using fuzzy logic and neural networks is presented.

As other practical areas of using data mining techniques in the literature, the possibilities of automatic detection and prediction of errors in program code were considered [10–12], system analysis of unstructured data [13], development and support of an economic data bank [14–16], meta-analysis of large amounts of data [17, 18].

Thus, the modern scientific approach to the organization of the search for scientific knowledge in intelligent systems is the implementation of complex IT solutions based on data mining technology and allowing the integration of analytical tools of decision support systems.

The aim of the study was to identify effective data mining tools for further use in the scientific field.

To achieve the goal of the study, the basic methods of data mining were studied, criteria for their comparison were determined, and analysis of the data obtained was carried out.

The process of applying data mining technology is to transform large volumes of raw data by collecting, processing, and analyzing it into scientific knowledge. The scheme is presented in Fig. 1.

With a more detailed consideration of data mining process, five successive stages can be distinguished that constituting the full implementation of data mining life circle (Fig. 2). The preliminary steps are the formulation of the purpose and main

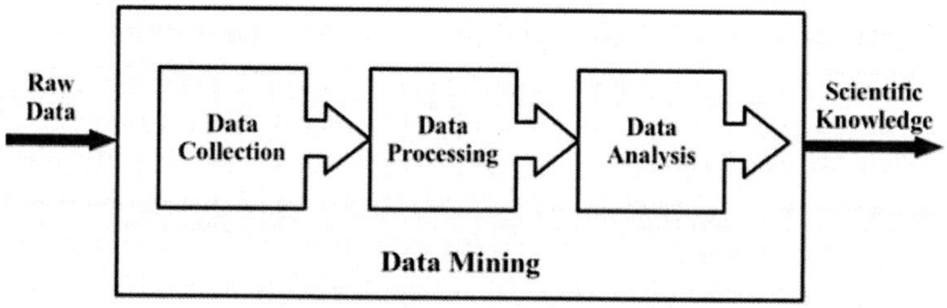

Fig. 1 Scheme for converting data into knowledge using Data Mining technology

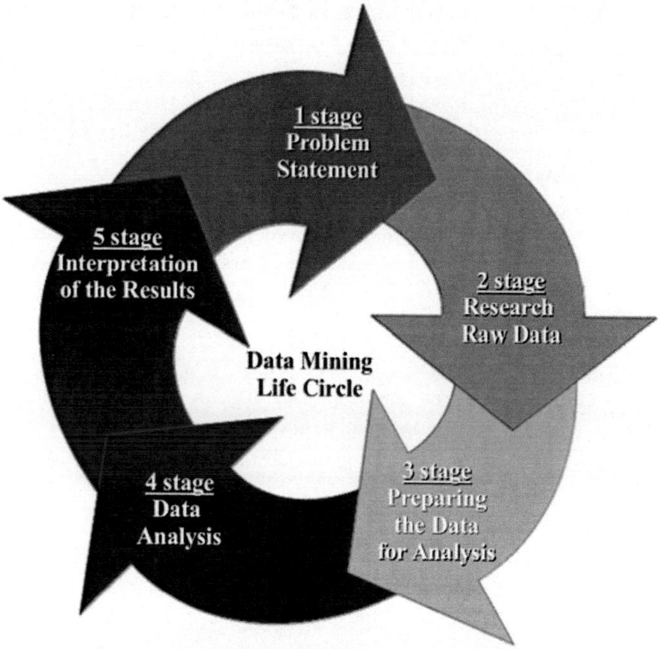

Fig. 2 Full implementation of data mining life circle

objectives of the study, followed by the initial collection and study of data. The result of the obtained data analysis and interpretation allows correcting the initial stage in the formulation of the problem and continuing the cycle until obtaining structured scientific knowledge.

3 Data Mining Tools

To implement the data mining technology, various information tools are used that differ in functional purpose and technical characteristics. The most common today are the following data mining methods, which have proven in projects of various kinds.

Data mining field tools (DMFT) are data mining tools focused on a specific applied scientific field of activity and allowing to solve specific highly specialized tasks, for example, related to the analysis and interpretation of audio, graphic, or text materials.

Data mining business tools (DMBT) have established as packages for business analytics, containing algorithms for solving statistical problems and generating analytical reports. A distinctive feature of this group of data mining tools is the ability to work with various, including remote data sources.

Data mining mat package (DMMP) is data mining applications that are implemented through built-in mathematical operators and functions, as well as its own integrated programming language.

Specialties data mining tools (SDMT) are specialized data mining tools used to intelligently process a specific family of methods or algorithms using associative rules or neural networks. Moreover, these tools are distinguished by the presence of a high-quality visualization system of the received data, capable of working with audio, video, and graphic data formats.

Data mining suite tools (DMST) are independent IT analytical platforms that include many algorithms and methods for modeling and analysis of intelligent systems. These applications can support work with both structured and unstructured data and also contain all the conditions for a high-quality visualization of the results.

In addition to these IT solutions, data mining tools also include applications that provide technical functionality, do not have their own interfaces and visualization tools, and however, develop separate modules and extension packages for other data mining methods. These tools include

1. Data mining extend package (DMEP)—modules for connecting table processors and mathematical applications.
2. Data mining library tools (DMLT)—sets of library modules that can use the API to connect to other data mining tools.
3. Research data mining tools (RDMT)—development and programming tools for data mining applications, including experimental methods and algorithms for importing and exporting data.
4. Integration data mining tool (IDMT)—applications that can be integrated into other data mining tools to create new software or extensions of importance.

Therefore, it is advisable to conduct a comparative analysis among the full-featured data mining tools, including many algorithms and modeling methods, as well as analysis of the source data. We selected the below tools for further comparison: DMFT, DMBT, DMMP, SDMT, and DMST.

4 Evaluation of Data Mining Tools

In order to compare the data mining tools in terms of data processing time and comparing the results with the manual mode, an array of data was tested. Manual mode in testing means data processing in the LibreOffice Calc table processor. The following were investigated as the main stages of the full implementation of data mining life circle:

- 2 stage. Research raw data.
- 3 stage. Preparing the data for analysis.
- 4 stage. Data analysis.

Figures 3 and 4 present the results of testing a data array manually and using various data mining tools.

According to the data obtained, the shortest processing time was 210 s when processing with DMST tools, which is almost 7 times faster than the processing speed in manual mode. DMBT tools also showed a good time, second only to the stages of data preparation and analysis. The types of curves for DMFT and SDMT are the same, since both tools spend approximately the same time preparing data for analysis and the analysis itself. This is due to the algorithms for applying the associative rules and interpretation of different data types. DMMP tools showed an average result at the stages of initial research and data preparation due to built-in mathematical operators and functions, but lost out at the data analysis stage.

The following criteria were selected as key criteria used for conducting a comprehensive comparative analysis of data mining tools:

1. support for import and export of data from various scientific fields;
2. ability to handle remote data through client–server architecture support;
3. the ability to generate reports according to specified criteria;

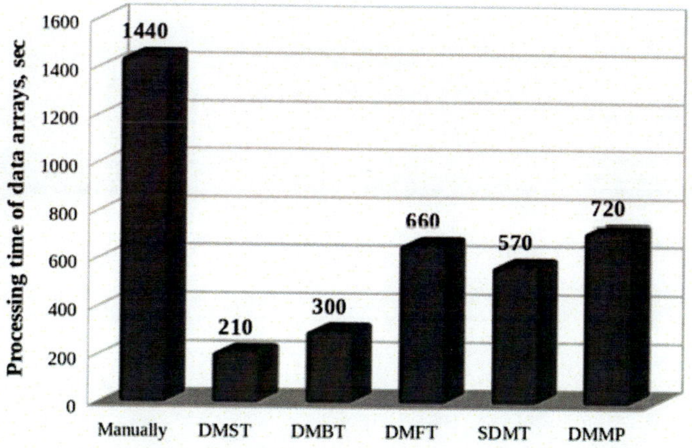

Fig. 3 Data mining tool processing time

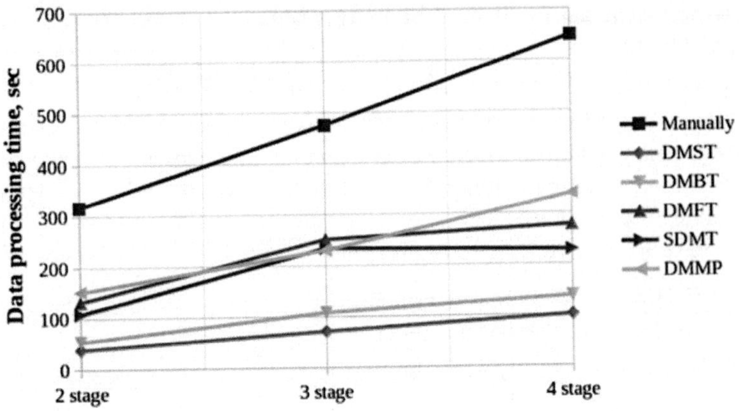

Fig. 4 Dependence of the data processing time on the data mining tool

4. support for various intelligent data processing algorithms;
5. a convenient user-friendly interface;
6. the availability of data visualization capabilities.

Figure 5 presents the results of a comparison of fully functional tools for implementing intelligent data processing. Of the presented methods, the leader is DMST, with the maximum capabilities for processing, storing, and visualizing data. DMBT takes the second position due to the support of clustering and implementation of analytical methods, second only to the DMST in graphical execution and next come the DMFT and SDMT data mining tools with the same set of features. The last position, according to a comprehensive assessment, is taken by the DMMP tool, which loses a lot due to the lack of visualization tools and its own designed interface.

Evaluation of Data Mining Tools by the following criteria

Comparison criteria	Data Mining Tools				
	DMFT	DMBT	DMMP	SDMT	DMST
Importing and Exporting Data	+	+	+	+	+
Supporting a Client-Server Architecture	-	+	-	-	+
Formation	-	+	-	-	+
Supporting of Various Algorithms	-	-	+	-	+
Availability of the Interface	+	+	-	+	+
Visualization	+	-	-	+	+

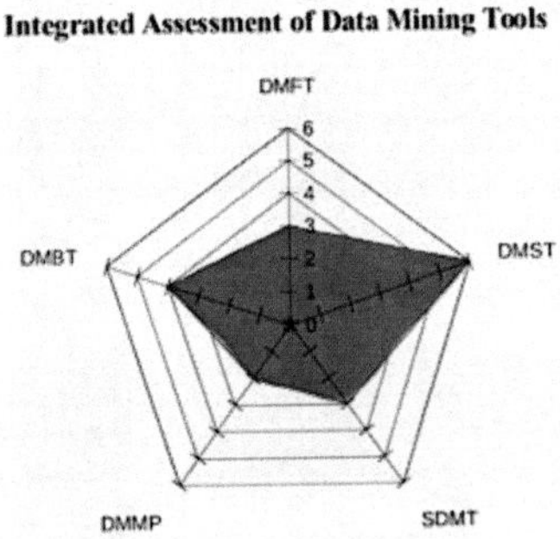

Fig. 5 Results of the data mining tools comparative analysis

5 Result and Conclusion

Thus, as a result of a comparative analysis of data mining tools, it was found that the most suitable method for the integrated implementation of research projects is the DMST toolkit, which has a complete set of analytical tools for organizing the processes of search, processing, analysis, visualization, storage, and replication of data. Additionally, DMST tools provide support for working with both structured and unstructured data through the implementation of modeling algorithms and analysis of intelligent systems. The DMST tools that are most common at present include Russian software products Deductor of Analytical Technologies Company, PolyAnalyst of Megaputer Intelligence Company, as well as international developments of Deductor Enterprise Miner, Data Detective, ISOFT Alice, etc.

References

1. D.V. Nechiporuk, Data mining technology features. Don State Technical University, Rostov-on-Don, Russian Federation **1**(4), 62–65 (2017)
2. A.A. Mitin, Methods and means of data mining. Inf. Syst. Technol. J. **1**(105), 34–38 (2018)
3. O.D. Sukhova, I.A. Popov, Data mining and the initiation of a project to introduce the technology. Electron. Sci. Pract. J. Youth Scientific Herald **8** (2018). URL: https://www.mnvnauka.ru/2018/08/Sukhova.pdf
4. Y.S. Krivenko, A.T. Minasyan, A.O. Razinkov, Research of technologies of data mining (data mining): actual problems of management in the electronic economy. Sat. Scientific tr - Kursk 182–184 (2018)
5. K.V. Mulukova, Comparative analysis of modern tools data mining. Young Sci. **1**, 19–21 (2019)
6. Y.I. Shokin, A.V. Yurchenko, On models of organizing the storage and use of scientific data: basic principles, processes and mechanisms. Inf. Control Syst. **3**, 45–54 (2019)
7. T.S. Umarov, I.Y. Bazhenova, Modern approaches to the mechanisms for extracting causal relationships from unstructured natural language text. Int. J. Open Inf. Technol. **7**, 81–89 (2019)
8. D.E. Prokudin, G.S. Levit, The use of methods for the quantitative processing of digital data in scientific research (using the example of studying the influence of scientific ideas by G. F. Gauze on the development of science). Int. J. Open Inf. Technol. **12**, 51–56 (2018)
9. V.V. Kukartsev, Z.A. Kolmakova, O.L. Melnikova, System analysis of possibilities for extracting named entities using text mining technology. Prospects Sci. **9**(120), 18–20 (2019)
10. D.E. Namiot, V.Y. Romanov, Data analysis for software repositories. Int. J. Open Inf. Technol. **4**, S. 18–23 (2018)
11. C. Rich, R.C. Waters, *Readings in Artificial Intelligence and Software Engineering* (Morgan Kaufmann, 2014)
12. M. White, Toward deep learning software repositories, in *Mining Software Repositories (MSR): IEEE/ACM 12th Working Conference* (IEEE, 2015), pp. 334–345
13. V.V. Kukartsev, O.L. Melnikova, System analysis of the field of tools for processing unstructured data. Prospects Sci. **7**(118), 22–24 (2019)
14. A.N. Oleinik, Content analysis of big quality data. Int. J. Open Inf. Technol. **10**, 36–49 (2019)
15. O.V. Kononova, S.K. Lyapin, D.E. Prokudin, The study of the terminology base of the interdisciplinary scientific field "digital economy" using context analysis tools. Int. J. Open Inf. Technol. **12**, 57–66 (2018)
16. E.E. Chekharin, Big data: big problems. Prospects Sci. Educ. **3**(21), 7–11 (2016)

17. M. Shepperd, D. Bowes, T. Hall, Researcher bias: the use of machine learning in software defect prediction. IEEE Trans. Softw. Eng. **40**(6), 603–616 (2014)
18. G. Nikiporets-Takigawa, «Socio-Political Insider» system: promises and limitations for the political analysis and prognosis. PolitBook (1), 6–20 (2018)

Apple Defects Detection Based on Average Principal Component Using Hyperspectral Imaging

MD. Nur Alam, Rakesh Thapamagar, Tilak Rasaili, Otabek Olimjonov, and Ahmed Abdulhakim Al-Absi

Abstract In this study, a detection method for skin defects in the red delicious apple with hyperspectral imaging (HSI) is presented. The method has been devised to process apple image samples in the spectral range between 403 and 998 nm. The purpose of the method is to detect the following apple skin defects: scabs, bruises, cracks, and cuts. The method detects these defects through a simple average principal component (Avg.PC) without apple image orientation adjustment. The average principal component was found by averaging a couple of PC-1 from the principal component analysis (PCA) of the visual spectral (VIS) region, and the near-infrared (NIR) wavelength region. Finally, a defect detection method was developed based on the Avg.PC that was coupled with a simple thresholding method. The average experimental apple skin detection rate of the method with the investigated apple image samples is 94.28%.

Keywords Apple · Bruise · Crack · Cut · Defect detection · Hyperspectral imaging · PCA · PC · Scab

1 Introduction

Indeed, the most common fruit in the world is an apple. For a long time, the apple industry has been establishing a quality fruit processing, time-consuming, provides better fruits for consumer, and reduces potential economic losses. Also, automatic apple defects the detection to decrease production costs and to compete with countries that have much lower production cost.

Hyperspectral imaging (HSI), like other spectral imaging, collects and processes information from across the electromagnetic spectrum. The goal of hyperspectral imaging is to obtain the spectrum for each pixel in the image of a scene, with the

MD. Nur Alam (✉) · R. Thapamagar · T. Rasaili · O. Olimjonov · A. A. Al-Absi
Department of Smart Computing, Kyungdong University, Global Campus (Gosung) 46 4-gil, Bongpo, Gosung, Gangwondo 24764, South Korea
e-mail: na@kduniv.ac.kr

P. K. Pattnaik et al. (eds.), *Proceedings of International Conference on Smart Computing and Cyber Security*, Lecture Notes in Networks and Systems 149,
https://doi.org/10.1007/978-981-15-7990-5_17

purpose of finding objects, identifying materials, or detecting processes [1, 2]. Much as the human eye sees visible light in three bands (red, green, and blue), spectral imaging divides the spectrum into many more bands. This technique of dividing images into bands can be extended beyond the visible. In hyperspectral imaging, the recorded spectra have fine wavelength resolution and cover a wide range of wavelengths. Different wavebands are visibilities different.

An unsupervised method for citrus fruit surface detection based on a multispectral image analysis and the PCA was introduced previously. The defects detection of it is 91.5%. Though the rate of defects detection is acceptable compared with an unsupervised method, the performance of the method can be susceptible to image orientation [3]. To detect the common defects on oranges, an HSI system has been built for acquiring the reflectance images of orange samples in the spectral region between 400 and 1000 nm. The hyperspectral images of samples were evaluated using PCA with the goal of selecting 3 or 6 wavelengths. The proposed method with two-band ratio and a thresholding showed a better recognition rate of 93.7%, and it could be effective to identify stem-ends from the common defects on oranges [4].

Fresh market fruits like apples are classified into quality categories according to their size, color, shape, and the presence of defects. The first two quality criteria are automated on industrial graders, but fruit classification according to the presence of defects is not yet efficient and consequently remains a manual operation, repetitive, expensive, and not reliable [5]. A classification according to apple skin defects (hereafter called defects) such as scab, bruise, crack, cut with considering stem-end, and calyx-end for which computer vision is employed is cumbersome. A feasible method is to classify into two types: good apples without defects and bad apples with at least one defect. Till now most advance technologies struggle to detect these kinds of defects [6, 7]. A near-infrared (NIR) rule-based vision system for the automation of the apple defect inspection was developed. This system used a monochrome CCD camera equipped with a 700-nm-long pass filter [8]. The results showed that stem-/calyx-ends were the primary cause of inspection errors for both defective and good apples. An asymmetric second difference method using chlorophyll-absorption wavelength at 685 nm and two wavelengths in the NIR band provided the best visual separation of the defective area from the sound area of Fuji apples, and so on [9]. But the method did not consider stem-/calyx-ends because it was a general-purpose imaging system not related to apple classification. The line-by-line defects detection algorithm with a threshold function was developed using three wavebands at 676, 714, and 779 nm [10]. The algorithm did neither consider calyx-ends nor differentiated normal skin on a stem-end from defects.

The main objective of this study is to find a method for the detection of defects on the apple surface. The proposed method considers apple defect detention without limit of apple image orientation adjustment and apple image masking using HSI. For this purpose, the research strategy is as follows: (1) to select the effective wavelength regions of the HSI for detecting the various skin defects; (2) to establish the average principal component (Avg.PC) way to detect the various skin defects on apple surfaces in full; (3) to develop a system to classify good/bad apples more accurately based on HSI.

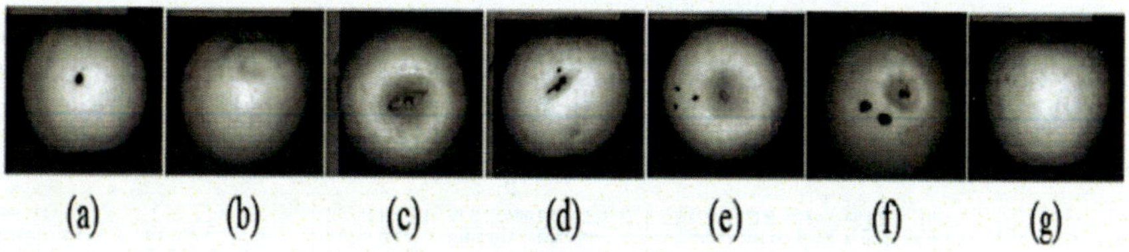

Fig. 1 Data set with different apple skin types: scab (**a**), bruise (**b**), crack (**c**), cut (**d**), calyx (**e**), defect with stem (**f**), normal (**g**). Images are from the 729.3 nm wavelength

The rest of the paper is organized as follows: Sect. 2 lists the sample data collection. Section 3 describes the proposed method. Section 4 displays experimental results and analysis. This paper is concluded in Sect. 5.

2 Sample Data Collection

2.1 Apple Collection

Red delicious apples were used in this study. The samples were provided by the Agro-Food Quality and Safety Sensing Lab from the Rural Development Administration (RDA) of the Republic of Korea. The apples were obtained from the apple orchard during the harvest period of 2015; the collection of apples was selected randomly from numerous "tree-runs" and classified visually. The apples were stored in a 4 °C cold room before the scanning. A total of 350 samples were separated into normal and defect apples using a visual inspection. The apple samples comprised classes with normal skin and 6 skin types: scab, bruise, crack, cut, stem-end, and calyx-end as shown in Fig. 1. A total of 350 red delicious apples were used to develop the algorithm, 40 were normal skin samples, and 310 were defect samples.

2.2 Hyperspectral Image Acquisition

For the hyperspectral image acquisition, the apples were dumped at the loading end, and the apples randomly moved with the conveyor belt as they were viewed by the overhead camera. To obtain more accurate data, some of the parameters were adjusted before the images were acquired [5]. The conveyor belt moved the apples across the field of view (FOV) of the camera at approximately four apples per second. The distance between the camera lens and the object viewpoint was arranged approximately at 45 cm, and the lighting angle is approximately 45°. The line-by-line scan data were saved in an image file that includes the spectral data along one axis and spatial data along the other axis. The EMCCD camera produced hyperspectral images with size 125 (spatial) × 125 (spectral). The spectrum for each pixel spanned the range from 403 to 998 nm across 125 wavebands (approximately 4.8 nm intervals). The pixel resolution of each image is 502 × 200 pixels, and the data type is two bytes.

3 The Proposed Method

The hyperspectral image data analysis and PCA processing were performed using MATLAB 2019a with the image processing toolbox. The basic algorithm of the proposed method can be subdivided into the following steps, as shown in Fig. 2.

3.1 Correction and Normalization of HSI

The device in the HSI system fully records the noise and the information due to the bump intensity of the different light source wavelengths and its illumination distribution, the different fruit shapes, and the existence of a dark current [11]. Therefore, the hyperspectral image needs to be corrected with white and dark references. The original hyperspectral images HSI_O should be corrected based on dark- and white-reference images, which will reduce the influence of the illumination and the dark current of the camera. In this study, the corrected relative reflectance R_λ is found

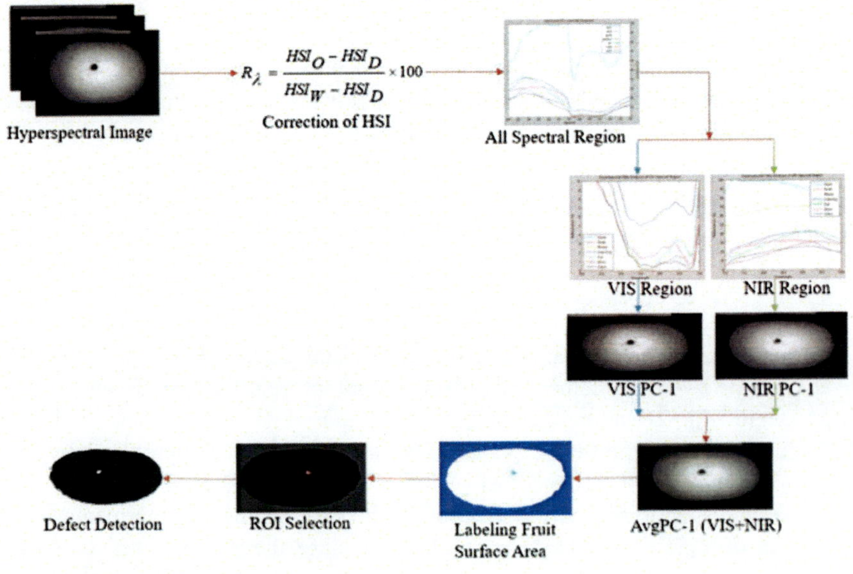

Fig. 2 Apple defect detection algorithm based on average principal component

from the acquired image HSI_O using the following equation [12]:

$$R_\lambda = \frac{HSI_O - HSI_D}{HSI_W - HSI_D} \times 100, \quad (1)$$

where HSI_W is the white-reference image for which a Teflon white board with a 99% reflectance was used, and HSI_D is the dark image with a 0% reflectance that was recorded by turning off the lighting source while the lens of the EMCCD camera were completely closed.

3.2 *Principal Component Analysis for Different Regions*

In the right-hand side of Fig. 3, PCA processing is carried out. In the proposed algorithm, PCA was used to differentiate the defects from an apple normal skin (hereafter called normal). According to the PCA, the mean and covariance of the hyperspectral image are found using the following equations:

$$m_x = \frac{1}{K}\sum_{k-1}^{K} X_k, \quad (2)$$

$$C_X = \frac{1}{K}\sum_{k-1}^{K} X_k X_k^T - m_x m_x^T, \quad (3)$$

where X_k, m_x, and C_X are the kth pixel value of a hyperspectral image X, the mean of the hyperspectral image X, and the covariance matrix of the hyperspectral images, respectively. In the process of the PCA image creation, an image covariance matrix was used to compute eigenvalues and eigenvectors. The PC images were obtained by

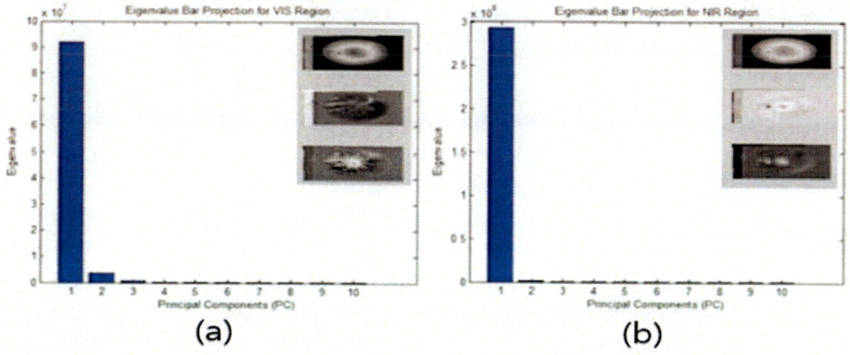

Fig. 3 Eigenvalue projection of PC in VIS-region (**a**). Eigenvalue projection of PC in NIR-region (**b**)

projecting the original images subtracted their mean to the eigenvector. The eigenvalues are equivalent to the variance of each PC image, and those PC images were decreasingly ordered according to the degree of the variance size, where the PC-1 accounts for the largest variance [13].

The PCA of the hyperspectral images can be performed for different wavelength regions: all band spectral (AB) region (403–998 nm), (2) visual spectral (VIS) region (403–695.7 nm), (3) near-infrared (NIR) region (700.5–998 nm), and (4) arbitrary selected several hyperspectral images.

3.3 PCs Selection from Different Region and Average

Figure 3 shows an eigenvalue projection of the original image on new image space defined by different principal components (PC). The PC images have different reflectance features. For example, the apple defects in PC-1 of the VIS-region are shown more clearly than that PC-2 of VIS-region (Fig. 6). And apple skin in the PC-1 of NIR-region has less noise than in the PC-3 of VIS-region (Fig. 7). To take advantage of those features, we propose a method to average PC images, which is called Avg.PC. In this method, the average PC image is obtained by selecting PCs from different regions: AB, VIS, and NIR and averaging them through the following equation:

$$\text{Avg.PC} = \frac{\text{PC-1} + \text{PC-2} \cdots \text{PC-}n}{n}, \tag{4}$$

where PC-1, PC-2, and PC-n are the PC images selected by arbitrary region. The number of the selected PC is n. Note that Avg.PC can be a single HS image or the average of arbitrary selected HS images, which depends on the implementation of the basic algorithm.

3.4 Defect Detection by Threshold

Thresholding is a fundamental method to convert a grayscale image into a binary mask so that the objects can be segmented from the background image. The threshold is very sensitive to the segmentation processing. If the threshold is low, both the objects region and a significant amount of background can be detected. In contrast, if the threshold is high, only a small subset of the object region can be detected. In our approach, the best simple global thresholding is Otsu's method.

After passing through the global threshold, the pixel difference can be a binary image that either contains a "0" (background) or a "1" (foreground). Then, the morphological operation will filter the image to reduce any undesired noise and to fill holes in the binary image. The morphological basic operations, based on a disk 2 kernel structuring element, are used to eliminate this error.

3.5 Labeling of Apple Surface Area and Region of Interest (ROI) Selection

A connected component is a set of pixels in which all pixels are connected to each other. Connected component labeling is a methodology to group all connected pixels into components based on pixel connectivity and mark each component with a different label. In a connected component, all pixels have similar values and are, in some manner, connected to each other [12].

In this study, after labeling of the apple, the fruit surface area is marked with false color indicating the label, and the centroid is calculated. If we found more than one label, we proceed to create the edges of the boundary. Those edges represent the ROI of the image. We consider an ROI always that its area is bigger than zero.

3.6 Labeling of Apple Surface Area and ROI Selection

Finally, in our study, we can define if an apple has defects (scab, bruise, crack, and cut) or if it is normal (good) using the algorithm we proposed. Subsequently, we compared the generation of different label ROIs with a human evaluation of possible defected areas. We proposed that any apple that has at least 4-pixel label of ROI is a defect apple. We have also experimented with different sizes of ROI; we have used values smaller than 4.

4 Experimental Result and Analysis

The average spectral reflectance of 125 wavebands in the range between 403 and 998 nm was collected from the ROIs of the red delicious apple samples, as shown in Fig. 4. This average spectral data, which is from each type of the defects and normal, was extracted to an HSI training data set. The HSI data set includes both the spectral and spatial information, and each pixel of an apple image corresponds to one spectrum. Each spectrum was obtained from a polygonal ROI with a size of approximately from 90 pixels to 120 pixels.

Typically, the spectral reflectance of every skin type in the NIR-region is higher than that in the VIS-region. The spectral reflectance of the normal shows a higher reflectance than that of the defects in the NIR-region. Also, the reflectance difference between the skin types in the NIR-region is larger than that in the VIS-region;

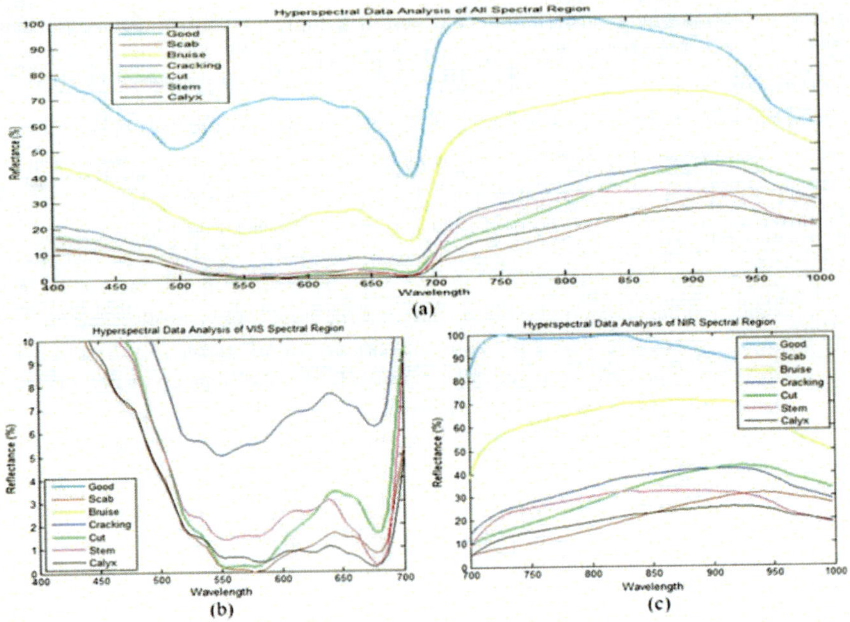

Fig. 4 The average representative ROIs reflectance spectra obtained from apple samples with different types of apple skin. **a** AB-region (403–998 nm), **b** VIS-region (403–695.7 nm), and **c** NIR-region (700–998 nm)

therefore, if only one image is used to identify defects, the one from the NIR-region must be more suitable.

In the NIR-region, the wavelength with the largest difference between the reflectance of the normal and that of the defects is 729.3 nm, and the wavelength has the largest reflectance of the normal as well. This wavelength could be a good candidate to identify defects with single wavelength image. Hereafter, this wavelength is called optimal wavelength to be used to compare it with the proposed method.

4.1 PC Analysis and Average Image

A PCA processing is performed in the three regions: AB, VIS, and NIR. The equivalent PCs to the reflectance of all the wavelength in the region was found. In this section, the first three images (denoted by "PC-1" to "PC-3") that were obtained from the PCA for the hyperspectral reflectance images of the apple training data set are analyzed. In the plot of the eigenvalues of Fig. 3, even though the first three PCs: PC-1, PC-2, PC-3 show relatively large eigenvalues, the images with relatively large eigenvalue did not always provide a clear correspondence to the original image as shown in Fig. 6a PC-3, and Fig. 7a PC-2. Hence, the individual PC image had to be visually evaluated to determine the PC images with the least variation to the original clear hyperspectral image and the best contrast between the apple defects and the normal skin. In general, as the PC number increases, the projected image of that PC becomes visually different from the original image. Note that the PC image analyses below were performed in view of not using apple image masking.

4.2 PC Image in the AB-Region

PC-1, PC-2, and PC-3 for the AB-region of wavelength 403–998 nm are shown in Fig. 5. The defect like the bruise which is not readily visible in the original single band image, Fig. 1b, is more apparent in these transformed PCA images: Fig. 5b: PC-1, PC-2, and PC-3. These images have the following features.

1. The PC-1 images have light gray normal skin, black background, and defects except bruise. Also, the boundary of the normal skin and the background is clear. In comparison, the images of PC-2 and PC-3 do not have a clear boundary among the normal skin, the background, and the defects. The PC-2 images have light gray normal skin and background, and in some cases, the PC-3 images have black normal skin and background; in other cases, it has light gray normal skin and background.
2. The PC-1 images have less noise on the normal skin and around the defects, but the images of PC-2 and PC-3 have a lot of noise on the normal skin and around the defects.

Thus, it is clear that the apple defects could be more easily identified in the images of PC-1 than in the PC-2 images and the PC-3 images in the AB-region.

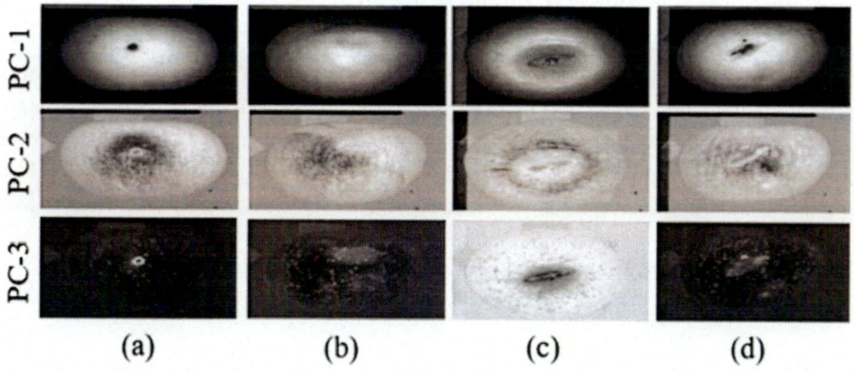

Fig. 5 First three principal component (PC) images obtained using the AB-region from 403 to 998 nm for: **a** scab, **b** bruise, **c** crack, and **d** cut. PC-1 to PC-3 are the first, second, and third PCs, respectively

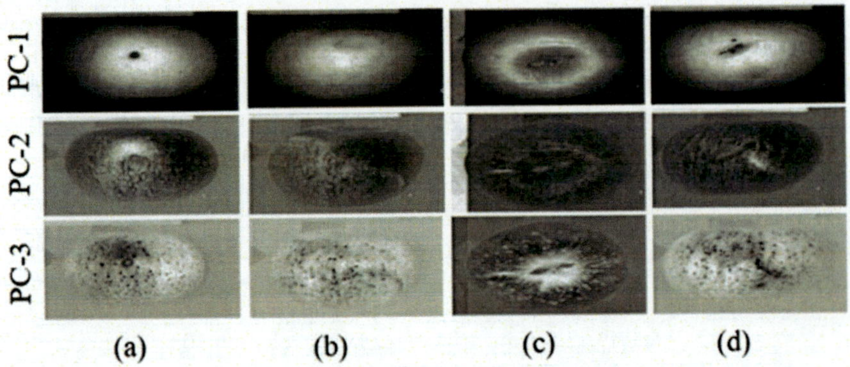

Fig. 6 First three principal component (PC) images obtained using the VIS-region from 403 to 695.7 nm for: **a** scab, **b** bruise, **c** crack, and **d** cut. PC-1 to PC-3 are the first, second, and third PCs, respectively

4.3 PC in the VIS Spectral Region

The first three images of PC-1, PC-2, and PC-3 for VIS-region of wavelength 403–695.7 are shown in Fig. 6.

Generally, the VIS-region spectra images are a little darker than the NIR-region spectral images. Especially the PC-2 images of the VIS-region are much darker than the PC-2 images of the NIR-region. In the PC-2 and PC-3 images, the images have noise on the normal skin and around the defects. In the VIR-Region, the PC-1 images have the best contrast and noisiness as well as that of AB-/NIR-region.

4.4 PC in the NIR Spectral Region

The first three images of PC-1, PC-2, and PC-3 for the NIR-region of wavelength 701–998 nm are shown in Fig. 7. Generally, the NIR-region spectra images are a little brighter than the VIS-region spectral image. Especially the PC-2 images of NIR-region are much brighter than the PC-2 images of the VIR-region. The PC-1, PC-2, and PC-3 images of the VIS-region are clear, and the noise is like that of the AB-/VIS-region. That is, PC-1 images depict the best contrast between the skin defects and the normal skin for the different types of skin.

4.5 Selection of Average PC Image in VIS-/NIR-Region

The PCA of the hyperspectral image was carried out for three regions: AB, VIS, and NIR. In selecting, the PC images to build the proposed method. There are six options:

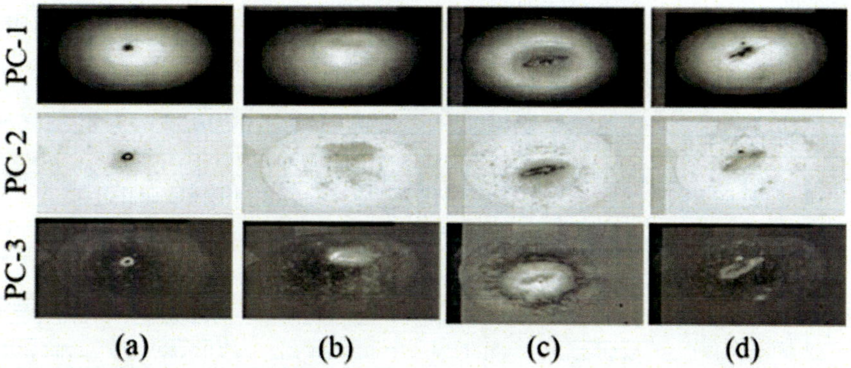

Fig. 7 First, three principal component (PC) images obtained using the NIR-region from 700.5 to 998 nm for **a** scab, **b** bruise, **c** crack, and **d** cut. PC-1 to PC-3 is first, second, and third PCs, respectively

AB, VIS, NIR, AB/VIS, AB/NIR, VIS/NIR, and AB/VIS/NIR. The evaluation of Sect. 4 showed that VIS/NIR has the best performance among them. The reason why VIS/NIR has the best performance can be explained as follows. The reflectance of the defects in the VIS-region is more sensitive than that in the NIR-region like the crack and the cut of Fig. 8. Namely, the size and the color of those defects in the VIS-region are larger and darker than that in the NIR-region. In contrast, the reflectance of the normal in the NIR-region is more sensitive than that of the VIS-region. The NIR-region has no noise in the normal skin region. Averaging these two PC-1 images could increase the possibility to take advantage of the clear defects of the VIS-region and the clear normal skin of NIR-region concurrently. Figure 8 Avg.PC-1 shows the averaged results of the VIS PC-1 and NIR PC-1.

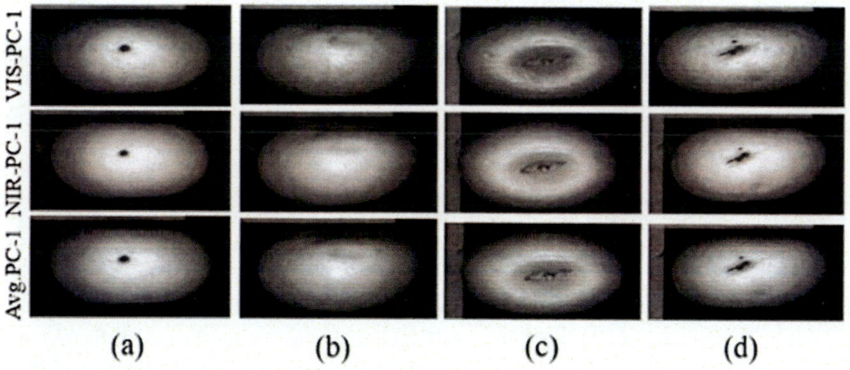

Fig. 8 Representative PC-1 images of the two groups of the spectral (VIS, NIR) regions and the average PC-1 for: **a** scab, **b** bruise, **c** crack, and **d** cut

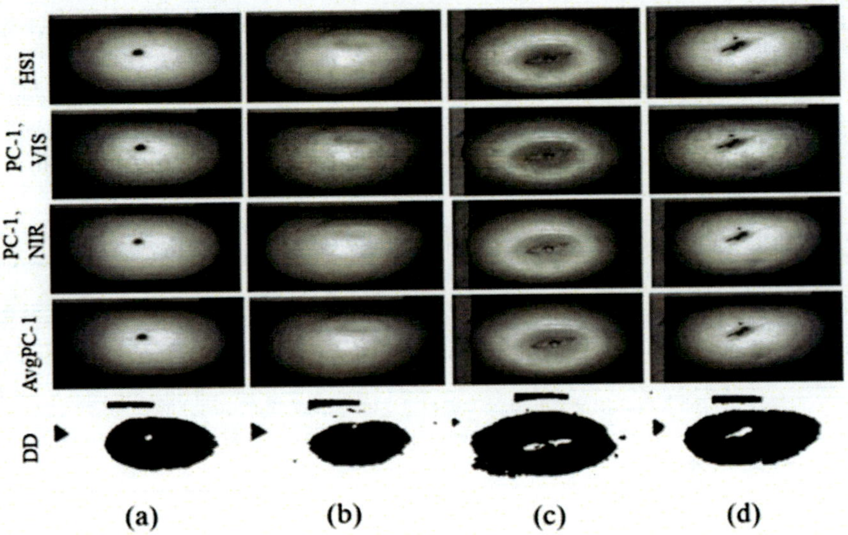

Fig. 9 Detection results of a defective apple sample for: **a** scab, **b** bruise, **c** crack, and **d** cut

4.6 Detection Result of Defective Apple Sample

Figure 9 shows the detection results of a defective apple sample for each skin type using the proposed method. The illustrated four defects were randomly chosen from the test data set. The first row shows the original apple images: HSI. The second row shows the PC-1 VIS ratio images. The third and fourth rows show the PC-1 NIR image and Avg.PC-1 generated from VIS and NIR (PC-1) images above, respectively. The fifth row corresponds to the binary images: DD getting from the developed method based on Avg.PC-1 images. Finally, represents the defects detection in binary image: DD.

4.7 Evaluation and Discussion

The apple detection accuracy of defective and normal apples of the proposed method was examined using 350 independent test data sets. The accuracy comparisons of the proposed method: Avg.PC-2R with the comparison group: Avg.PC-1 3R, AB PC-1, NIR PC-1, VIS PC-1, and optimal band constructed from the basic apple defects detection algorithm are listed in Table 1. For example, in the optimal band of the comparison group, optimal band (729.3 nm) image was used instead of Avg.PC-2R image.

The test data set was composed of the following two classes: first, the defected class having 285 samples with six types of skin defects and second, the normal class

Table 1 Apple skin type detection of the proposed method and comparison group

Class	Skin type	Number of samples	Number of detection (skin type of detection rate)					
			[t1]Avg.PC-1 2R	[t2]Avg.PC-1 3R	[t3]AB PC-1	[t4]NIR PC-1	[t5]VIS PC-1	Optimal band (729.3 nm)
Defects	Scab	85	79	77	75	79	72	79
	Bruise	55	48	43	42	42	43	39
	Crack	75	73	73	61	59	61	64
	Cut	70	68	68	64	65	64	67
Normal	Normal	65	62	61	58	59	52	59
Total (%)	[a]*D* (4) + [b]*N* (1) = 5	350	94.28	92.00	85.71	86.85	83.45	88.00

[t1]Avg.PC-1 2R: average of principal component 1 from 2 regions (VIS and NIR)
[t2]Avg.PC-1 3R: average of principal component 1 from 3 regions (AB, VIS, and NIR)
[t3]AB PC-1: principal component 1 from AB-region
[t4]NIR PC-1: principal component 1 from NIR-region
[t5]VIS PC-1: principal component 1 from VIS-region
[a]*D*: defect sample type
[b]*N*: normal sample type

having 65 samples with one type. The defected class has the following number of samples: (1) 85 samples of scab, (2) 55 samples of bruise, (3) 75 samples of crack, (4) 70 samples of cut.

The average apples detection rate of the proposed method is 94.28%. And that of the comparison group: Avg.PC-1 3R, AB PC-1, NIRPC-1, VIS PC-1, and optimal band are 92.00, 85.71, 86.85, 83.45, and 88.00, respectively.

5 Conclusion

In this study, a method to classify apples into normal ones or defected ones using hyperspectral apple image was proposed. The hyperspectral bandwidth was divided into three regions: VIS, NIR, and AB. The reflectance of each region was analyzed, and PCA for each region was carried out. Finally, an apple classifying method based on Avg.PC-1 was designed and evaluated.

In the reflectance analysis, the overall reflectance image of VIS-region had a dark gray color with noise. Therefore, in the VIS-region, an apple image with noise could be classified as a defected apple. And that of NIR-region had a bright gray color, some bruises were like the normal skin. These features of the hyperspectral data were the same as that of the PC-1s obtained from PCA processing. It required that both regions should be used tactfully when making an apple classifying algorithm. The evaluation results showed that the proposed method, with Avg.PC-1 2R: averaging the PC-1 of VIS-/NIR-region among the methods constructed from the basic apple defects detection algorithm, has the best detection rate of apple skin, the rate is 94.28%. Note that the proposed method is simple because it does not need apple orientation adjustment and apple image masking.

The proposed method could be feasible for practical applications of detection of various skin defects on the red delicious apple surfaces. However, because of the limited sample size for each skin type of apples, it could not indicate the best fruit classifying wavelengths for the development of an algorithm for all the skin types of apples.

Further research will be required to focus on the following matters: (1) Development of an efficient HSI system with the proposed method. (2) Increasing the sample size and collect samples with other types of skin defect. (3) Application of this procedure to other kinds of apples such as Fuji, Jonagold, Granny Smith, and Cox's Orange Piping.

References

1. C.-I. Chang, *Hyperspectral Imaging: Techniques for Spectral Detection and Classification* (Springer Science & Business Media, Berlin, 2003). ISBN 978-0-306-47483-5
2. H. Grahn, P. Geladi, *Techniques and Applications of Hyperspectral Image Analysis* (Wiley, Hoboken, 2007). ISBN 978-0-470-01087-7
3. F. López-García, G. Andreu-García, J. Blasco, N. Aleixos, J. Valiente, Automatic detection of skin defects in citrus fruits using a multivariate image analysis approach. Comput. Electron. Agric. [Online] **71**(2), 189–197 (2010). Available: https://doi.org/10.1016/j.compag.2010.02.001
4. J. Li, X. Rao, Y. Ying, Detection of common defects on oranges using hyperspectral reflectance imaging. Comput. Electron. Agric. [Online] **78**(1), 38–48 (2011). Available: https://dx.doi.org/10.1016/j.compag.2011.05.010
5. V. Leemans, M.F. Destain, A real-time grading method of apples based on features extracted from defects. J. Food Eng. [Online] **61**(1), 83–89 (2004). Available: https://dx.doi.org/10.1016/S0260-8774(03)00189-4
6. J. Blasco, N. Aleixos, E. Moltó, Computer vision detection of peel defects in citrus by means of a region-oriented segmentation algorithm. J. Food Eng. [Online] **81**(3), 535–543 (2007). Available: https://dx.doi.org/10.1016/j.jfoodeng.2006.12.007
7. N. Aleixos, J. Blasco, F. Navarrón, E. Moltó, Multispectral inspection of citrus in real-time using machine vision and digital signal processors. Comput. Electron. Agric. [Online] **33**(2), 121–137 (2002). Available: https://dx.doi.org/10.1016/S0168-1699(02)00002-9
8. Z. Wen, Y. Tao, Building a rule-based machine-vision system for defect inspection on apple sorting and packing lines. Expert Syst. Appl. [Online] **16**(3), 307–313 (1999). Available: https://doi.org/10.1016/S0957-4174(98)00079-7
9. P.M. Mehl, Y.R. Chen, M.S. Kim, D.E. Chan, Development of hyperspectral imaging technique for the detection of apple surface. J. Food Eng. [Online] **61**(1), 67–81 (2004). Available: https://doi.org/10.1016/S0260-8774(03)00188-2
10. H. Lee, C.-C. Yang, M.S. Kim, J. Lim, B.-K. Cho, A. Lefcourt, K. Chao, C.D. Everard, A simple multispectral imaging algorithm for detection of defects on red delicious apples. J. Biosyst. Eng. [Online] **39**(2), 142–149. Available: https://dx.doi.org/10.5307/JBE.2014.39.2.142
11. G. Polder, W. Gerie, V.D. Heijden, Calibration and characterization of imaging spectrographs. J. Near Infrared Spectro. [Online] **11**, 193–210 (2003). Available: https://journals.sagepub.com/doi/abs/10.1255/jnirs.366

12. Md. N. Alam, I. Pineda, J.G. Lim, O. Gwun, Apple defects detection using principal component features of multispectral reflectance imaging. Sci. Adv. Mater. **10**, 1051–1062 (2018)
13. Y. Lia, Y.R. Chen, M.S. Kim, D.E. Chan, A.M. Lecourt, Development of simple algorithms for the detection of fecal contaminants on apple from visible/near infrared hyperspectral reflectance imaging. J. Food Eng. [Online] **81**(2), 412–418 (2007). Available: https://doi.org/10.1016/j.jfoodeng.2006.11.018

Development of an Information System for the Collection and Processing of Big Data in Construction

Eugene Istratova, Dina Sin, and Konstantin Strokin

Abstract The article presents the development results and research of an information system for the collection and big data processing in construction. The designed information system allows you to collect both structured and unstructured data from a large number of different Internet information resources, store them, process, and analyze. Furthermore, the system provides the ability to flexibly configure the frequency of data monitoring, visualize data analysis, and output processes.

Keywords Big data · Unstructured data · House-building plants · Construction · Apache Hadoop · MapReduce · Data analysis

1 Introduction

Recently, information technologies in construction have been used to optimize a lot of technical and management-related business processes. Moreover, along with the use of traditional computer-aided design (CAD) and building information modeling (BIM) systems, the use of big data technologies becomes necessary to collect and process the huge amount of data in such dynamic environment which exceeds the existing traditional volume [1, 2].

As one of the most indicative areas of applying big data in construction is the use of these technologies for information and decision support of the work of house-building

E. Istratova
Department of Automated Control Systems, Novosibirsk State Technical University, K. Marksa pr. 20, 630073 Novosibirsk, Russia
e-mail: istratova@mail.ru

D. Sin (✉) · K. Strokin
Department of Construction, Technical Oil and Gas Institute, Sakhalin State University, Pogranichnaya Street 2, 693007 Yuzhno-Sakhalinsk, Russia
e-mail: shin.dina16.01@gmail.com

K. Strokin
e-mail: strokin07@rambler.ru

P. K. Pattnaik et al. (eds.), *Proceedings of International Conference on Smart Computing and Cyber Security*, Lecture Notes in Networks and Systems 149,
https://doi.org/10.1007/978-981-15-7990-5_18

plants. Construction organizations deal with large modules of construction processes such as manufacturing components, products and structures; conducting transport, construction and installation works; design, and preparing designed-estimated documentation. Therefore, house-building plants form an integrated constantly updated information system. In management-related business processes of house-building plants, the data is complex, huge in size, frequently updated, with high different variety of data types. The most used data operations are searching, processing, and storage of unstructured data, interpretation and conversion of data from one format to another [3].

Thus, the purpose of this study was to develop an information system to optimize the collection, processing, and storage of big data which are required for the main operation of house-building plants.

To achieve the goal of the study, it was necessary to collect data from various Web sites to support our research on the operation of house-building plants. Data such as prices of certain building materials, construction, and assembly works which later on can be done by various companies to determine average market value. Since majority of Web sites data are available in unstructured form, and using the application programming interface (API) to automate the data collection process is impractical [4, 5].

2 Information System Modeling

The initial stage of the developed information system is the data model design with context diagram of the distribution and interaction of data flows. During the implementation of this stage, the key data streams were identified and associated with the process of obtaining data from the external environment along with the flows of internal data circulating within the framework of the information system. The information system context data flow diagram was depicted in Fig. 1.

The context data flow diagram includes the following key processes: definition of search objectives and criteria data, data collection, data upload, data management and analysis, data output, and verification.

The first process is initiated by the user based on data criteria, resulting inputting data in the proper formation of the target, objectives, and basic requirements for the data collection. During this process, keywords are determined, possible search queries are formed, and thematic content is revealed. The user determines the format of the data (text, audio recordings, fragments of graphic images, video recordings, etc.) along with the frequency of the searched data on information resources on the Internet. In addition to the search criteria, the addresses of certain information resources, and the time of the search run can be indicated in order to set the frequency of the user data collection mode. At the end of this process, objectives and criteria for data collection are formed.

The second process is the data collection from information resources on the Internet. In the course of this process, new and periodic checks are automatically

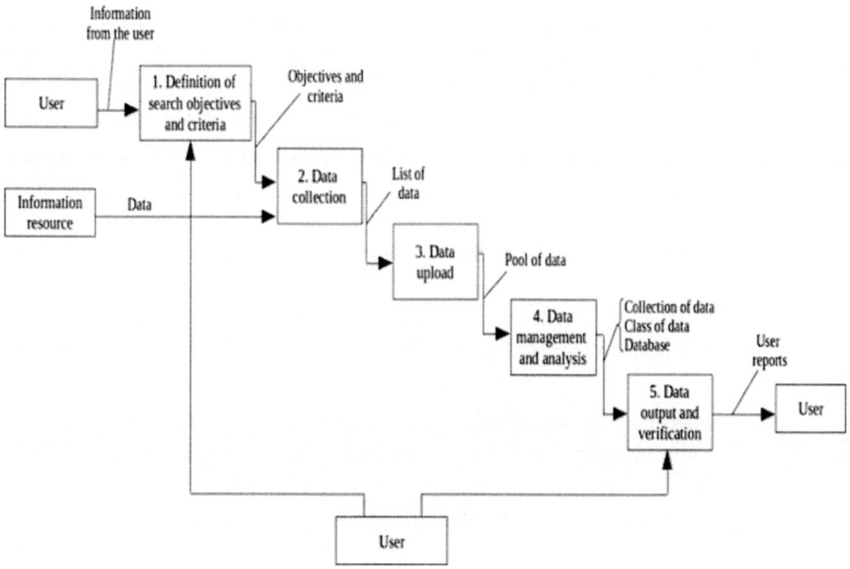

Fig. 1 Context data flow diagram of the developed information system

launched. The process is carried on a sequential order fashion: generating a request for data collection, sending a request, waiting for a response from an information resource, receiving and finally processing a response. The process is based on data obtained from information sources according to the parameters specified by the user. The output is a list of data for further uploading into the information system.

The next logical process is to upload data into the information system for further data cleaning, normalization, and processing. This process not only provides loading the list of data but also their preliminary processing by comparing the downloaded data with its previous version in order to detect the presence or absence of changes.

After the uploading into the information system, the data pool is processed and analyzed. Data management and analysis are made by extracting and structuring objects from the data pool. Data analysis is carried out automatically and identifies the existence of relationships within the pool, calculates the probability of the relationship to a certain group structure or object. Structured data collections with defined classes and database are formed at the output.

The final process of the context data flow diagram is the output with data verification, provided in the form of reports to the user. During this process, the following actions are carried out: selection of data presentation templates, preparation of data for display, generation of reports on key operations in accordance with established filters and user preferences. Data verification is conducted by the user via confirming the results of the analysis.

Therefore, based on the given context data flow diagram, the information system structure was designed with the following set of the modules:

- module for providing storage and access to found data;
- data collection module for implementing a user-defined algorithm of work;
- module for analysis of found data (for pool data);
- interaction module for providing access to all information resources according to the established parameters, as well as launch and formatting options for these parameters;
- graphical user interface with the set of tools for viewing data in the system and the results of the processing.

3 Big Data Processing

The listed set of the mentioned modules and the search and data collection functionality can be provided through carrying out a context-based search and parsing technology implemented via the hadoop software package.

Apache Hadoop is a framework designed for efficient work with big data technologies and an open source code, which significantly reduces the cost of developing the information system. The main distinctive feature of hadoop is processing data selectivity with minimum time delay of data processing and synchronized capability with multiple big data sources in hadoop external data storages.

Apache Hadoop provides a great infrastructure for processing data in parallel within a distributed storage for large amounts of constant and updated data that have different degrees of structure (be structured, semi-structured, or unstructured).

In Apache Hadoop, data is stored as document-oriented databases in Hadoop Distributed File System (HDFS). The data processing and analysis are carried out using Hadoop MapReduce. In order to form queries and data retrieval from the HDFS, Pig and Hive similar to SQL models are used. A simple MapReduce model is shown in Fig. 2.

The next logical stage in the development of the information system was the selection of the elements for data processing (MapReduce), normalization and analysis of the collected data (Apache Pig and Apache Hive), storage (HDFS) and data visualization (Cloudera Hue).

MapReduce is a distributed data processing model that is used to process large amounts of data in computer clusters. The principle of MapReduce includes two key phases: processing (Map) and data aggregation (Reduce). At the first phase, the preliminary data processing is carried out. As a result, data is transferred to the master node for further separation into separate flows and distribution to worker nodes for processing. The operation of the stage is controlled by the higher-order function (by the map). At the second phase, the data generated in the first phase is aggregated. As a result, the pre-processed data is transferred to the master node from the working nodes to generate a report that is, solutions to the initially formulated problem.

One of the distinctive features of MapReduce is handling the distributed data processing, which effectively implements the processes of preparing large amounts of data. The ability to perform processing and aggregated data operations in parallel

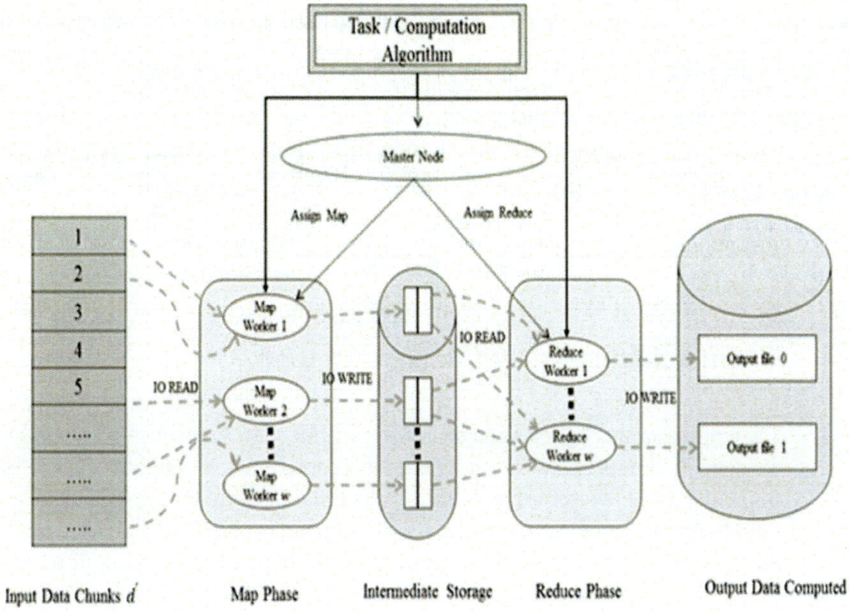

Fig. 2 Hadoop MapReduce model [6]

also facilitates the process of data recovery in case of possible server failures due to load balancing between work nodes.

Apache Pig is a high-level software platform used to normalize big data and is one of Apache Hadoop framework tools. This platform uses the high-level programming language Pig Latin and process data for further analysis, applying, and storage [7].

Apache Hive is an SQL-interface for accessing data, which enables the formation of queries, as well as the aggregation and analysis of data stored on the HDFS file system in the Apache Hadoop framework. Generated using the HiveQL language, queries are converted using Apache Hive in the MapReduce task sequence [8].

HDFS is an independent file system framework used to store files grouped in separate blocks. These files are distributed between cluster nodes, which ensure high stability of the data to individual node failures. HDFS is playing an important role in supporting MapReduce with the data processing.

Unlike Apache Hive, Cloudera Hue is a graphical interface for data visualization and analysis, which includes a set of applications for access to cluster's modules and for application development. Due to this, Cloudera Hue accesses the file storage system from the browser and can be installed on any machine in the Hadoop-cluster.

The next fundamental stage in the development of the information system, after the selection of all its elements, was the implementation of a structure for storing data. As a result, the corresponding database tables were developed. However, due to the fact that the incoming input data is poorly structured, it requires preliminary processing before being entered into the database, using the high-level Pig Latin language.

The main purpose of interaction with the user's Internet resources necessary is to obtain the initial information carried out in the request-response mode. The algorithm of this process includes the search for information on a specific Web site, uploading it to the information system, comparing with existing data from the same Internet resource and determines the frequency of updating information based on the time it was updated.

4 Information System Implementation

To implement the information search algorithm, a special script was developed that accepts input data and converts them into a format that is transferred later for entries to the database. Moreover, this script takes into account the frequency of accessing online services to collect data. The script algorithm was depicted in Fig. 3 and includes the following main working stages: running a script, establishment of baseline data, entering the raw data to the HDFS file system framework, normalizing and subsequent processing the data using Apache Pig and Apache Hive, uploading the prepared data into the corresponding database tables, checking the frequency of launch. If the

Fig. 3 Script algorithm

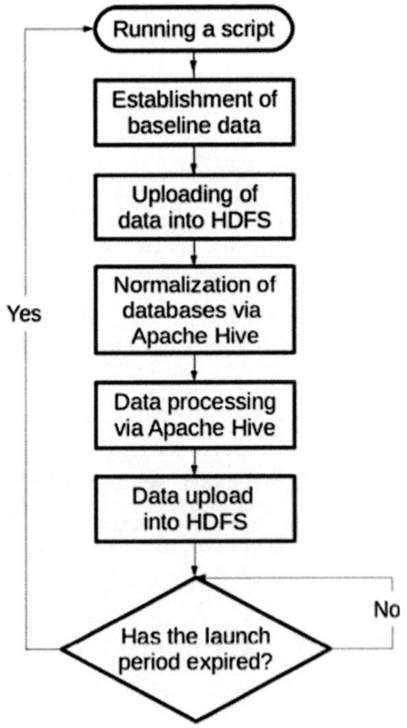

launch period has expired, the script is repeated, and if not, the script is in standby mode.

The designed information system is a Web application for the data collection and processing of the house-building plant's operation and management, as well as other construction organizations. The main distinctive characteristics of the developed information system are the following:

- The ability to collect, process, and further analyze unstructured and weakly structured data due to the context-based search system and technologies for cleaning, normalizing, and preprocessing Apache Pig and Apache Hive data;
- Flexible setting of data monitoring frequency on specified online services and the ability to manually add or remove Web site addresses;
- Visualization of data analysis and output processes through the use of the Cloudera Hue graphical interface, which provides the user interaction with data directly through a user-friendly interface, and not through the console;
- The ability to form criteria for users to select and refine data by using the HiveQL language that underlies Apache Hive and organizes the process of searching and selecting data using SQL queries.

To check the operability and effectiveness of the designed information system, a study was conducted consisting in comparing the results obtained manually and automatically. For this, eighteen thousand thematic links were processed with about 5.6 thousand unique documents uploaded into the information system, of which only 20.7% were structured documents.

As a result, the data collection processing and analysis revealed that the most active document management is typical for such structural units of house-building plants as designed-estimated contract services, manufacturing bases, and integrated teams.

In the overall structure of the house-building plants data, their proportions were 23%, 21% and 18%, respectively. The structure of the house-building plants data was depicted in Fig. 4.

5 Conclusion

As a result, the trial operation of the developed information system revealed reducing the complexity of the data collection and processing and the appearance regularity of certain data in reports depending on the degree and nature of the formation of search criteria. These results were also confirmed by the manual data collection.

In the future, it is planned to compare the results of data collection and processing using various methods, furthermore, to increase the information system's efficiency by reducing the response time of the system to user input by amending its graphical interface.

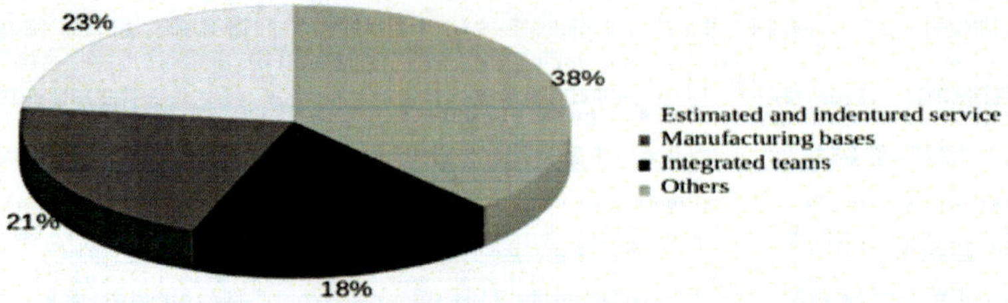

Units of house-building plants	Data types	Share	
		Structured data	Unstructured data
Estimated and indentured service	Data on documents traffic Data on the cost of product Cost data	21 % 25 % 24 %	79 % 75 % 76 %
Manufacturing bases	Data on documents traffic Data on the cost of product Cost data	19 % 23 % 22 %	81 % 77 % 78 %
Integrated teams	Data on documents traffic Data on the cost of product Cost data	20 % 18 % 17 %	80 % 82 % 83 %

Fig. 4 Structure of the house-building plants data

References

1. P.B. Kagan, Analytical studies of large data arrays in construction. Ind. Civ. Constr. **3**, 80–84 (2018)
2. Introduction to Big Data [Electronic resource]. Accessed from: https://www.coursera.org/learn/big-data-introduction
3. Y. Leskovets, A. Rajaraman, J. Ulman, *Analysis of Large Data Sets* (M.: DMK Press, 2016)
4. N. Martz, J. Warren, *Big Data. The Principles and Practice of Building Scalable Real-Time Data Processing Systems* (M: Williams, 2017)
5. D. Silen, A. Meisman, *Fundamentals of Data Science and Big Data. Python and Data Science* (Peter, St. Petersburg, 2018)
6. A.A. Al-Absi, D. Kang, A novel parallel computation model with efficient local memory management for data-intensive applications, in *Proceedings of the 2015 IEEE 8th International Conference on Cloud Computing (CLOUD)*, New York City, NY, USA, June 2015, pp. 958–963
7. Apache Pig [Electronic resource]. Accessed from: https://pig.apache.org
8. Apache Hive [Electronic resource]. Accessed from: https://hive.apache.org
9. MapReduce [Electronic resource]. Accessed from: https://ru.wikipedia.org/wiki/MapReduce

Genetic Algorithm for Decrypting User's Personal Information

Fu Rui, Mohammed Abdulhakim Al-Absi, Ki-Hwan Kim, Ahmed Abdulhakim Al-Absi, and Hoon Jae Lee

Abstract With the information system has been greatly improved, how to fast and effective crack the passwords can be of great significance to detect unknown information, obtain important information, prevent hacker attacks, and protect information security in detective work. Genetic algorithm is a kind of evolutionary algorithm. It searches for the optimal solution by mimicking the choice of nature and the mechanism of genetics. Genetic algorithm has three basic operators: selection, crossover, and mutation. The main method of numerical method for solving NP problem is an iterative operation. This paper studies the possible consequences of password cracking and how to quickly and effectively enter the system to decrypt the personal information. The main work is study the algorithmic process of genetic algorithm and solve the problem. First introduces the specific genetic process of genetic algorithm and then uses an example to show how genetic algorithm works. In the analysis of how genetic algorithms works, it is better we develop better method to avoid the information has been stolen.

F. Rui · M. A. Al-Absi · K.-H. Kim
Department of Computer Engineering, Dongseo University, 47 Jurye-ro, Sasang-gu, Busan 47011, Republic of Korea
e-mail: furui.qilianteng@gmail.com

M. A. Al-Absi
e-mail: Mohammed.a.absi@gmail.com

K.-H. Kim
e-mail: ghksdl90@naver.com

A. A. Al-Absi
Department of Smart Computing, Kyungdong University, 46 4-gil, Bongpo, Gosung, Gangwon-do 24764, Republic of Korea
e-mail: absiahmed@kduniv.ac.kr

H. J. Lee (✉)
Division of Information and Communication Engineering, Dongseo University, 47 Jurye-ro, Sasang-gu, Busan 47011, Republic of Korea
e-mail: hjlee@dongseo.ac.kr

P. K. Pattnaik et al. (eds.), *Proceedings of International Conference on Smart Computing and Cyber Security*, Lecture Notes in Networks and Systems 149,
https://doi.org/10.1007/978-981-15-7990-5_19

Keywords Crack passwords · Information security · Genetic algorithm · NP problem

1 Introduction

With the continuous development of the network, the openness of information systems has greatly increased and global informatization has become a trend of human development. For example, cloud services and mobile office have brought about the demand for open and interconnected information systems in e-commerce. At the same time, key data and personal privacy of enterprises also put forward higher requirements for the security of information systems. Due to the open and interconnected characteristics of the network, the information system is vulnerable to attacks by computer viruses, hackers, malware, and other misbehaviors. Taking customer information theft in China as an example, RSA announced in March 2011. In the attack, in April, hackers stole 77 million customer information, including credit card accounts from Sony's online broadcast station network. Subsequently, Sony stated in May that the attack caused it to lose $1.7 billion. In June, Citibank of the USA confirmed that the banking system was recently hacked the names, accounts as well as email addresses of 210,000 North American bank card users. In December, hackers published the user database of the CSDN Web site online, resulting in the replacement of more than six million user registration information. Many Web site user databases have been published one after another. So we try to use genetic algorithm to explore how hackers invade multiple Web site databases to obtain user information.

Genetic algorithm is a heuristic search algorithm based on the principle of natural selection and natural genetic mechanism [1]. The genetic algorithm draws on the natural selection process of survival of the fittest in the biological world and shows the excellent selection process of the natural world. Darwin called the survival of the fittest in the process of selecting good individuals and keeping bad individuals from being eliminated [2] (Fig. 1).

By simulating the natural mechanism (selection, crossover and mutation operations) of biological genetic evolution in nature, the algorithm continuously inherits

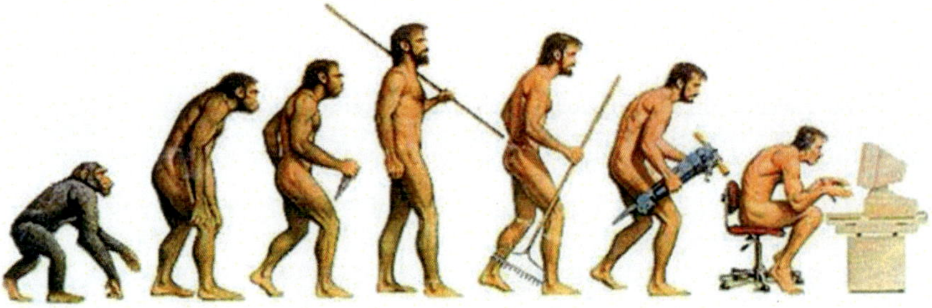

Fig. 1 Process of genetic algorithm

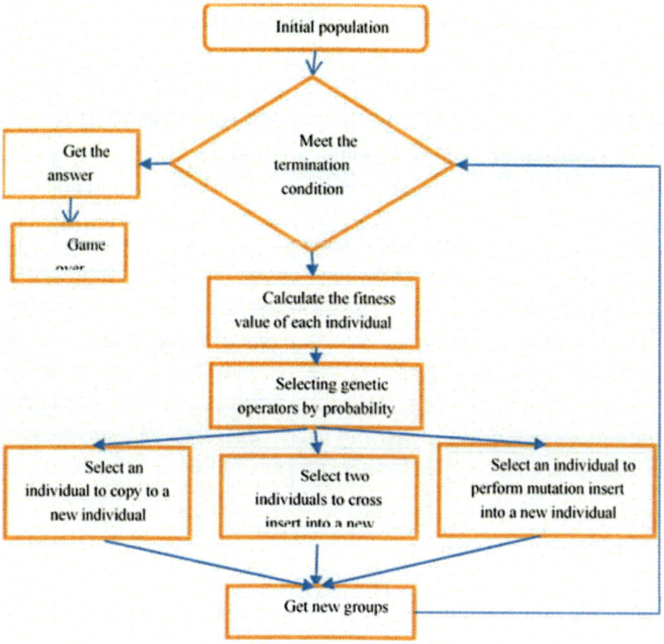

Fig. 2 Flow of genetic algorithm

good genetic genes (optimal goals) to the offspring, resulting in an increased probability that the offspring will produce the optimal solution (the offspring will still There are some poor results) [3]. Its entire genetic algorithm flow is as follow Fig. 2.

2 Process of Genetic Algorithm

In the algorithm of GA, the passcode itself would be called a chromosome. Each digit in our passcode would be considered a gene.

A batch of attempted solutions for the passcode would be a population. The first population is made up of randomly created chromosomes and subsequent populations are made from the evolutionary rules. Each iteration of a new population is called a generation.

For each chromosome in our population, we should test how well it matches with our actual passcode. Our metric for this is called fitness. In our specific case, each attempted solution will be compared against the actual passcode to see how many digits are correct.

Once we have fitness scores for all chromosomes in the current population, we decide that we only want to keep a few of the best ones (the ones with the highest fitness scores) for use in the future generations.

The ones we decide to keep are known as parents. These parents then breed in a process called crossover to create children. There are lots of ways this can be done, in the example we will take a randomly chosen section of genes from one parent and swap them with that same section of genes from another parent. Each time, we do this, we are creating new children.

These children form the population for the next generation.

Something else we can do in the breeding process is called Elitism. This is where we move a small number of the highest scoring parents to the next generation, in their current state. This ensures that some of the best possible solutions persist through generations.

Another useful step in the process is called mutation. This helps avoid local convergence by introducing some level of randomness to a select number of solutions. This is often only applied on a probabilistic basis and only 10% of children are altered. In the experiment, for one in every 10 children, we will change a randomly selected gene to a new, randomly selecting integer (Table 1).

1. Fitness Scoring
   For each chromosome in the population, we are checking the number of correct digits that appear in the correct position. Each chromosome just gets a single number to represent this score. The output is a list of paired chromosomes and their scores [4].
2. Selecting Parents
   In this step, we are sorting the fitness scores in descending order and selecting the top N of them. We are then finding the specific chromosomes that achieved those scores and appending those to a new list [5].
3. Breeding to create new Children
   The first function is the logic for breeding. It takes two parents and randomly selects start and end cut point positions. These positions are used to combine the genes of the two parents to create a new child.

Table 1 Basic concept of genetic algorithm

Concept	Function
Individual	Feasible solution
Chromosome	Encoding of solutions represented by strings and vectors
Gene	Elements in the chromosome
Fitness	Individual adaptability to environment
Population	The selected number is a set of solutions of group size
Reproduction	A set of solutions selected according to the fitness function value
Choose	Eliminate inferior individuals and choose excellent ones
Crossover	Gene interaction in chromosomes
Mutation	Genetic changes in chromosomes

I have visualized this below, each time breeding happens, the cut points will be different as they are randomly selected meaning that the influence of Parent A and B will change.

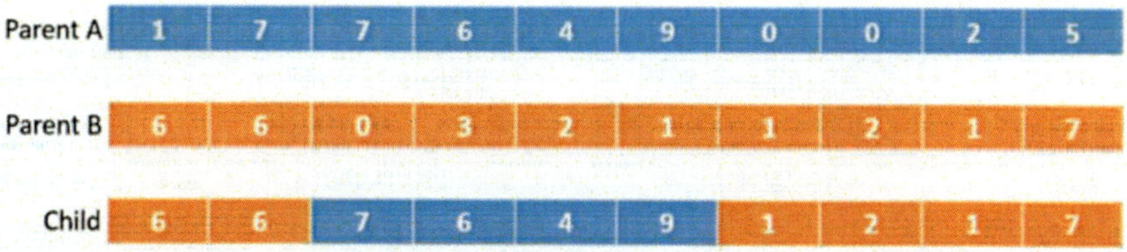

4. Mutation
   Mutating children is not acceptable subject matter in most places. This function runs through each child in our new population and for one in every 10, it will swap out one randomly selected gene for a randomly selected new digit. The output from this is our new population, ready to run through the whole process again. One of the children that has been randomly selected for mutation. The position of the digit to be mutated is random as is the digit that replaces it [6].

3 Experimental Result

1. Initialize the population
   First, convert individual DNA to ASCII code. Secondly, initialize parameters using init_population methods, including desirable substitution of independent variables, re-insertion, expansion size, crossover rate, mutation rate, and breeding algebra [7].

```
 def init_population(self):
 population          =          np.random.randint(low=self.ascii_bounder[0],
high=self.ascii_bounder[1],
                    size=(self.n_population,
self.password_size)).astype(np.int8)
 return population
 def translateDNA(self, DNA): # convert to readable string
 return DNA.tostring().decode('ascii')
```

2. Calculate fitness
   Calculate the fitness of each individual in the group, choose the higher the fitness as the better the individual's genes [7, 8].

```
 def fitness(self, population):
 match_num = (population == self.password_ascii).sum(axis=1)
 return match_num
```

3. Selecting
   By sampling the population according to fitness, individuals with high fitness will be selected with a higher probability and the best individuals in the contemporary population will be selected as historical records.

```
def select(self, population):
    fitness = self.fitness(population) + 1e-4  # add a small amount to avoid all zero fitness
    idx = np.random.choice(np.arange(self.n_population), size=self.n_population, replace=True,
                           p=fitness / fitness.sum())
    return population[idx]
```

4. Crossover
   This is one of the cores of the genetic algorithm and its role is to select some excellent individuals from the original group to form the initial group. First, the random original individuals to be screened are stored in the original group variables and then the fitness of each individual is calculated. Also the individuals that meet certain requirements are selected, that means the fitness value is above a certain threshold. It can be adjusted according to the needs and then qualified individuals are stored in the initial group. If the initial group is not full, a group of original individuals is regenerated to continue screening until the initial group is filled.

```
def create_child(self, parent, pop):
    if np.random.rand() < self.cross_rate:
        index = np.random.randint(0, self.n_population, size=1)  # select another individual from pop
        cross_points = np.random.randint(0, 2, self.password_size).astype(np.bool)  # choose crossover points
        parent[cross_points] = pop[index, cross_points]  # mating and produce one child
        # child = parent
    return parent
```

5. Mutation
   The function mutate_child is mainly used to make a mutation judgment for each individual in a certain generation group according to the generated random value with a certain probability and then randomly generate mutation points for individuals that need to be mutated. The specific operation of mutation is based on the random value that determines the mutation to increase or decrease the value of the current mutation point by plus or minus, it is actually to change the letter on the mutation point to the letter before or after it.

```
    def mutate_child(self, child):
    for point in range(self.password_size):
        if np.random.rand() < self.mutate_rate:
            child[point] = np.random.randint(*self.ascii_bounder)  # choose a ran-
dom ASCII index
    return child
```

```
"D:\untitled\Genetic algorithm\venv\Scripts\python.exe" "D:/untitled/Genetic algorithm/GA4.py"
After generation about 0    times, the best result is:   CZ__dQhTmE|mxrsN:e
After generation about 10   times, the best result is:   DZ#gsLo#UniNersi~y
After generation about 20   times, the best result is:   DZvgseocUniversity
After generation about 30   times, the best result is:   Dongseo+UnivUrsity
After generation about 40   times, the best result is:   DongsLoZUniversity
After generation about 50   times, the best result is:   Dongseo)University
After generation about 60   times, the best result is:   Dongseo#University
After generation about 70   times, the best result is:   Dongseo<University
After generation about 80   times, the best result is:   Dongseo<University
After generation about 90   times, the best result is:   Dongseo~University
After generation about 99   times,find the secret code:      Dongseo University
```

4 Conclusion

The genetic algorithm has a good global search ability and can quickly search out all solutions in the solution space without falling into the rapid decline trap of the local optimal solution. Using its inherent parallelism, it is easy to perform distributed computing. The main work of this experiment is to decipher the user's personal information stored in the system, including user name, user password, and other personal information. Using genetic algorithm can greatly shorten the running time and evolutionary algebra. Using this method can quickly decipher lots of Web site's personal information.

Acknowledgements This work was supported by Basic Science Research Program through the National Research Foundation of Korea (NRF) funded by the Ministry of Education, Science and Technology (grant number: NRF-2016R1D1A1B01011908).

References

1. X. Binglei, S. Yi, L. Rongxi, Genetic algorithm for solving the problem of collecting travel salesman. J. Shaanxi Inst. Technol. **18**(1), 70–75 (2002)
2. P. Lijun, F. Zhuo, Solving genetic algorithm with time window for delivery problem. Syst. Eng. Theory Pract. **32**(1), 120–126 (2012)
3. J. Branke, C. Schmidt, H. Schmeck, Efficient fitness estimation in noisy environments. Proc. Genet. Evol. Comput. **11**(2), 243–250 (2011)
4. E. Cantu-Paz, On random numbers and the performance of genetic algorithms, in *Proceedings of the Genetic and Evolutionary Computation Conference*, Morgan Kaufnann, **2**(2), 311–318 (2012)
5. M.S. Withall, The evolution of complete software systems. Ph.D. Thesis, Department of Computer Science, Loughborough University (2013)
6. G. Xiechao, Study on genetic algorithm application in information security. Comput. Eng. Appl. **48**(25), 127–131 (2012)
7. R. Spillman, Cryptanalysis of Knapsack Ciphers using genetic algorithms. Cryptologia **1**(17), 367–377 (1993)
8. A.J. Bagnall, The applications of genetic algorithms in cryptanalysis. Master of Science Thesis, University of East Anglia (1996)

Text File Protection Using Least Significant Bit (LSB) Steganography and Rijndael Algorithm

Apri Siswanto, Yudhi Arta, Evizal Abdul Kadir, and Bimantara

Abstract Nowadays, thousands of kilobytes personal data are transmitted every day through insecure communication media (such as the Internet, computer networks, communication systems, etc.). This makes data vulnerable to information theft, especially for fraud, illegal trade, and so on. So, there is a need for protecting the information in its storage and transmission. To improve data and information security, in this study, we propose a Least Significant Bit (LSB) steganography to insert message information in a 24-bit jpg image and Rijndael cryptography that is used to encrypt jpg images so that message information can be secured from unauthorized parties.

Keywords Encryption · Cryptography · LSB steganography · Rijndael · Information hiding

1 Introduction

The rapid development of computer technology has triggered crimes that exploit the weaknesses of computer network transmission systems. One form of crime is hackers try to retrieve data and information through the transmission of computer networks or known as a man-in-the-middle attack [1]. Transfer of essential data on companies, agencies, or the military is vulnerable to attack if it only relies on

A. Siswanto (✉) · Y. Arta · E. A. Kadir · Bimantara
Department of Informatics Engineering, Faculty of Engineering, Universitas Islam Riau, Pekanbaru, Indonesia
e-mail: aprisiswanto@eng.uir.ac.id

Y. Arta
e-mail: yudhiarta@eng.uir.ac.id

E. A. Kadir
e-mail: evizal@eng.uir.ac.id

Bimantara
e-mail: bimbimjabrikz@gmail.com

P. K. Pattnaik et al. (eds.), *Proceedings of International Conference on Smart Computing and Cyber Security*, Lecture Notes in Networks and Systems 149,
https://doi.org/10.1007/978-981-15-7990-5_20

a standard security system [2]. Confidential information can be taken and used by irresponsible parties. So, this must be given special attention by the parties concerned. Some ways to overcome this problem is to secure the message using the information hiding technique. Information hiding is a field of science that studies how to hide messages so that they cannot be perceived (both visually and audial). There are two ways techniques used in information hiding, i.e., cryptography and steganography [3].

Cryptography is the study of mathematical techniques related to aspects of information security such as confidentiality, data integrity, and authentication [4]. While steganography is the science that studies, researches, and develops the art of hiding information. Steganography can be classified as one part of communication science [5]. In the digital information era, steganography is a technique and art of hiding information and digital data behind other digital data, so that digital information is invisible.

Recently, some methods can carry out attacks on steganography by utilizing the weaknesses of steganography. These methods are visual attacks and statistical attacks [6]. Visual attacks explain the difference between noise and visual patterns, while statistical attacks to detect the steganography method used. Because the method of attack on steganography has been found, problems arise how to provide security for data so that data can be hidden. Besides, confidentiality can also be maintained from the parties who are not authorized to access it. Therefore, to increase data and information security, in this study, we implement message encryption (cryptography) while hiding data and information in image files.

This paper is organized as follows. Section 2 describes Rijndael and LSB steganography theory. Then, Sect. 3 introduced literature review where different methods of hiding information are discussed. Next, Sect. 4 discusses the research method of this paper. After that, Sect. 5 explained results and discussion. Finally, Sect. 6 presents conclusions and references used at the end.

2 LSB Steganography and Rijndael Algorithm

LSB is a technique commonly used to encrypt confidential information and to decrypt it. The way the LSB method works is to change the cover image's redundant bits which have no significant effect on the bits of the secret message. Figure 1 showed the LSB method mechanism in 8-bit images by using 4 bits of LSB [7].

Figure 1 showed LSB application using the 8-bit pixel-based image media (gray value). Each 8-bit pixel is divided into two parts, namely 4 MSB bits (most significant bits) and 4 LSB bits (lowest significant bits). The LSB part of the message to be inserted is changed to the value. After each pixel has been sprinkled with a secret message, it is reconstructed into a complete image that resembles the original media. In human eyes, the advantages of LSB are less suspicious, easy to enforce, and high eternal transparency. On the other hand, LSB's drawbacks include robustness and

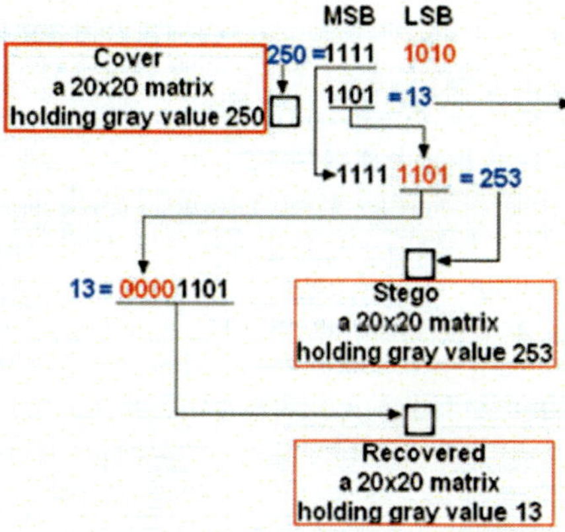

Fig. 1 LSB mechanism

sensitivity to filtering, and scaling, rotation, adding noise to the image, and cropping can damage confidential messages [8].

The Rijndael algorithm used substitution, permutation, and a number of rounds. Each round used a different internal key. The key of each round is called around key. However, unlike DES operates bit-oriented, Rijndael operates in byte orientation. The goal is to minimize software and hardware resources. The Rijndael algorithm works on 128-bit blocks with 128-bit keys with the AddRoundKey process. AddRoundKey is to do XOR between the initial state (plaintext) and the cipher key [9]. This stage is also called initial round. The process carried out in each round is:

1. SubBytes: byte substitution using a substitution table (S-box).
2. ShiftRows: shifting array state lines in wrapping.
3. MixColumns: scrambles data in each state array column.
4. AddRoundKey: perform XOR between the current state of the round key.

The Rijndael algorithm has three parameters [10]:

1. Plaintext: a 16-byte array, which contains input data.
2. Ciphertext: an array of 16-byte size, which included the results of encryption.
3. Key: an array of 16-byte size, which contains a ciphering key (also called a cipher key). With 16 bytes, both the data block and the 128-bit key can be stored in all three arrays ($128 = 16 \times 8$).

3 Related Research

Data security and confidentiality are essential aspects needed in the process of exchanging data on the Internet network. Two techniques can be used for data protection, namely cryptography and steganography. Several studies related to cryptography and steganography, for example, Syawal et al. [11], proposed text message encryption using Vigenere cipher algorithm and LSB technique for inserting messages into images. The proposed encryption was programmed in MATLAB 2014b. The object of research is to enter text into the image to produce hidden files and cannot be accessed by unauthorized parties.

Then, Purba et al. [12] has conducted a study Implementation of Text Message Steganography into Sound Files (.Wav) with Byte Distance Modification in the Least Significant Bit (LSB) Algorithm. The purpose of this study is to hide files with the extension .txt and files ending in .Wav. Data bits are hidden or secured using LSB into the audio media. The result of the study found that the bit values are inserted into the audio media and are still looks like normal so as not to arouse suspicion of the listener. Then, if extracted, it will get back the whole bit values that have been inserted. Therefore, the results of the research show that the resulting wav stego file has a good level of imperceptibility, fidelity, and recovery.

Utomo and Purnama [13] have proposed image steganography with the Least Significant Bit Method for Protection of Communication in Online Media. In this study, a message is inserted in the image file to be extracted again into a message. This method is done to secure the message and avoid unauthorized parties from utilizing the message.

The research conducted by Utomo and Purnama, Purba et al., Syawal et al., and the research that the authors did together secure data by hiding the data into other data. The difference is in the object under study, the research method, and the programming language used in developing the system. Like Purba, hide the.txt file into the file extension .Wav. Syawal used a different algorithm. And Utomo securing the message on the image file can then be extracted again into a message.

4 Research Methodology

The LSB steganography and Rijndael algorithm are implemented using the Visual Basic Net programming language. We used modified LSB steganography method as a medium that will hide text file information in the form of each bit data value into the image media bit values. Data bits that will be hidden or secured with LSB into the jpg image media. The proposed encryption scheme is like Fig. 2.

In the encryption and decryption process, users must input object image files that will be steganography with text files that will be encrypted. Then, the data is encrypted with the Rijndael algorithm. The Rijndael algorithm did the encryption process using substitution and permutation process. For the decryption process, the user enters the steganographic image file and then decomposes it with the Rijndael algorithm so that the ciphertext file returns to the original text file. See details in Fig. 3 the decryption process.

Fig. 2 Encryption process

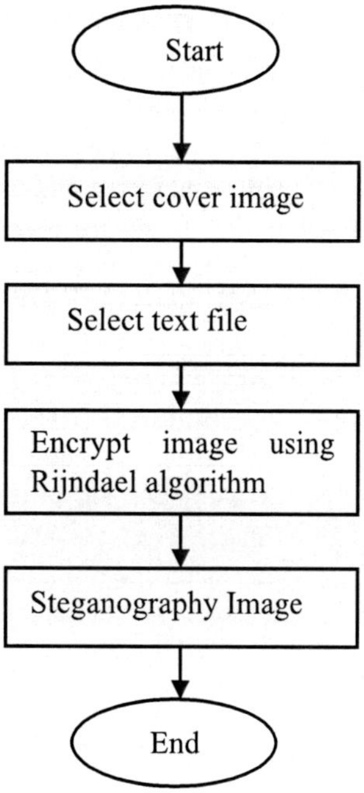

Fig. 3 Decryption process

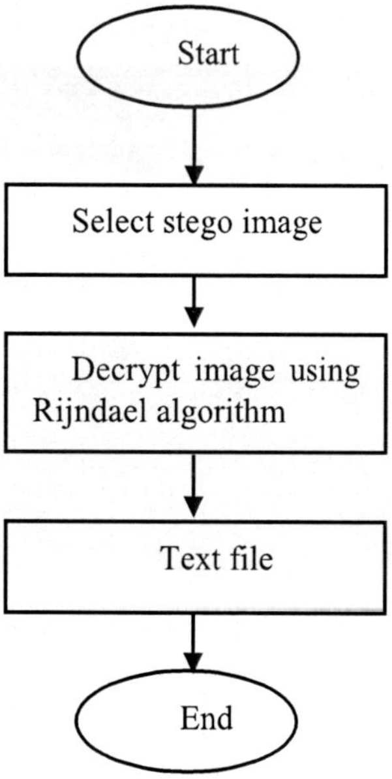

5 Result and Discussion

This research output is an encryption scheme to secure text file. In simple application, the process steps are insert text files as a hidden message into a digital image. It is built using Visual Basic Net programming language, which has several supports for digital image programming. To accommodate the image when the process of hiding and reading the message, it used picture box control. The interface display of the application is like Fig. 4.

The first evaluation conducted was a histogram analysis. We have compared histogram analysis of the original image and stego image that has been inserted with the text file. The result is in Table 1.

The peak signal-to-noise ratio (PSNR) method is used to determine image quality as a comparison of the quality of the stego image with the original image (cover image). The term peak signal-to-noise ratio (PSNR) is a term in the engineering field that states the ratio between a digital signal's maximum possible signal strength and the noise power that affects the signal's correctness. Since many signals have a broad range of dynamics, PSNR is usually expressed on a logarithmic decibel scale [14].

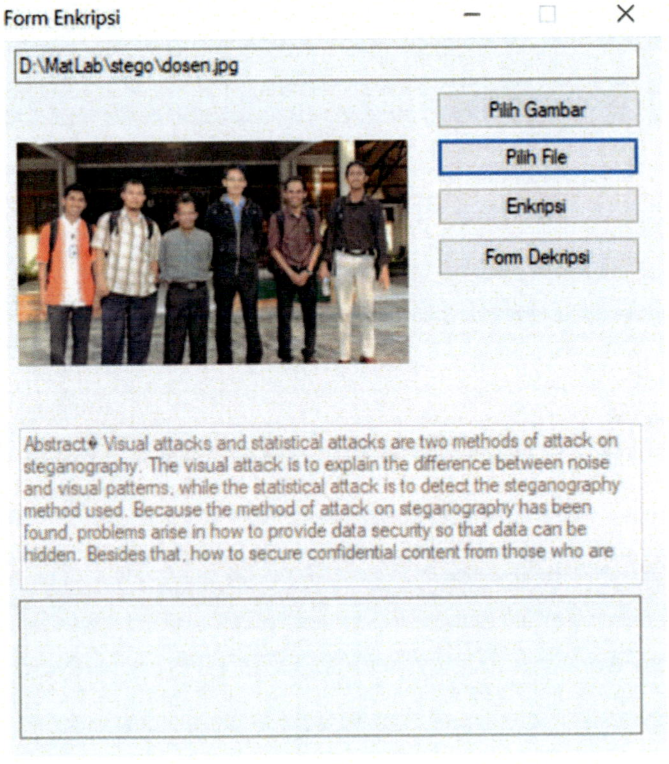

Fig. 4 Encryption/decryption interface application

Table 1 Histogram analysis

Original image (A)	Stego image (B)	Histogram A	Histogram B
		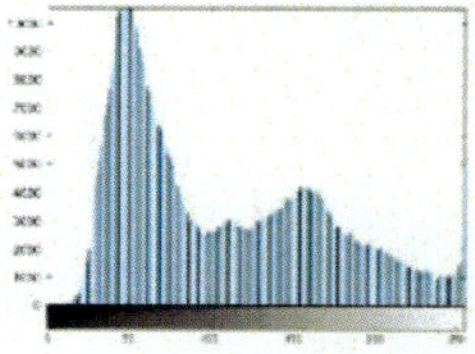	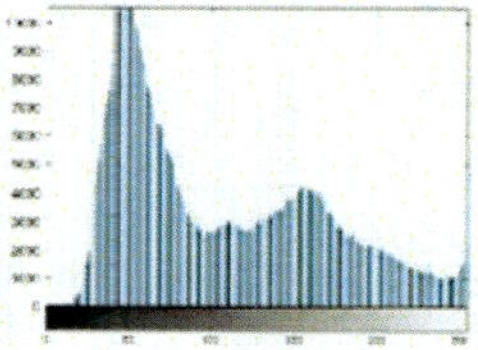
		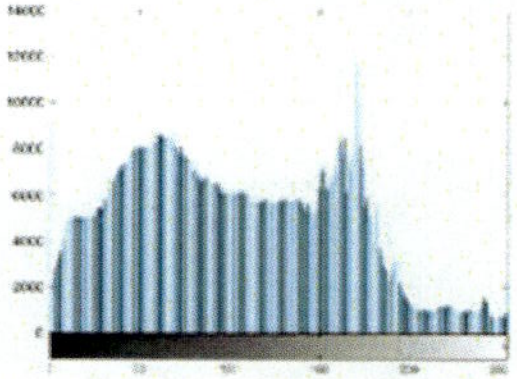	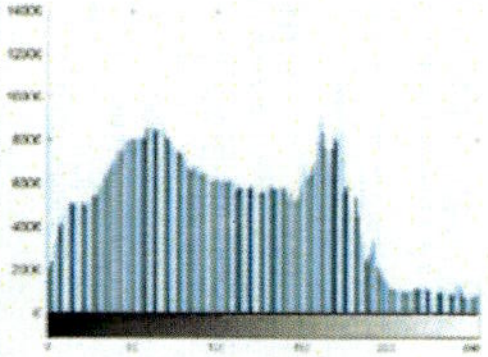

Table 2 MSE, PSNE, and MSE results

Image		MSE	PSNR	NSR
Cover image	Dosen.jpg	414.9138	22.0142	15.7063
Stego image	Dosen1.jpg	410.4317	21.9629	15.6550
Cover image	Kolam.jpg	408.2399	22.0304	15.3456
Stego image	Kolam1.jpg	404.2999	21.9541	15.2694

The formula for calculating PSNR is as follows:

$$\mathrm{MSE} = \frac{1}{mn}\sum_{i=0}^{m-1}\sum_{j=0}^{n-1}[I(i, j) - K(i, j)]^2 \quad (1)$$

$$\mathrm{PSNR} = 10 \cdot \log_{10}\left(\frac{\mathrm{MAX}_1^2}{\mathrm{MSE}}\right) \quad (2)$$

PSNR was defined through the signal-to-noise ratio (SNR). SNR is used to measure the level of signal quality. This value is calculated based on the comparison between the signal and the noise value. Signal quality is directly proportional to the SNR value. The higher the SNR value, the better the quality of the signal produced. Table 2 showed the results of calculation of values PSNR which is represented on a decibel scale (dB) [15].

As the calculation results in Table 2 show, inserting a text message with different sizes will yield different MSE and PSNR values. The greater the size of the message file, the higher the value of MSE and the smaller the value of PSNR, the smaller the size of the message file, the smaller the value of MSE and the higher the value of PSNR. If the PSNR value is low, it can be said that the image quality is getting worse, which means the quality of the image is physically poor. Whereas the image quality is still good if the PSR value is large, which means that the damage to the image is relatively small.

6 Conclusion

From the research that has been done, it can be concluded several things, namely steganography is a very efficient and powerful technique that allows to send text files safely and hidden. The LSB method that is applied to the message hiding process does not significantly affect the quality of the cover image.

References

1. B.A. Forouzan, D. Mukhopadhyay, *Cryptography and Network Security (Sie)* (McGraw-Hill Education, New York, 2011)
2. A. Siswanto, A. Syukur, I. Husna, Perbandingan metode data encryption standard (DES) dan advanced encryption standard (AES) Pada Steganografi File Citra, in *Seminar Nasional Teknologi Informasi dan Komunikasi 2018* (2018), pp. 190–197
3. K. Challita, H. Farhat, Combining steganography and cryptography: new directions. Int. J. New Comput. Archit. Appl. (IJNCAA) **1**, 199–208 (2011)
4. A.J. Menezes, P.C. Van Oorschot, S.A. Vanstone, *Handbook of Applied Cryptography* (CRC Press, Boca Raton, 1996)
5. R. Rahim, H. Nurdiyanto, R. Hidayat, A.S. Ahmar, D. Siregar, A.P.U. Siahaan et al., Combination Base64 algorithm and EOF technique for steganography. J. Phys. Conf. Ser. 012003 (2018)
6. A. Westfeld, A. Pfitzmann, Attacks on steganographic systems, in *International Workshop on Information Hiding* (1999), pp. 61–76
7. M. Pelosi, N. Poudel, P. Lamichhane, D. Lam, G. Kessler, J. MacMonagle, Positive Identification of LSB Image Steganography Using Cover Image Comparisons (2018)
8. A. Cheddad, J. Condell, K. Curran, P. Mc Kevitt, Digital image steganography: survey and analysis of current methods. Sig. Proc. **90**, 727–752 (2010)
9. R. Munir, Pengantar Kriptografi, in *Penerbit Informatika Bandung* (2010)
10. R. Munir, Steganografi dan Watermarking, in *Departemen Teknik Informatika, Institut Teknologi Bandung*. Diakses dari https://informatika.stei.itb.ac.id/~rinaldi.munir/Kriptografi/Steganografi%20dan%20Watermarking.pdf (2004)
11. M.F. Syawal, D.C. Fikriansyah, N.A.-U.B. Luhur, Implementasi Teknik Steganografi Menggunakan Algoritma Vigenere Cipher Dan Metode LSB. Jurnal TICom **4** (2016)
12. J.V. Purba, M. Situmorang, D. Arisandi, Implementasi steganografi pesan text ke dalam file sound (.wav) dengan modifikasi jarak byte pada algoritma least significant bit (LSB). Dunia Teknologi Informasi-Jurnal Online **1** (2012)
13. P. Utomo, B.E. Purnama, Pengembangan Jaringan Komputer Universitas Surakarta Berdasarkan Perbandingan Protokol Routing Information Protokol (RIP) Dan Protokol Open Shortest Path First (OSPF). IJNS-Indonesian J. Networking Secur. **1** (2012)
14. D.F. Alfatwa, Watermarking Pada Citra Digital Menggunakan discrete wavelet transform, in *Bandung: Institut Teknologi Bandung* (2005)
15. M.M. Amin, Image steganography dengan metode least significant bit (SLB). J. Comput. Sci. Res. Dev. (CSRID) **6** (2014)

Apple Defect Detection Based on Deep Convolutional Neural Network

MD. Nur Alam, Shahi Saugat, Dahit Santosh, Mohammad Ibrahim Sarkar, and Ahmed Abdulhakim Al-Absi

Abstract Fruit detection and classification are a challenging task in image processing. This paper presents a novel approach to fruit detection using deep convolutional neural networks. The aim is to build an accurate, fast, and reliable fruit detection system, which is a vital element of an autonomous agricultural artificial intelligence platform; it is a key element for fruits yield detection and automated fruits processing industry. Recent work in deep neural network has led to the development of a state-of-the-art object detection and classification. This paper demonstrates the design and implementation of deep learning-based automated categorization of the apple images captured from fruits processing industry. The system is based on a convolutional neural network (CNN) followed by the selection of proposed regions. The training of the classifier is performed, with a dataset derived from the set of images taken from the agricultural-based industrial fruit sorting process. In the preparation of the CNN architecture model, initializing the parameter configuration accelerates the network training process. The results of the experiments using CNN algorithm showed the performance of defect detection on the apple fruit of 96%.

Keywords Apple detection · Apple classification · Deep learning · Convolutional neural network (CNN)

MD. Nur Alam (✉) · S. Saugat · D. Santosh · A. A. Al-Absi
Department of Smart Computing, Kyungdong University, Global Campus (Gosung) 46 4-gil, Bongpo, Gosung, Gangwondo 24764, Republic of Korea
e-mail: na@kduniv.ac.kr

M. I. Sarkar
Department of Electronic Engineering, Chonbuk National University, Baekje-daero, Deokjin-gu, Jeonju-si, Jeollabuk-do 54896, Republic of Korea

P. K. Pattnaik et al. (eds.), *Proceedings of International Conference on Smart Computing and Cyber Security*, Lecture Notes in Networks and Systems 149,
https://doi.org/10.1007/978-981-15-7990-5_21

1 Introduction

Nowadays, in fruit agricultural production, most of the farming works rely on the manual labor of fruit planters. A great quantity of simple and repetitive labors not only consumes time and energy and increases production costs but also brings more uncertainties to agricultural production. Fruit farmers may make mistakes in the judgment of farming due to the lack of knowledge and experience. The wrong judgment will directly lead to the wrong implementation of farming, which could have a serious impact on crop yield. With the continuous progress of precision agricultural technology [1], sensors have become the prime sources of crop information. As one of the major components of sensor information, image data plays a significant role in obtaining crop growth status and judging crop health states [2]. With the development of vision sensors, automation and intelligence of agricultural production have been promoted, and various image processing approaches have been applied in agricultural production [3, 4].

While moving ahead with machine learning technology, the traditional image processing methods are gradually replaced by methods such as neural network and support vector machine [5]. Arribas et al. adopted General Softmax Perceptron neural network to train the extracted image features and then used the features to classify sunflower leaves [6]. In our previous work, a BP neural network improved by genetic algorithm was applied to realize multithreshold image processing. The region of green apple in the image was segmented, and the lesion area was extracted by subsequent SVM method. The diseased apple was further identified [7]. This method has been well applied in the images collected in orchards but could not achieve real-time image processing.

Nowadays, the deep learning technology gains more popularity. Deep learning achieves desirable performance in computer vision since it takes the advantage of mass amount of data and does not need to extract the image feature manually. In agriculture, deep learning techniques [8] have also been widely applied in crop detection [9–11] and classification [12], pest and disease identification [13], and so on. In the classification of plant diseases, Tan et al. used convolutional neural networks (CNN) to identify and diagnose the surface lesions of fruits [14]. Mohanty et al. used AlexNet, GoogleNet, and some other deep learning models to classify 26 plant diseases. The highest classification accuracy reached 99.35% [15]. Ferentinos compared the classification performance of five deep learning methods such as AlexNet and VGG on 58 different plant diseases, and the highest accuracy reached 99.53% [16].

In this paper, convolution neural networks (CNN) have been exploited to distinguish the defects from apple fruits in the fruits processing industry. CNNs act as an advanced feature extractor as it creates many small features from a huge feature set representing the characteristics of the image. The performance and flexibility of CNNs reveal a clear improvement over the classic machine learning techniques. Our main contribution of this paper is a vision-based classification and detection system to distinguish apple fruits and defects using convolutional neural networks (CNNs). To build up our classification and detection system, we provide an end-to-end approach, in which there are no handcrafted features used.

The rest of the paper is organized as follows: Sect. 2 presents the sample data collection information. Section 3 explains the proposed architecture and methodology. Section 4 discusses the experimental results, and Sect. 5 concludes the paper.

2 Sample Data Collection

2.1 Apple Collection

We used red delicious apples in this study. The image acquisition and samples were provided by the Agro-Food Quality and Safety Sensing Laboratory from the Rural Development Administration (RDA) in the Republic of Korea (ROK). The samples were separated into normal and defect apples using a visual inspection. The apple samples comprise groups with normal skin and skin defect conditions. Figure 1 shows a sample with the common defects scab, bruise, crack, cut, defect with stem, and calyx. The red delicious apples normal skin samples and defect apple samples were used to develop the algorithm.

The training data must be provided to train these deep learning models. In this case, the image regions containing apple are labeled as normal samples; whereas, the fruit surface such as the clean and more than one label of fruits surface area is considered as defect samples.

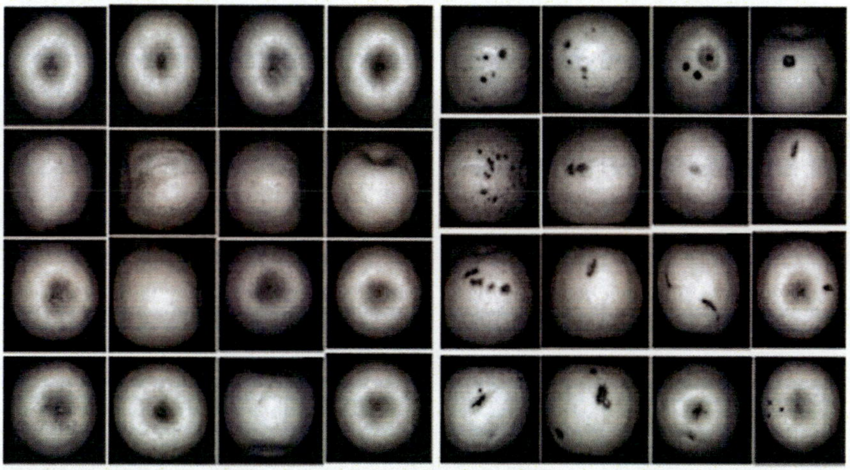

(a) (b)

Fig. 1 **a** Normal skin apple samples, **b** defect skin apple samples

To formulate a sample, the original image is subjected to a sliding window of fixed size. The sub-images are assigned as normal or defects label according to the presence or absence of defects. They are then fed into the neural network for the purpose of training.

3 The Proposed Method

3.1 Construction of the Dataset

The construction of the input data, to make the system computationally efficient and effective, is concisely presented here.

The proposed classifier is trained with the features extracted from the supplied images, captured from agricultural sectors, but affected in different areas by defects. These images are segmented by visual inspection, with sectors with "defects". These images are cropped into size 64 × 64. The size of the input image is chosen as 64 × 64 as this reduces not only the computational complexity during training but also is sufficient for providing distinguishability between the defects and normal apples.

The following steps are performed to analyze the dataset and decide on the number of pixels that can be moved by the sliding window at a time. Primarily, it is checked if the object lies within the image bounds.

The classifier is trained with our own dataset, where images are captured from different agricultural industry. The captured images are cropped (64 × 64 size) by visual inspection to produce new images. These input images that are cropped are labeled as positive samples if the presence of defects is observed. Similarly, clippings or images are generated, and regions with the absence of defects (of equal size) are labeled accordingly as negative samples. Special emphasis is placed on positive samples being densely filled with defects (the object sought). This is to ensure that, in the event of classification, there would be a high probability of match between the model learned by the classifier in training and the image segment under exploration. The dataset considered is in the same area and yielded 7200 total samples with each class containing 3600 positive and 3600 negative samples.

Having provided details and justification on the construction of the dataset, the next section explains the construction of the classifier.

3.2 Construction of the Classifier

Convolution neural network (CNN) architecture is exploited to extract the features from the input images rather than adopting the conventional time-consuming feature extraction process. CNN, also believed to be effective in distinguishing defects from apple fruits, is utilized to learn the representation of the image. Convolution and

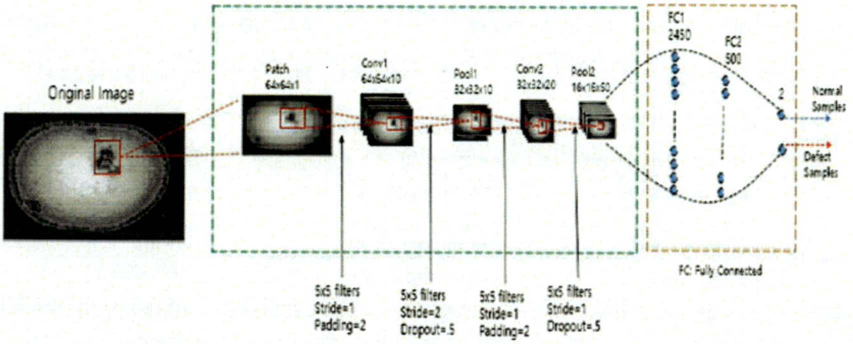

Fig. 2 Apple classification network

pooling layers are used alternatively in the proposed architecture. Convolution layer is a set of trainable modules in implementing complex nonlinear function stacked together to capture the underlying patterns in unlabeled data. This process is a kind of feature learning in which the learning algorithm will generate filters, thereby extracting feature information such as color information, edge information, and texture information. Then, the subsequent pooling layer is utilized to downsample the feature maps. Pooling is done as an operation to divide a feature into regions and collect statistics for each region. Figure 2 depicts the classification network used in this work.

The input layer is of the convolutional type made up of 20 filters of 3 × 3 dimensions with RELU type activation function. This layer receives an input image of size 64 × 64 pixels (64 × 64 × 1) and performs 5 × 5 filtering at stride set to 1 and padding set to 2 to yield a feature map of 64 × 64 × 20 dimensions. Subsequently, subsampling or "Mamax pooling" is done, and the feature map is reduced by a scale of 2 (32 × 32 × 20). The pooling layer reduces the size of the feature map in the view of requiring fewer calculations with the network architecture. This strategy is incorporated to minimize over-fitting through the "dropout" technique that drops out 50% of neurons of the layer in hand. The previous arrangement ("Conv + Activation + Max-Pooling") is repeated causing the feature maps to be reduced further by a factor of 2, thereby a reduction of 4 from the original image.

Following, there are two fully connected layers. The output images from the previous structure are "flattened" to incorporate them as input to a neural network with a hidden layer of 128 neurons (with dropout). The final layer of the neural network forms the output layer with two neurons. There are two neurons as the problem in hand is a binary classification problem, in which the output is positive (presence of defect) or negative (absence of defect). Hence, each neuron in the output layer incorporates a cost function "Softmax" that allows the estimation of probability of the predicted class.

The CNN-based proposed classifier is trained using a set of 7200 samples of size 64 × 64 pixels. Dividing into two sets of 3600 images each of positive and negative samples, the presence of defects is considered as positive samples while that of its absence is regarded as the negative sample. For training the classifier, 80% of the positive sample set (80% of 3600) is used. The remaining 20% is used for testing the learned model.

This classifier, in preliminary tests, is applied through the test image using the sliding window technique (of the same 64 × 64 format), and in the event where the object is detected, it is necessary to register the coordinates of its location and append it to a list created to gather these events. For each sliding window, the network gives two outputs, one is the probability a defect is present in the input image, and another is the probability of the defects in each position.

Having presented the construction of the proposed classification network, the following section presents on the selection of the candidate defect regions.

3.3 Selection of Proposed Regions

Once the classifier has been trained with the proposed architecture, test images are fed as input to trained model. The input is provided in a sliding window fashion (of the same 64 × 64 format) to determine the presence of the desired objects (defects in this case). The classifier output returns a binary response according to the presence or absence of the defects. According to the binary output response, the coordinates of the defect locations are registered and thereby a bounding box is created. For each sliding window, the network yields two outputs; one is the probability that a defect is present in the input image, and second is the probability of the defects in each position.

The framework depicted in Fig. 3 initially receives an image of size 502 × 200 × 1. Then it is repositioned or resized to 100 × 100 × 1 to reduce computationally the

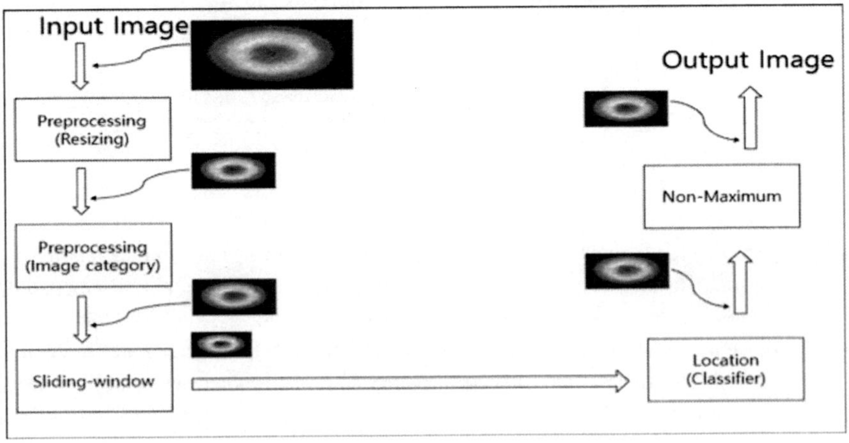

Fig. 3 Localization network architecture

complexity. The sliding window procedure is adopted by sliding a 64×64 window along the height and width of the original image. The size is 64×64 to match the format used during training. The image patches are then fed into the proposed classifier network, where the convolution–activation–pooling architecture yields the 50 features representing the 64×64 input patch. The features are provided into the neural network to yield a probability for both the classes, namely presence or absence of defects. If the probability for presence of defect is more than 99%, then the output is considered as 1, else 0. The outputs are then concatenated to predict the defects in the original image.

Having explained the localization of the predicted defect regions and hence the entire methodology, the next section discusses the experimental results.

4 Experimental Result

There are various deep learning frameworks capable of creating, training, and using CNN. These frameworks vary in speed, abstraction level, modifiability, and creation of new layers. Keras libraries have been utilized for the convolutional network implementation in this work owing to its high abstraction level, which makes network architecture definition, solver, and application relatively easy. The conducted experiments and its results justify the accuracy and efficiency of the proposed system by proving that the proposed system can learn feature representations and can be implemented quickly by using previously learned features. Stochastic gradient descent (SGD) is used with a mini-batch size of 50 and an epoch size of 1000. The parameters, namely the weight decay is set to 0.000001, momentum is set to 0.9, and learning rate is tuned as 0.01.

The performance of the classifier during training and validation is justifiable. The metrics such as precision, recall, F1-score, and support are computed when the number of epochs is 4000 and tabulated in Table 1.

Analysis on Table 1 reveals that all the metrics yield high values around 0.95 and 0.97 (precision, recall, F1-score) for 4000 epochs, thus reaching 96% accuracy.

It depicts four conditions, namely the true positives, true negatives, false positives, and false negatives. True positives denote the number of positive samples correctly detected as positive while false positive denotes the negative samples being detected as positive. Similarly, true negative refers to the corrected identified negative samples as negative while false negative signifies the negative sample being wrongly recognized as positive sample.

Table 1 Average accuracy of input image

Class	Precision	Recall	F1-score
Normal Skin	0.97	0.97	0.96
Defect	0.96	0.95	0.95
Average	**0.96**	**0.96**	**0.955**

5 Conclusion

In this paper, the defect detection method is proposed based on deep learning. Firstly, in view of the problem of image dataset due to the random occurrence of apple defects, a deep learning method is adopted to extract the features of normal apples and defect apples and to produce defects on the surface of normal apple images. Convolutional neural networks were proved to yield better results in visual object recognition. Hence, the proposed system incorporated CNN to classify defects from apple fruits. The input image was initially processed to make it computationally effective and efficient. Then the processed image was subjected to convolution filtering, activation function, and pooling twice, thereby resulting in 50 feature maps representing the characteristic of the image that would differentiate it into a defect or an apple fruit. The features were fed into a neural network with 128 neurons. The output probabilities resulted in a binary map. Non-maximum suppression was then applied to derive at the boundary of the defects in the entire original image. The formulated architecture can be viewed as a light neural network and hence is memory efficient. The experimental results clearly prove the effectiveness of the proposed system in helping the fruit sorting industry.

References

1. M. Paustian, L. Theuvsen, Adoption of precision agriculture technologies by German crop farmers. Precis. Agric. **18**(5), 701–716 (2017)
2. Y. Zhao, L. Gong, Y. Huang, C. Liu, A review of key techniques of vision-based control for harvesting robot. Comput. Electron. Agric. **127**, 311–323 (2016)
3. B. Zion, The use of computer vision technologies in aquaculture a review. Comput. Electron. Agric. **88**(88), 125–132 (2012)
4. Q. Wang, S. Nuske, M. Bergerman, S. Singh, Automated crop yield estimation for apple orchards, in *Experimental Robotics*, vol. 88 of *Springer Tracts in Advanced Robotics* (Springer, Berlin, 2013), pp. 745–758
5. M.A. Ebrahimi, M.H. Khoshtaghaza, S. Minaei, B. Jamshidi, Vision-based pest detection based on SVM classification method. Comput. Electron. Agric. **137**, 52–58 (2017)
6. J.I. Arribas, G.V. Sánchez-Ferrero, G. Ruiz-Ruiz, J. Gómez-Gil, Leaf classification in sunflower crops by computer vision and neural networks. Comput. Electron. Agric. **78**(1), 9–18 (2011)
7. Y. Tian, E. Li, L. Yang, Z. Liang, An image processing method for green apple lesion detection in natural environment based on GA-BPNN and SVM, in *2018 IEEE International Conference on Mechatronics and Automation (ICMA)*, Changchun, China (2018), pp. 1210–1215
8. A. Kamilaris, F.X. Prenafeta-Boldú, Deep learning in agriculture: a survey. Comput. Electron. Agric. **147**, 70–90 (2018)
9. S. Bargoti, J. Underwood, Deep fruit detection in orchards, in *2017 IEEE International Conference on Robotics and Automation (ICRA)*, Singapore (2017), pp. 3626–3633
10. S.W. Chen, S.S. Shivakumar, S. Dcunha et al., Counting apples and oranges with deep learning: a data-driven approach. IEEE Robot. Autom. Lett. **2**(2), 781–788 (2017)
11. P.A. Dias, A. Tabb, H. Medeiros, Apple flower detection using deep convolutional networks. Comput. Ind. **99**, 17–28 (2018)
12. S.H. Lee, C.S. Chan, S.J. Mayo, P. Remagnino, How deep learning extracts and learns leaf features for plant classification. Pattern Recogn. **71**, 1–13 (2017)

13. S. Sladojevic, M. Arsenovic, A. Anderla, D. Culibrk, D. Stefanovic, Deep neural networks based recognition of plant diseases by leaf image classification. Comput. Intell. Neurosci. **2016**, Article ID 3289801, 11 p (2016)
14. W. Tan, C. Zhao, H. Wu, Intelligent alerting for fruitmelon lesion image based on momentum deep learning. Multimedia Tools Appl. **75**(24), article 16741 (2016)
15. S.P. Mohanty, D.P. Hughes, M. Salathé, Using deep learning for image-based plant disease detection. Front. Plant Sci. **7**, 1419 (2016)
16. K.P. Ferentinos, Deep learning models for plant disease detection and diagnosis. Comput. Electron. Agric. **145**, 311–318 (2018)

Satellite Image Segmentation and Classification Using Fuzzy C-Means Clustering and Support Vector Machine Classifier

P. Manjula, Ojasvita Muyal, and Ahmed A. Al-Absi

Abstract Feature extraction and classification are important areas of research in image processing and computer vision with an extreme great number of applications in science and industry. One of the applications is satellite image classification. The main objective of this research work is to study image segmentation, feature extraction, and image classification algorithm, apply it on satellite images to classify residential, mountain, forest, desert, and river region. Automated segmentation is performed using fuzzy C-means technique and study and identify technique for classification of regions in satellite images. Features are extracted from the segmented output image using grey level co-occurrence matrix (GLCM) and Gabor filter. Then, the classification is used to classify the region using SVM classifier based on feature extracted. We can propose work based on colour and texture features with the help of Gabor filter and GLCM which are more appropriate in extracting colour and texture features. As a result, the number of features will increase accordingly classification accuracy will also get increased. We can also add one more phase which is image segmentation for more accurate results. We can segment the input image using fuzzy C-means clustering technique and after that we can perform the feature extraction on the segmented image, in this way, we will get the results which are more optimized and takes less time to extract feature than the previous work. This work can be enhanced in context of classification accuracy and enhance feature extraction phase.

P. Manjula (✉)
SRM University, Sonepat, India
e-mail: gpmanjula28@gmail.com

O. Muyal
VIT University, Vellore, India
e-mail: ojasvita.41@wipro.com

A. A. Al-Absi
International Faculty (Academic), Global Academic Collaboration Centers, Kyungdong University, Global Campus, Wonju-si, South Korea
e-mail: absiahmed@kduniv.ac.kr

P. K. Pattnaik et al. (eds.), *Proceedings of International Conference on Smart Computing and Cyber Security*, Lecture Notes in Networks and Systems 149,
https://doi.org/10.1007/978-981-15-7990-5_22

Keywords Satellite image · Segmentation · Classification · Feature extraction · Gabor filter · Fuzzy C-means clustering

1 Introduction

The planet is continually being observed and imaged by satellites. Before 1972, satellites were not designed to study or monitor Earth's surface. Instead, they were mainly used for military missions, for example, threat monitoring and assessment and were exceptionally costly. Imagery was commercialized in 1984 but faced many funding issues. This led to the passing of the Land Remote Sensing Policy Act of 1992. The new law promoted research, public sector applications and allowed commercial companies to launch satellites and sell imagery. As a result, Earth observation satellites were designed. As few business satellites have been launched, such as IKONOS, QUICK BIRD, and IRS and so on, they give worldwide, precise, high-resolution pictures to people, associations, and governments. Along these lines, satellite images have turned out to be more affordable and the area of utilization has expanded enormously. Over 2200 satellites orbit around Earth today.

The primary applications have been for weather prediction, to monitor global environment conditions and geographical and geological applications. Satellite images provide detail information about Earth's surface. The advantages of satellite imagery include covering huge zones; visit to any part of the globe, regardless of its remoteness; the capacity to gather information unhindered by local air traffic.

Satellite images of various spatial resolutions are commercially available. Pictures with high-resolution information with ground pixel sizes of under 5 m give detail data about the Earth's surface and small objects, for example, structures, lanes, and trees can be shown in extraordinary points of interest. High-resolution pictures are utilized for applications, for example, transportation organizes mapping, calamity readiness, urban arranging, exactness cultivating, and broadcast communications. Then again, low-resolution satellite pictures, with ground determination more prominent than 10 m, are utilized for applications like environmental assessment, local mapping, forestry management, across the board catastrophe evaluation, and urban monitoring which is shown in Fig. 1.

In this paper [1], the fuzzy-based clustering approaches fuzzy C-means clustering (FCM), possibilistic C-means (PCM), and possibilistic fuzzy C-means (PFCM) are compared, and the performance of these algorithms was tested with number of satellite images. In this article [2], the problem of fuzzy clustering is posed as one of searching for some suitable number of cluster centres so that some measures of validity of the obtained partitions should be optimized. In this paper, a recently developed multi-objective simulated annealing-based technique and archived multi-objective simulated annealing technique (AMOSA) are used to perform clustering, taking two validity measures as two objective functions.

In this paper [3], author proposed a new approach to Gabor filter bank design, by incorporating feature selection, i.e., filter selection, into the design process. The

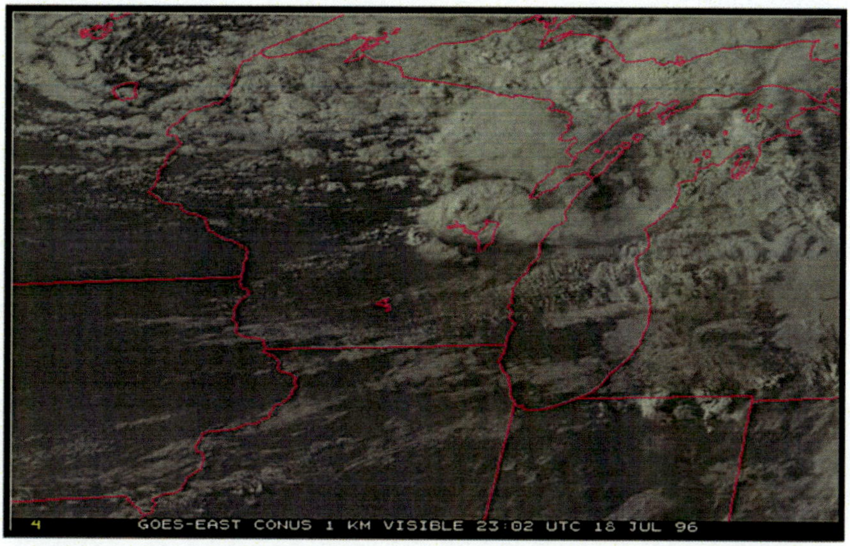

Fig. 1 Example of visible satellite image

merits of incorporating filter selection in filter bank design are twofold. Firstly, filter selection produces a compact Gabor filter bank and hence reduces computational complexity of texture feature extraction. Secondly, Gabor filter bank thus designed produces low-dimensional feature representation with improved sample-to-feature ratio, and this in turn leads to improved performance of texture classification. This paper [4] introduces a classification system for remote sensing ASTER satellite imagery using SVM and particle swarm optimization (PSO) algorithm.

This paper [5] describes a texture segmentation algorithm to segment satellite images using Gabor filter bank and neural networks. In the proposed method, feature vectors are extracted by multi-channel decomposition. The spatial/spatial-frequency features of the input satellite image are extracted by optimized Gabor filter bank. Some important considerations about filter parameters, filter bank coverage in frequency domain and the reduction of feature dimensions are discussed. In this article [6], in order to make the most of deep CNN and Gabor filtering, a new strategy, which combines Gabor filters with convolutional filters, is proposed for hyperspectral image classification to mitigate the problem of overfitting. The obtained results reveal that the proposed model provides competitive results in terms of classification accuracy, especially when only a limited number of training samples are available.

This study [7] presents a novel object-based remote sensing image texture extraction method to aid the classification of mountain economic forest. This paper [8] investigated texture feature extraction using 2D Gabor filter to extract the texture features of inverse fast Fourier transform (IFFT), texture energy and transformed IFFT. The Gabor filter bank experimented on seventy-two collected samples of skull-stripped T1-weighted, T2-weighted, and FLAIR MRI brain images utilizing four frequencies and four orientations. Results showed that texture feature extractions of two highest frequencies with all four orientations produced the highest acceptance rate. This paper [9] presents a new method for satellite image classification. Specifically, we make two main contributions: (1) we introduce the sparse coding method for high-resolution satellite image classification; (2) we effectively combine a set of diverse and complementary features-SIFT, colour histogram, and Gabor to further improve the performance. The retrieval efficiency [10] is further increased by using the SVM classifier by classifying the satellite images based on urban area, water body, and vegetation. The experimental results show that the fusion technique gives better result and more accuracy can be obtained by classifying the dataset using SVM. The goal of this research [11] is to provide the efficiency in classification of satellite images using the object-based image analysis.

The aim of this paper [12] is to study and compare the different texture-based approaches for object recognition and feature extraction. GLCM and Hear wavelet transform are the most primitive methods for texture analysis. In this paper, two more techniques based on their fusion have been considered. These techniques have been tested on sample images and their detailed experimental results along with the method of implementation have been discussed.

In this paper [13], the Gabor texture features combined with original bands of image, PCA, and NDVI were adopted as the characteristic vector of training samples for SVM and decision tree classification. Finally, traditional classification schemes of maximum likelihood were comparatively studied. For most of the cases, the SVM method gave the highest correct classification rate within these three methodologies. Decision tree and SVM have their superiority, respectively. In minimum time and evaluate the better accuracy based on the support vector machine algorithm [14]. The objective of the study [15] is to use SVM technique for classifying multi-spectral satellite image dataset and compare the overall accuracy with the conventional image classification method. LISS-3 and AWIFS sensors data from Resourcesat-1, Indian Remote Sensing (IRS) platform were used for this analysis.

The primary objective of this research is to use high-resolution satellite images to distinguish:

- River images
- Forest images
- Residential area
- Desert images
- Mountain image.

To accomplish this objective, we reviewed and analysed various image segmentation algorithms. This paper proposed an approach for classification of satellite images and the techniques utilized as a part of the proposed framework are fuzzy C-means, Gabor filter, GLCM, and support vector machines. Also, to accomplish this objective, we have studied and analysed image segmentation, feature extraction, and classification algorithms. We used the MATLAB image processing tool kit for developing the algorithms.

2 Proposed Work

Step 1: Load Image

A satellite image is chosen from dataset for classification.

Step 2: Image Pre-processing

Gaussian filter is used to filter the unnecessary information and remove various types of noises from the images using image processing. Gaussian filter utilizing its Gaussian function alters the input images and the equation of Gaussian function can be defined as:

$$g(x, y) = \frac{1}{2\pi\sigma^2} \cdot e^{-\frac{x^2+y^2}{2\sigma^2}} \tag{1}$$

Here, x describes the distance to horizontal axis from origin; y describes the distance to vertical axis from origin, and the gamma represents the deviation value of Gaussian distribution. Gaussian filters mostly used in image processing for noise removal and it also improve the image signal. This pre-processing step makes satellite image more useable for classification processes.

Step 3: Image Segmentation

Segmentation is basically used for identifying contextual or non-contextual area from a satellite image. These contexts can be defined as important attribute in any image processing phase. The goal of this phase is to alter the image such that the representation of image is more meaningful and usually locate the specific objects or boundaries. In this paper, image segmentation is performed as colour-based utilizing FCM. In colour-based segmentation, utilizing a satellite image perform a task to find or locate objects and differentiate them.

$$d(x, y) = \sqrt{(R(x, y) - R_0)^2 + (G(x, y) - G_0)^2 + (B(x, y) - B_0)^2} \tag{2}$$

This step assigns a label to each pixel in a satellite image such that it shares similar statistical characteristics and help the feature extraction phase. Colour-based segmentation is said to be more correct due to the information in pixel level is greater than grey scale images. After this step, the satellite image is segmented into different regions containing every pixel with similar statistical characteristics; thus, image regions will strongly be interrelated to feature of interest or objects.

Step 4: Feature Extraction

GLCM and Gabor filter extract feature vectors from input satellite pictures like texture. Texture element is extracted from the RGB colour picture. The GLCM functions characterize these texture feature of an image by calculating how often pairs of pixels with specific values and in a specified spatial relationship occur in an image and global colour histograms in extricating the colour features of satellite pictures.

The texture features utilized most widely includes—homogeneity, correlation, and entropy. Gabor feature is utilized here in classification process and provides more accurate outcomes. Gabor function can be defended as:

$$g(x, y) = \frac{1}{2\pi\sigma_x\sigma_y} exp + \left[-\frac{1}{2}\frac{x^2}{\sigma^2 x} + \frac{y^2}{\sigma^2 y}\right] + 2\pi j W x \tag{3}$$

Gabor feature can be described by its parameters which includes frequency. Gabor filter bank decrease the complexity of feature extraction by incorporating low-dimensional pixel representation. After this phase, a feature vector is generated utilizing extracted features and utilized for classification in next phase.

Step 5: Training and Testing Framework

Support vector machine algorithm utilizes these element vectors (colour and texture) to prepare and train our proposed structure. The features colour and texture of each satellite image are stored in database and these features will be used for the next stage of classification. In light of these component vectors, colour and texture, this proposed structure using SVM will group the satellite pictures into five various classes which include desert, mountain, residential area, river, and forest. For the effective classification of the image with images, various distance metrics are used to measure similarities of features. Here, the similarity evaluation using SVM classifiers achieved between the features of the query image and the features of the database images.

The SVM classifier will compute the feature value of input image and the feature value of database images, based on these values, the SVM classifier will classify the input image belong to which class.

Let A and B be the two datasets, A as input and B is output.

$$B = f(A, \alpha) \tag{4}$$

Here, A is the trained dataset contains feature vectors and α is parameter for SVM kernel function contains features of input satellite image that need to be tuned for better classification accuracy. Here, a kernel function, radial basis function is used for calculating satellite image classification.

Step 6: Classified Image

Input image is classified as either from five categories or other than these five categories. Satellite image classification architecture is shown in Fig. 2.

3 Results and Discussion

Load a satellite image to be classified into any of the five categories which is shown in Fig. 3. Next, image pre-processing is applied on the input image. Gaussian filter is

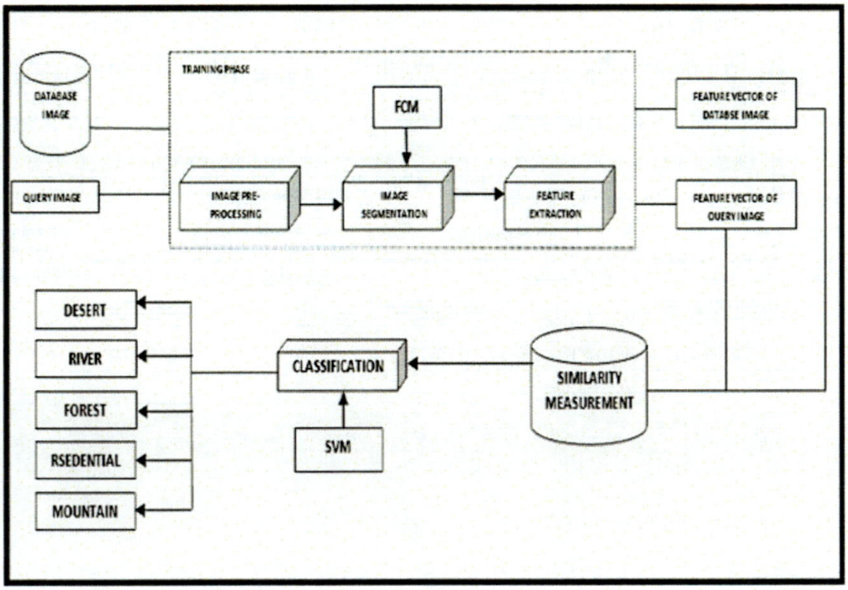

Fig. 2 Satellite image classification architecture (data flow diagram)

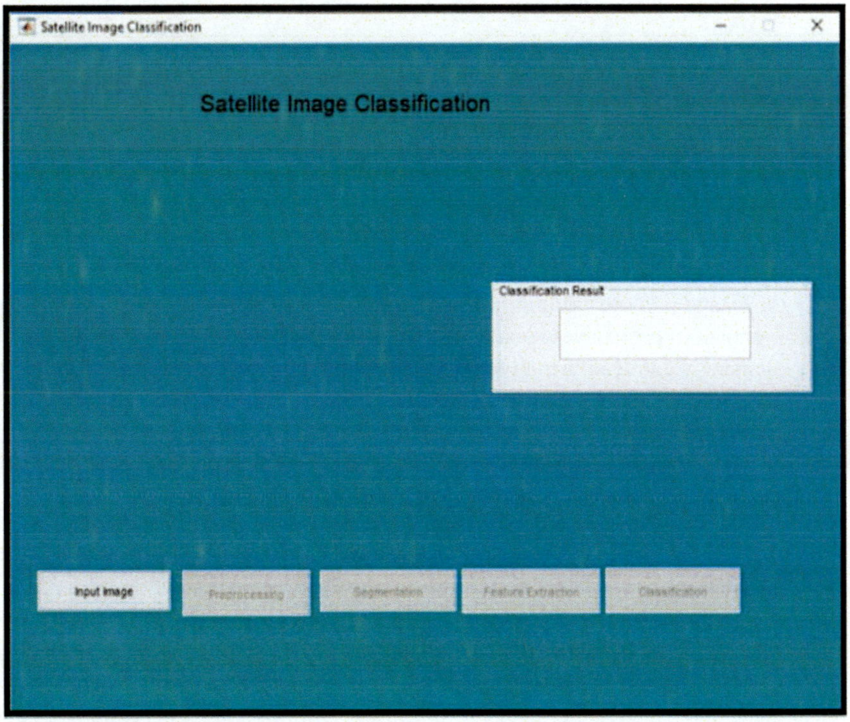

Fig. 3 Load input image

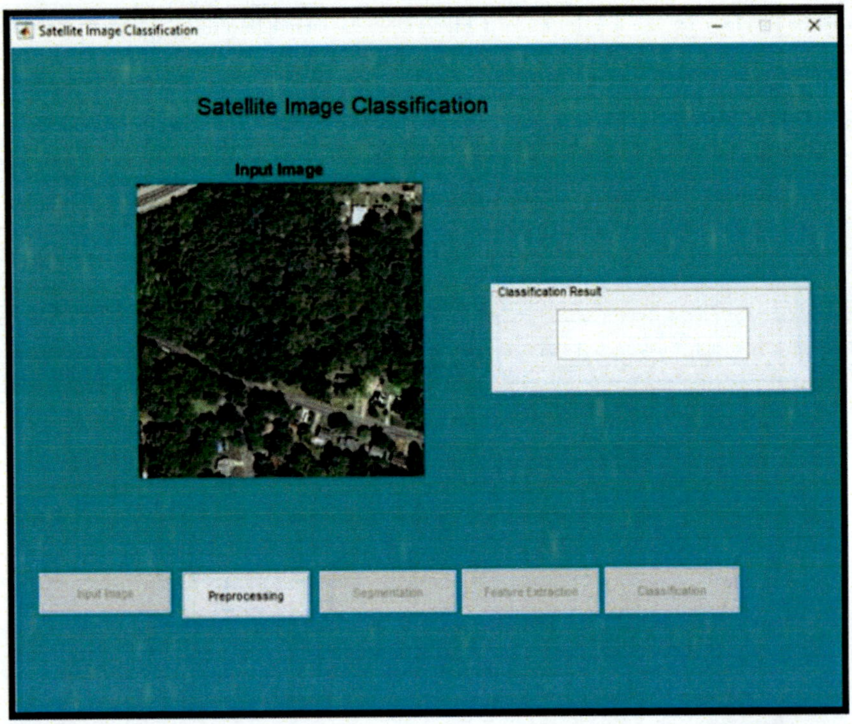

Fig. 4 Image pre-processing

used to filter the unnecessary information and remove various types of noises from the images using image processing which is shown in Fig. 4.

Image Segmentation is used to identify the contextual and non-contextual areas from a satellite image. It alters the image in such a way that the representation of the image is more meaningful and represents specific objects or boundaries which is shown in Fig. 5. Next phase is the **feature extraction** for satellite image. Texture element is extracted from the RGB colour picture which is shown in Fig. 6.

Support vector machine algorithm utilizes these element vectors (colour and texture) to prepare and train our proposed structure. Image is **processed** (Fig. 7).

Input **image is classified** as either from five categories or other than these five categories (Forest Area, in the given test scenario) (Fig. 8).

This paper presented the result executing feature extracting and classification process in terms of precision, recall, and accuracy. The accurate classification depends upon how effectively the image features extracted and every classification algorithm relies on a good accuracy to understand. The proposed framework is trained and tested with 100 satellite images in which 50 images are taken for testing purpose and 50 images are used to classify against trained framework. The dataset is generated from taking images from Google Earth satellite. This work is implemented in MATLAB\Simulink 2016a.

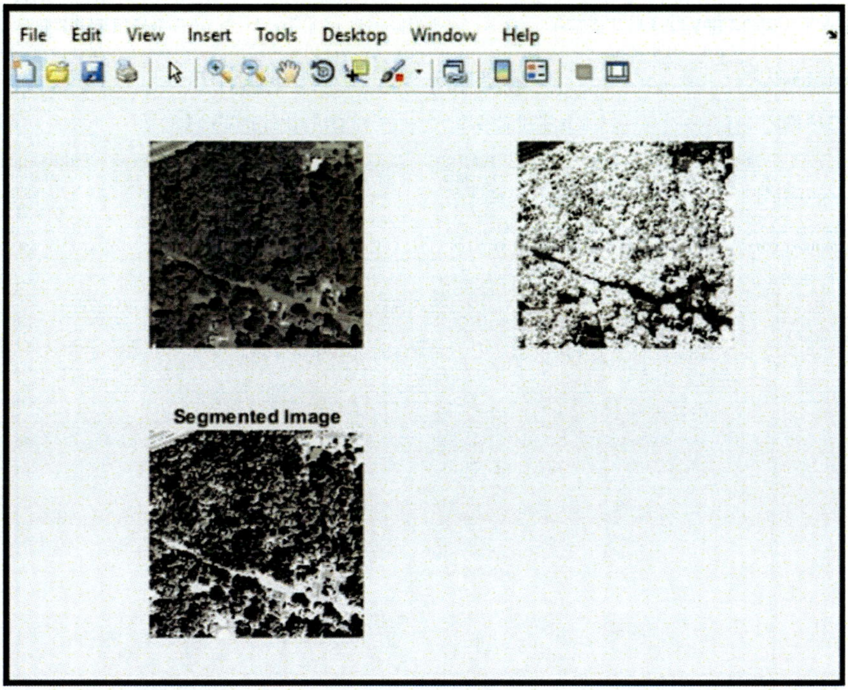

Fig. 5 Image segmentation

We use precision and recall improving the classification process. Precision is the ratio between measurements of number of relevant classified image to sum of relevant classified image and irrelevant classified image. Recall is the ratio between measurements of number of relevant classified image to total number of relevant classified image present in collection. The precision and recall is calculated as (Table 1)

$$\text{Precision} = \frac{\text{number of relevant classified images}}{\text{total number of classified images}} \tag{5}$$

$$\text{Recall} = \frac{\text{number of relevant classified images}}{\text{total no of images in collection}} \tag{6}$$

Table 2 and Fig. 9 demonstrate the comparison analysis carried out after successful implementation of proposed classification technique and described below. In Table 2, we have presented the accuracy (%) obtained after execution of our proposed framework. The execution results demonstrate the effectiveness of our proposed framework in satellite image classification application.

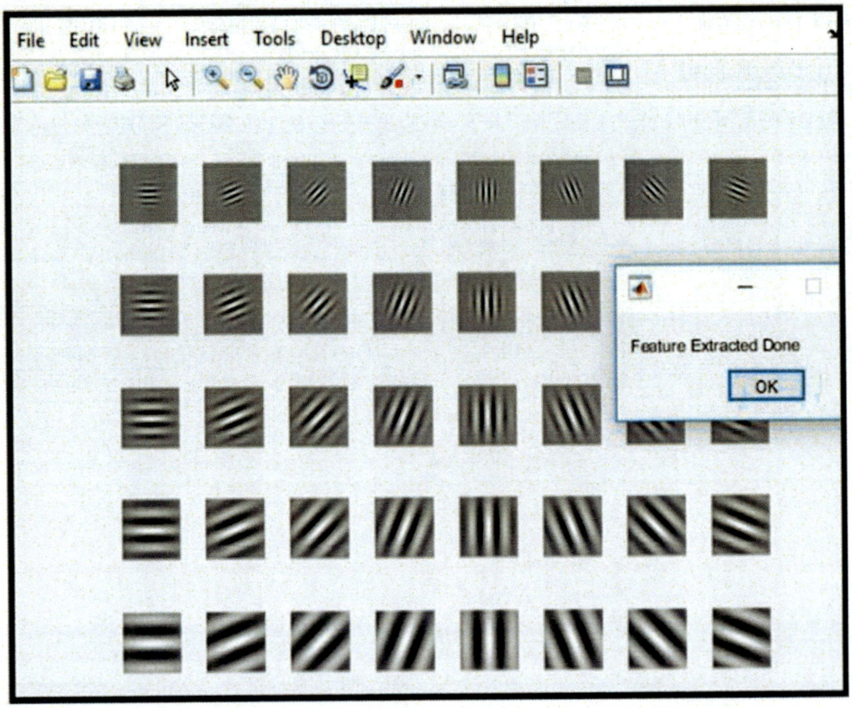

Fig. 6 Feature extraction

4 Conclusion

This paper proposed a hybrid approach for classification of satellite images into five classes which are residential area, river, desert, mountain, and forest image by utilizing fuzzy C-means for image segmentation and clustering, GLCM and Gabor filter for colour and texture feature and SVM classification for image classification based on feature value of feature vectors. Fuzzy C-means segments images into different clusters. Gabor filter is designed to extract the features from an image. Gabor filter is most appropriate method for texture representation and discrimination. SVM classifier utilizing Gabor feature vectors then used for the image classification into different classes. Further, the execution of proposed framework shows that one can be certain of high classification results utilizing the proposed approach and the framework yields better classification accuracies.

Fig. 7 Image processing

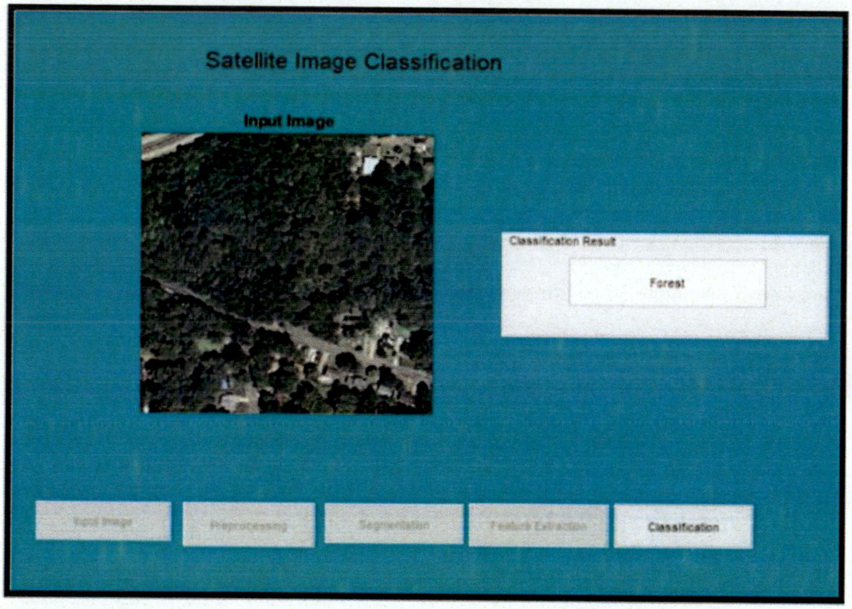

Fig. 8 Classified image

Table 1 Evaluation of performance of the proposed framework based on precision and recall

Techniques/no. of queried images	10	20	30	40	50	100
Precision	1.00	0.41	0.32	0.40	0.23	0.10
Recall	0.15	0.33	0.52	0.68	0.85	1.00

Table 2 Table of accuracy between existing approach and proposed approach

Techniques	No. of queried images	Classification accuracy
GLCM and gabor filter, FCM, and SVM (proposed)	100	99.6
GLCM, LWT, and SVM (existing)	100	99.4

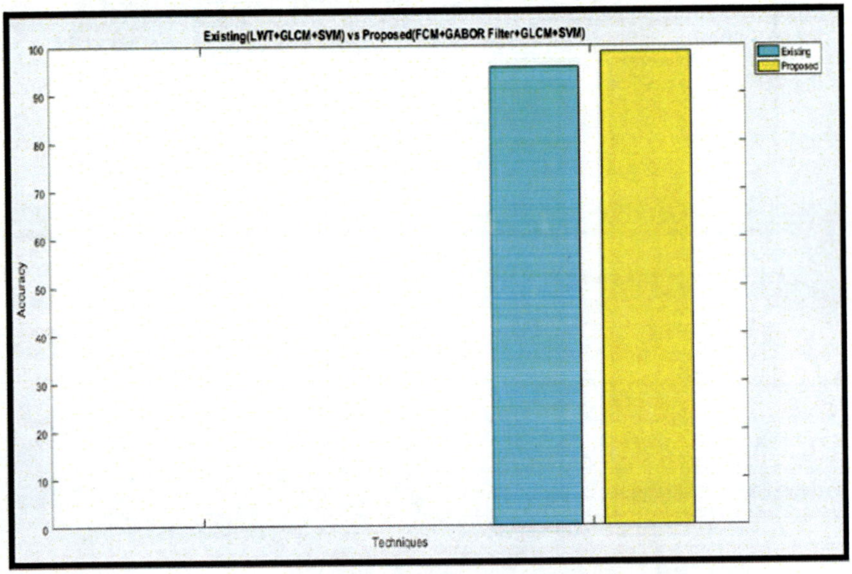

Fig. 9 Accuracy graph between existing approach and proposed approach

References

1. S. Saha, S. Bandyopadhyay, A multiobjective simulated annealing based fuzzy-clustering technique with symmetry for pixel classification in remote sensing imagery, in *19th International Conference on Pattern Recognition, 2008. ICPR 2008*, 23 Jan 2009. https://doi.org/10.1109/ICPR.2008.4761723. ISSN: 1051-4651
2. G. Sheng, W. Yang, L. Chen, H. Sun, Satellite image classification using sparse codes of multiple features, in *2010 IEEE 10th International Conference on Signal Processing (ICSP)*, 03 Dec 2010. https://doi.org/10.1109/ICOSP.2010.5655718. ISBN: 978-1-4244-5897-4
3. J.-T. Hwang, K.-T. Chang, H.-C. Chiang, Satellite image classification based on Gabor texture features and SVM, in *2011 19th International Conference on Geoinformatics*, 12 Aug 2011. https://doi.org/10.1109/GeoInformatics.2011.5980774. ISSN: 2161-024X

4. S.M. Moorthi, I. Misra, Kernel based learning approach for satellite image classification using support vector machine, in *2011 IEEE Recent Advances in Intelligent Computational Systems (RAICS)*, 03 Nov 2011. https://doi.org/10.1109/RAICS.2011.6069282. ISBN: 978-1-4244-9478-1
5. O.S. Soliman, A.S. Mahmoud, S.M. Hassan, Remote sensing satellite images classification using support vector machine and particle swarm optimization, in *2012 Third International Conference on Innovations in Bio-Inspired Computing and Applications*, 25 Oct 2012. https://doi.org/10.1109/IBICA.2012.61. ISBN: 978-1-4673-2838-8
6. F. Mirzapour, H. Ghassemian, Using GLCM and gabor filters for classification of PAN images, in *Electrical Engineering (ICEE), 2013 21st Iranian Conference*, 14–16 May 2013. https://doi.org/10.1109/IranianCEE.2013.6599565. ISSN: 2164-7054
7. A.K. MansiSaraswat, A.T. Goswami, Object recognition using texture based analysis. Int. J. Comput. Sci. Inf. Technol. **4**(6), 775–782 (2013)
8. N. Sharma, A. Verma, Performance comparison of texture based approach for identification of regions in satellite image. Int. J. Comput. Appl. (0975-8887) **74**(2) (2013)
9. S.V.S. Prasad, T. Satyasavithri, I.V. Murali Krishna, Object classification of satellite images using cluster repulsion based kernel Fcm and Svm classifier. IOSR J. Electron. Commun. Eng. (IOSR-JECE) **7**(3), 25–35 (2013). e-ISSN: 2278-2834, p-ISSN: 2278-8735
10. G. Usha, A.N. Valliyappan, Gabor features based classification of multispectral images using clustering with SVM classifier. Int. J. Trends Eng. Technol. **13**(1) (2016). ISSN: 2349-9303
11. R. Kaur, D. Sharma, A. Verma, Enhance satellite image classification based on fuzzy clustering and Marr-Hildreth algorithm, in *2017 4th International Conference on Signal Processing, Computing and Control (ISPCC)*, 25 Jan 2018. https://doi.org/10.1109/ISPCC.2017.8269663. ISBN: 978-1-5090-5838-9
12. Y. Chen, L. Zhu, P. Ghamisi, Hyperspectral images classification with gabor filtering and convolutional neural network. IEEE Geosci. Remote Sens. Lett. **14**(12), 2355–2359 (2017). https://doi.org/10.1109/LGRS.2017.2764915
13. A. Dixit, N. Hedge, B. Eswar Reddy, Texture feature based satellite image classification scheme using SVM. Int. J. Appl. Eng. Res. **12**(13), 3996–4003 (2017). ISSN 0973-4562
14. P. Ganesan, K. Palanivel, B.S. Sathish, V. Kalist, K.B. Shaik, Performance of fuzzy based clustering algorithms for the segmentation of satellite images—a comparative study, in *2015 IEEE Seventh National Conference on Computing, Communication and Information Systems (NCCCIS)*, 12 Oct 2015. https://doi.org/10.1109/NCCCIS.2015.7295906. ISBN: 978-1-4799-8990-4
15. P. Yang, Z. Hou, X. Liu, Z. Shi, Texture feature extraction of mountain economic forest using high spatial resolution remote sensing images, in *2016 IEEE International Geoscience and Remote Sensing Symposium (IGARSS)*, 03 Nov 2016. https://doi.org/10.1109/IGARSS.2016.7729816. ISSN: 2153-7003

The Determinants of Internet Financial Reporting for Investor Decision Making: Evidence from Indonesia Companies

Kurnia Rina Ariani and Gustita Arnawati Putri

Abstract This research presents the latest finding the effect of profitability, age of the company, and audit quality to Internet financial reporting with the board of commissioners as a moderating variable in manufacturing companies listed in the Indonesia Stock Exchange (IDX) for year-end 2018. Adding a moderating variable is expected to find out whether the board of commissioners could moderate the relationship between profitability, age of the company, and audit quality to IFR. Using moderated regression analysis (MRA), this research could not empirically prove that there is a correlation between profitability, age of the company, and audit quality to IFR moderated by the board of commissioners level due to lack of use of technology in developing countries.

Keywords Profitability · Companies age · Audit quality · The board of commissioners · IFR

1 Introduction

In this digital era, the rapid expansion of technology has led to some changes not only in the social mindset but also the way of how companies run their business and how their information is exchanged [1–8].

Internet usage globally for the dispersion of financial information purposes is a common practice for many publicly traded companies worldwide. The Internet can create a new reporting environment for companies to constantly give information to their shareholders and attract their future investors. It is also possible to facilitate

K. R. Ariani (✉)
Muhammadiyah Surakarta University, Surakarta, Indonesia
e-mail: kra123@ums.ac.id

G. A. Putri
Veteran Bangun Nusantara Sukoharjo University, Kabupaten Sukoharjo, Indonesia
e-mail: gustita.ap@gmail.com

P. K. Pattnaik et al. (eds.), *Proceedings of International Conference on Smart Computing and Cyber Security*, Lecture Notes in Networks and Systems 149,
https://doi.org/10.1007/978-981-15-7990-5_23

access to information and to proffer cost savings associated with printing and sending physical reports for acceleration data analysis.

Nowadays, IFR is commonly used in the company's business operations cycle in providing easy access to investors' financial statements. IFR has emerged as the fastest media to inform matters related to the company and changed the reporting system to the paper-less reporting system [2, 9–11].

Prior research stated that more than half the companies sampled in their research have a Web site that presents financial reports [12]; this indicates companies consider the Internet to be a crucial medium for transferring financial information. Some prior studies found that the Internet is currently the main tool used in publishing and updating company financial information promptly [1, 4, 12–15].

There is no rule governing disclosure since financial reporting through the Internet is voluntary. Although it is not compulsory, companies can increase the level of transparency of the company with IFR. A high level of transparency will reduce the level of information asymmetry (mismatch) between managers and stakeholders. Increasing information transparency will ultimately improve corporate governance [3, 9, 10, 13, 16, 17]. Another advantage of using IFR is that companies can disclose information about company performance in real time.

This paper focused on testing the determinants of financial information reporting via Web site listed companies in Indonesia. The selection of samples based in Indonesia is intended because of Indonesia's characteristics, including developing countries so that the Web paper-based application is still relatively new, no more than a decade, so that it can predict IFR determinants in developing countries.

2 Literature Review and Hypothesis Development

One theory that supports the company's business activities is agency theory. It states the working relationship between the party that gives the authority (the principal) or the investor and the party that receives the authority (agency), or the manager, in the form of a cooperation contract. Agency theory arises because of differences in interests between interested parties; each party tries to increase profits for themselves. In this case, there is a tendency for agents to convey information about their performance in a certain way to maximize the wishes of the principals, in this case, the owner, creditor, and the government.

Prior research [18] conducted a survey of large companies in Europe. The survey was conducted on 1000 companies. The survey results show that as many as 536 companies already have Web sites and have disclosed their financial reports on the Internet [12]. Examined the correlation between IFR and stocks revealed that if a company had implemented IFR and had a high level of information disclosure, it would have a stock price whose movements tended to be high.

Khan [15] conducted a study of companies in Egypt. As many as 56% of companies in Egypt are known to have revealed most of the data information on the company's Web site. This study uses the variables of profitability, liquidity, leverage,

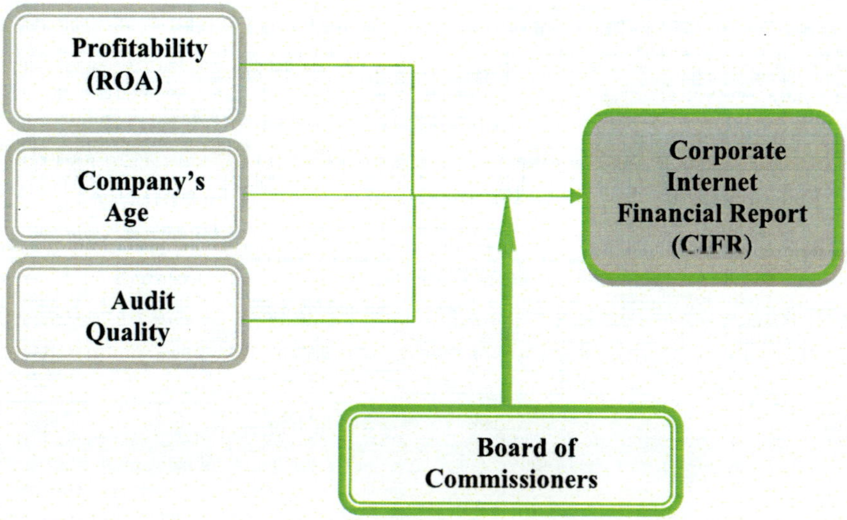

Fig. 1 Hypothesis framework

type of industry, public share ownership, company age, foreign share ownership, and auditor size. The research of [2] shows that company size variables significantly affect the delivery of financial information on the Internet of companies in France (Fig. 1) [19].

3 Research Methods

3.1 Population and Sample

Manufacturing companies listed in IDX for year-end 2018 was used as the population in this research. The sampling technique in this study was determined using a purposive sampling technique.

3.2 Operational Definition and Variable Measurement

Dependent Variable

IFR measurement using the index of reporting via the Web site [10]. Measurement of this variable is also divided into several criteria, as follows:

Criteria	Assessment (%)	Measurement
Content	40	If financial information is disclosed in HTML format, the score obtained is 2 If financial information is disclosed in PDF format, the score obtained is 1
Timeliness	20	The assessment of this criterion uses a scale of 0–3. The existence of disclosure of press releases, stock prices, the latest quarterly audit report will add a score for the company
Technology index	20	The availability of items that cannot be provided by printed paper reports include download plug-ins, online feedback, presentation slides, uses of multimedia technology (audio and video), analysis tools, and advanced features such as XBRL
User support	20	Availability of navigation (FAQ, links to the main page, links to the top, site maps, search sites) and the number of clicks to get financial information (scale 0–3)

$$\text{IFR-DS} = \left(\frac{\text{Score}}{\text{Max}}\%\text{CONT}\right) + \left(\frac{\text{Score}}{\text{Max}}\%\text{TIME}\right) + \left(\frac{\text{Score}}{\text{Max}}\%\text{TECH}\right) + \left(\frac{\text{Score}}{\text{Max}}\%\text{SUPP}\right)$$

where:

Score Total score for each disclosure component
Max Maximum score of each disclosure component
% CONT Proportion of criteria for evaluating the contents of financial statements by 40%
% TIME Proportion of financial time reporting criteria by 20%
% TECH Proportion of technology assessment criteria of 20%
% SUPP Proportion of user support assessment criteria by 20%.

Independent variable

1. Profitability	(ROA) $= \frac{\text{Earning after Tax}}{\text{Total Asset}} \times 100\%$
2. Company's age	Year of observation—year established
3. Audit quality	"0" if the company does not use the Big4 Affiliated Public Accounting Firm "1" if the company uses Big4 Affiliate Public Accounting Firm

Moderating variable

Board of commissioners	Total of board commissioners [2]

4 Results and Discussion

There are 154 manufacturing companies listed on the IDX in 2018 that will be sampled in this study, by eliminating the following predetermined sample criteria, so that in the final process. Table 1 shows the results of regression tests with an adjusted R square value of 0.056 or 5.6%. It shows that a 5.6% internet financial reporting can be explained by variables of profitability, age of the company, audit quality and the variable moderating the board of commissioners, while 94.4% is explained by other factors outside the research model. Judging from the significance value of the t-test shows that there is no supported hypothesis.

The results of testing hypothesis in this research cannot prove that there is an interaction between profitability, age of the company, and audit quality that is moderated by good corporate governance to CIFR [1, 2, 4, 10, 12, 13, 17, 20, 21].

It gives a signal even though companies in Indonesia have high profits, sound control systems, and reliable auditors, but instead not to overly utilize technology to attract potential investors to invest in their companies. It is understandable because investors themselves are more confident to visit the local authority's Web site such as the Indonesian Stock Exchange to obtain information about the financial statements of an audited company, thereby reducing the company's efforts to optimize the current technology to facilitate access to the availability of financial reports on their Web site.

Table 1 Hypothesis test

Variables	Regression coefficients	*t*-values
ROA	0.184	0.508
CA	0.001	0.576
AQ	−0.128	−1.801
AC	0.015	−0.701
ROA * AC	0.006	0.091
CA * AC	0.0001	−0.587
AQ * AC	0.025	1.658
Adj R square = 0.056		
F statistic = 0.039		

5 Conclusion

IFR is a tool used for companies to share financial and non-financial information with their stakeholders. This study failed to find evidence that there is a relationship between profitability, company age, and audit quality that is moderated by good corporate governance on IFR. This research has limitations on the sample and research period used, i.e., only manufacturing companies and uses a year period 2018, so the results only represent conditions in that period. However, this was chosen based on relevant considerations; manufacturing companies are the majority of Indonesia, and the company's Web site is always changing. Based on these limitations, future research can use all companies listed on IDX and use more recent periods to examine the effect of corporate governance mechanisms on IFR levels and to understand the development of IFR practices from period to period, especially in an Indonesian context company.

References

1. D. Aly, J. Simon, K. Hussainey, Determinants of corporate internet reporting: evidence from Egypt. Manag. Auditing J. **25**(2), 182–202 (2010)
2. F. Laswad, R. Fisher, P. Oyelere, Determinants of voluntary internet financial reporting by local government authorities. J. Acc. Public Policy **24**(2), 101–121 (2005)
3. S.-C. Lai, C. Lin, H.-C. Li, F.H. Wu, An empirical study of the impact of internet financial reporting on stock prices. Int. J. Digit. Account. Res. **10**, 1–26 (2010)
4. M. Tarmidi, R.A. Roni, An international comparison of the determinants and financial information quality in XBRL reporting environment. Procedia Soc. Behav. Sci. **164**, 135–140 (2014)
5. K.A. Al Daoud, K.N.I.K. Ismail, N.A. Lode, The timeliness of financial reporting among Jordanian companies: do company and board characteristics, and audit opinion matter? Asian Soc. Sci. **10**(13) (2014)
6. P. Oyelere, N. Kuruppu, Voluntary internet financial reporting practices of listed companies in the United Arab Emirates. J. Appl. Acc. Res. **13**(3), 298–315 (2012)
7. K. Ojah, T. Mokoaleli-Mokoteli, Internet financial reporting, infrastructures and corporate governance: an international analysis. Rev. Dev. Financ. **2**(2), 69–83 (2012)
8. R. Fisher, P. Oyelere, F. Laswad, Corporate reporting on the internet: audit issues and content analysis of practices. Manag. Auditing J. **19**(3), 412–439 (2004)
9. L. Botti, S. Boubaker, A. Hamrouni, B. Solonandrasana, Corporate governance efficiency and internet financial reporting quality. Rev. Acc. Financ. **13**(1), 43–64 (2014)
10. O. Abdelsalam, A. El-Masry, The impact of board independence and ownership structure on the timeliness of corporate internet reporting of Irish-listed companies. Manag. Financ. **34**(12), 907–918 (2008)
11. S. AlMatrooshi, A.M. Al-Sartawi, Z. Sanad, Do audit committee characteristics of Bahraini listed companies have an effect on the level of internet financial reporting? Corp. Ownership Control J. **13**(2), 130–146 (2016)
12. A.S. Kelton, R.R. Pennington, Internet financial reporting: the effects of information presentation format and content differences on investor decision making. Comput. Hum. Behav. **28**(4), 1178—1185 (2012)
13. H. Akhiruddin, The effect of financial reporting on the internet on market reaction (Empirical study of companies registered in the Kompas 100 Index 2011 period) (Universitas Brawijaya, 2013)

14. J. Bananuka, S. Night, M. Ngoma, G.M. Najjemba, Internet financial reporting adoption. J. Econ. Financ. Adm. Sci. **24**(48), 266–287 (2019)
15. T. Khan, Financial reporting disclosure on the internet: an international perspective (Victoria University, 2006)
16. D. Puspitaningrum, S. Atmini, Corporate governance mechanism and the level of internet financial reporting: evidence from Indonesian companies. Procedia Econ. Financ. **2**, 157–166 (2012)
17. T. Dolinšek, A. Lutar-Skerbinjek, Voluntary disclosure of financial information on the internet by large companies in Slovenia. Kybernetes **47**(3), 458–473 (2018)
18. A. Saleh Al Arussi, M. Hisyam Selamat, M. Mohd Hanefah, Determinants of financial and environmental disclosures through the internet by Malaysian companies. Asian Rev. Acc. **17**(1), 59–76 (2009)
19. S. Boubaker, F. Lakhal, M. Nekhili, The determinants of web-based corporate reporting in France. Manag. Auditing J. **27**(2), 126–155 (2012)
20. I. Khadaroo, Corporate reporting on the internet: some implications for the auditing profession. Manag. Auditing J. **20**(6), 578–591 (2005)
21. R. Debreceny, G.L. Gray, A. Rahman, The determinants of Internet financial reporting. J. Account. Public Policy **21**(4–5), 371–394 (2002)

Resource Allocation in the Integration of IoT, Fog, and Cloud Computing: State-of-the-Art and Open Challenges

Baseem Al-athwari and Hossain Md Azam

Abstract In recent years, the number of smart devices has a tremendous growth and brought, along with the high Internet connectivity, about the emergence of the IoTs as a new computing paradigm that resulted in a dramatic change in many areas of life, including Smart Home, Smart Cities, Smart Industry, Smart Traffic, Smart Grids, etc. However, to a great extent, this rapid growth of the smart devices and communication technologies have not resulted in significant implementation of IoTs. There are many challenges associated with the IoTs yet to be addressed to pave the transformation to the smart world. One obstacle is the lack of computing resources of the IoT devices to collect, process, store, and transmit the huge amount of generated data which demands high speed, large capacity, and high reliability, and availability. This paper presents a survey on the integration of IoT, fog, and cloud with a focus on the state-of-the-art of resource allocation. The paper also discusses open challenges yet to be addressed in this field.

Keywords Cloud computing · Fog computing · IoT · Resource allocation

1 Introduction

The IoTs is a new paradigm comprises things that are intelligently connected with a variety of communication technologies such as Wi-Fi, wireless sensor networks (WANs), radio-frequency identification (RFID), to name a few [1]. IoT connected devices are highly heterogeneous ranging from smartphones, laptops, home appliances, sensors, and wearable devices, to name a few. As per HIS, the IoT market will grow from an installed base of 15.4 billion devices in 2015 to 30.7 billion devices in 2020 and 75.4 billion in 2025. This unprecedented number of devices results in a huge amount of data to be stored, processed, and transmitted. Such activities imply

B. Al-athwari (✉) · H. M. Azam
Smart Computing Department, Kyungdong University, Goseong-gun, Gangwon-do, South Korea
e-mail: Baseem_cs@kduniv.ac.kr

P. K. Pattnaik et al. (eds.), *Proceedings of International Conference on Smart Computing and Cyber Security*, Lecture Notes in Networks and Systems 149,
https://doi.org/10.1007/978-981-15-7990-5_24

a heavy workload on IoTs devices and results on significant resource consumptions (e.g., memory, processing, communication, and energy). However, due to their computing resources constraints, these devices may not be able to store and process the generated data. As a solution, such IoT devices can rely on cloud computing to store, process, and analyze their data where computing resources are provided as Internet-based or they can leverage the computing capacity of fog computing where computing devices are located nearby.

From various researches, it has been well acknowledged that the integration of cloud and/or fog computing and IoT can be as a feasible solution to overcome the computing resource constraints of IoT devices. By leveraging the cloud and fog resources, it will not only extend the computing resources of IoTs devices but also to support and enhance the interoperability between the heterogeneous IoT devices and their applications. In this regard, resource allocation has become a hot topic and has gain significant attention from the researchers and industry. Frameworks related to the resource allocation in cloud computing have been widely exploited, jointly with other emerging computing paradigm such as cloudlet, fog computing, and edge computing, to fulfill the requirements of IoT ecosystem. However, to the best of our knowledge, the integration of IoTs with such computing paradigm has not received an adequate attention.

This paper is organized as follow. The high level IoT-fog-cloud architecture that supports the integration will be described in Sect. 2. Section 3 discusses the start-of-the-art of the resource allocation in the integration of IoT, fog, and cloud. Section 4 discusses the open challenges of the resource allocation in IoT-fog-cloud environment.

2 IoT-Fog-Cloud Architecture

In this section, we discuss the integrated IoT-fog-cloud architecture that is covered by this study, as shown in Fig. 1. It consists of three layers, and each layer has different computing resources [2]. The IoT layer comprises things or end-users' devices; the fog computing layer where computing resources are located nearby the IoT devices; and cloud computing which provides an Internet-based computing resources that can be used to extend the computing resources of fog and IoT.

2.1 IoT

IoT has very short history of 20 years. The term of IoT was firstly proposed by Ashton in 1999 [3]. The concept itself has its definition which means things will be connected via Internet. The basic idea of IoT is to connect anything, anyone anywhere, and anytime. It emerged as a new platform in which devices become smarter. Various researchers have described IoT in multitude forms. However, for

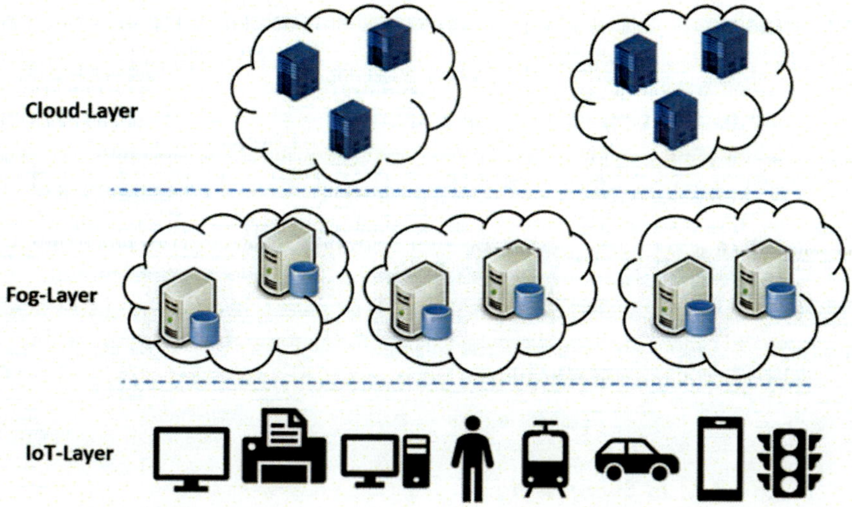

Fig. 1 IoT-fog-cloud architecture

the purpose of our work, we will consider the definition given by the ITU as cited in [4]:

"A global infrastructure for the information society enabling advanced services by interconnecting (physical and virtual) things based on, existing and evolving, interoperable information, and communication technologies."

The IoT architecture can be viewed as a physical system, virtual system, or a hybrid of the two systems. It consists of the collection of numerous active physical things, sensors, actuators, communication technologies, protocols, and users [4]. There is a significant amount of survey studies on IoT architecture, taxonomy, applications, and trends [4–11].

IoT can benefit from the integration of fog and cloud computing to compensate its resources constraints (e.g., storage, processing, communication). However, the IoT-fog-cloud continuum entails many challenges that should be addressed to make it suitable for all kinds of applications and services leveraged by the IoT paradigm. Such challenges are (but not limited to) resource allocation and optimization, protocols, business models, security, privacy, etc.

2.2 Fog Computing

Despite the rapid growth of the cloud computing paradigm, the adoption of cloud computing has exposed some limitations in fulfilling the requirements of some classes of applications such as delay-sensitive applications, privacy-sensitive applications [12], and security-sensitive applications. In order to overcome the limitations of cloud computing, fog computing has been developed in such way that it is located between the IoT devices and the cloud to provide cloud-like computing resources close to the IoT device.

Fog computing is expected to full the requirements that cloud are not able to [10, 13–15]. Considering the characteristics of the fog such as low latency [16] and location awareness, wide-spread geographical distribution, mobility, very large number of nodes, predominant role of wireless access, strong presence of streaming and real-time applications, and heterogeneity, [17] argued that these characteristics make the fog appropriate platform for a number of critical Internet of things (IoT) services and applications. In this sense, fog computing has been proposed recently to be integrated with IoTs and cloud [18] to create a new genre of ecosystem which would benefit the overall growth of the information and communication technology-based development [19].

2.3 Cloud Computing

Cloud computing is a paradigm which provides on-demand Internet-based computing resources including storage, processing, applications, services, and networks. The basic idea of cloud computing is to provide all the infrastructure needed to run applications over the Internet. Therefore, IoT can leverage the computing power of clouds by offloading the computationally intensive applications onto clouds instead of running them locally. That is referred to as computing offloading. The concept of offloading the computation and data to the cloud is used to address the limitation of mobile devices by using the resources provided by the cloud rather than the mobile device itself to run the mobile application. In the recent years, several computational offloading frameworks have been proposed with different objectives such minimizing energy consumption [20–24], reduction in execution time [25–27], minimizing the monetary cost [27–29]. In addition to those, there are also combinations of two or more objectives such as minimizing energy and execution time [30–34], and even some of them combines more than two criteria such as minimizing execution time, cost, and network latency [34–37].

With the emergence of IoT paradigm, many studies empathized on the strong need of integrating IoT and cloud [4, 38–42]. The following section presents the state-of-the-art of the IoT integration with cloud and/or fog with the focus on the resource allocation aspect as it is one of the main challenges in the IoT-fog-cloud infrastructure and becoming a hot topic that has been attracting attention in the academic community.

3 State-of-the-Art of Resource Allocation in IoT-Fog-Cloud

In the last few years, the IoTs and cloud have gained popularity and the integration of both is expected to disrupt both current and future Internet [43–45]. Many works in literature have been done on cloud, fog and IoT separately. However, several studies on the integration of IoTs and cloud computing have been conducted recently [11, 46]. For instance [46], proposed a framework that incorporates and supports adaptive interaction of the user with the IoT cloud architecture based on the quality of context information and quality of services. Another platform framework by Li et al. [47] called IoT platform as a service (IoTPaaS) was proposed to provide essential platform services for IoT solutions providers by leveraging computing resources and middleware services on cloud to provide efficient delivery and continuous extension of their services. However, these platforms did not consider the fog which can provide the intermediate layer between the IoT device and the cloud to support IoT applications, especially delay-sensitive services and to fulfill the requirement for mobility.

Despite the significant amount of surveys and review papers works, which has been carried out on the resource allocation related to cloud, there is a lack of survey papers that refer directly to the resource allocation problem in the IoT-fog-cloud integrated environment. One of the most recent and important surveys is carried out by Ghanbari et al. [48] in which they conducted a systematic literature review (SLR) to investigate the resources allocation methods in the IoT. In their paper, the resource allocation techniques are classified into different categories based on different parameters including cost-aware, context-aware, efficiency-aware, load balancing-aware, power-aware, QoS-aware, SLA-based, and utilization-aware. Despite the large number of cited resource allocation mechanisms in this paper, however, most of the works focus on either cloud, fog, or IoT environment. Although some of the works considered the resource allocation in IoT and cloud/fog, however, only one work considered the resource allocation in IoT-fog-cloud environment [49] in which the authors proposed an architecture of IoT service delegation and resource allocation based on collaboration between fog and cloud computing.

In IoT-fog-cloud environment, IoT applications could choose the cloud or fog to execute the computation. In this regard, load balancing should be considered for an effective resource allocation strategy as it is the main goal of resource allocation. Hence, computing resources of the integrated IoT-fog-cloud environment should be allocated in such way that achieve balanced utilization across all resources [48]. In this regards, the authors of [50] proposed a dynamic resource allocation method, named DRAM, for load balancing in fog environment. Although their proposed method is designed through static resource allocation and dynamic service migration. However, only fog devices are used to achieve the load balance.

A recent systematic review of the load balancing mechanisms in the Internet of things is carried out in [51]. The paper examined the challenges relevant to the load balancing issue in IoT considering many parameters such as energy consumption, fault-tolerant, heterogeneity, overhead, reliability, scalability, and others. However, according to the authors' conclusion, there is not any single research considered all the load balancing parameters. In addition, this work did not consider the integration of IoT, fog, and cloud.

As the IoT-fog-cloud environment comprises a huge number of heterogeneous devices that run heterogeneous applications, optimization in resource allocation becomes a vital issue. In this regards, the optimization approach for resource allocation on cloud computing for IoT is proposed in [52]. In their approach, they used combinatorial auction which is a popular approach for resource allocation in cloud computing. Their objective is to maintain the quality of the service (QoS) and maximize the profits for the provider. For that, they consider the execution time constraint as SLA constraint. In order to efficiently allocate the resources, the winners at each bidding round are determined according to the job's urgency based on the execution time deadline. The other QoS factors are not included, and also their proposed system is only for cloud.

Many other researchers have proposed dynamic resource allocation methods in the context of cloud and/or edge computing [28, 53–60] but none of these works considers the integration of Io, fog, and cloud. In addition, it is worthy to mention that, although there are many frameworks have been proposed to build an IoT-fog-cloud ecosystem, there is no common methodology to integrate the IoT, fog, and cloud. As many IoT applications have common requirements, more efforts should be spent to come up with a generic and flexile platform that could be the starting point for defining a standardized workflow [40].

4 Open Challenges and Future Directions

There are many challenges regarding the resource allocation relevant to the integration of IoT, fog, and cloud that need to be investigated comprehensively. Hence, this section discusses various open challenges and future directions relevant to the resource allocation in the IoT-fog-cloud environment.

4.1 IoT Device/User's Context

For effective resource allocation decision making in the IoT-fog-cloud environment, IoT device/user's context such as mobility and his available resources should be considered with an attempt to consider all the factors that influence the QoS. For instance, as for the case of the mobility, the IoT user/device may lose connectivity to the cloud or even the fog computing resources because of changing his location but might reconnect again later after short time or might not reconnect for a long time. Therefore, the resource allocation framework should ensure the QoS in such way that the IoT user/device will be connected to the available resource.

4.2 Cost

In addition to the user context, the resource allocation decision making should consider different costs associated with the offloading the computation from the IoT device to the IoT-fog-cloud environment (i.e., monetary cost, time cost, energy cost, communication cost, and computation cost). Therefore, more efforts should be done to develop a multi-criteria framework that should consider all the cost incurred.

4.3 Business Model

As there will be different entities involved in the integrated IoT-fog-cloud environment, each of them has its own resources, customers, pricing and billing systems, there is strong need to develop a business model that can help in setting the price, allocate the resources among the involved entities in such way that ensures the fairness of sharing the revenue among all the entities involved in the IoT-fog-cloud ecosystem.

4.4 Dynamic Changes of the IoT-Fog-Cloud Environment

Making adaptive resource allocation decisions at runtime requires consideration of the dynamic changes in the IoT-fog-cloud environment where the IoT device/user, resources availability either in the fog or the cloud, network conditions, application requirements, and user context may change. Therefore, there is a need for a multi-objective resource allocation platform that can help in resources discovery, allocation, and monitoring which can result in making optimal resource allocation decision.

The effectiveness of the decision-making process for the resource allocation highly depends on how the resource allocation mechanism deals with dynamic nature of IoT-fog-cloud environment considering the optimization objectives. Therefore, the resource allocation framework should consider the IoT device/user requirements and application requirements. It should also consider the available resources at the edge or/and cloud with their specifications/features as well as their cost.

References

1. M. Díaz, C. Martín, B. Rubio, State-of-the-art, challenges, and open issues in the integration of internet of things and cloud computing. J. Netw. Comput. Appl. **67**, 99–117 (2016). https://doi.org/10.1016/j.jnca.2016.01.010
2. B. Liu, X. Chang, B. Liu, Z. Chen, Performance analysis model for fog services under multiple resource types, in *Proceedings of 4th International Conference on Dependable Systems and*

Their Applications, DSA 2017, vol. 2018, Janua (2017), pp. 110–117. https://doi.org/10.1109/DSA.2017.26
3. K. Asthon, That internet of things–the real world, things matter more than ideas. RFID J. 4986 (2010). https://doi.org/10.1038/nature03475
4. P.P. Ray, A survey on internet of things architectures. J. King Saud Univ. Comput. Inf. Sci. **30**(3), 291–319 (2018). https://doi.org/10.1016/j.jksuci.2016.10.003
5. A. Čolaković, M. Hadžialić, Internet of things (IoT): a review of enabling technologies, challenges, and open research issues. Comput. Netw. **144**, 17–39 (2018). https://doi.org/10.1016/j.comnet.2018.07.017
6. A. Alreshidi, A. Ahmad, Architecting software for the internet of thing based systems. Future Internet **11**(7) (2019). https://doi.org/10.3390/fi11070153
7. M.U. Farooq, M. Waseem, S. Mazhar, A. Khairi, T. Kamal, A review on internet of things (IoT). Int. J. Comput. Appl. **113**(1), 1–7 (2015). https://doi.org/10.5120/19787-1571
8. P.P. Ray, A survey of IoT cloud platforms. Future Comput. Inform. J. **1**(1–2), 35–46 (2016). https://doi.org/10.1016/j.fcij.2017.02.001
9. A. Whitmore, A. Agarwal, L. Da Xu, The internet of things—a survey of topics and trends. Inf. Syst. Front. **17**(2), 261–274 (2015). https://doi.org/10.1007/s10796-014-9489-2
10. I. Ud Din et al., The internet of things: a review of enabled technologies and future challenges. IEEE Access **7**, 7606–7640 (2019). https://doi.org/10.1109/ACCESS.2018.2886601
11. M.M. Gomes, R. Da Rosa Righi, C.A. Da Costa, Future directions for providing better IoT infrastructure, in *UbiComp 2014, Adjunct Proceedings 2014, ACM International Joint Conference on Pervasive and Ubiquitous Computing* (2014), pp. 51–54. https://doi.org/10.1145/2638728.2638752
12. L. Bittencourt et al., The internet of things, fog and cloud continuum: integration and challenges. Internet of Things **3–4**, 134–155 (2018). https://doi.org/10.1016/j.iot.2018.09.005
13. A.V. Dastjerdi, H. Gupta, R.N. Calheiros, S.K. Ghosh, R. Buyya, Fog computing: principles, architectures, and applications, in *Internet of Things* (2016), pp. 61–75. https://doi.org/10.1016/B978-0-12-805395-9.00004-6
14. S. Mostafavi, W. Shafik, Fog computing architectures, privacy and security solutions. J. Commun. Technol. Electron. Comput. Sci. **24**, 1–14 (2019). https://doi.org/10.22385/JCTECS.V24I0.292
15. C. Puliafito, E. Mingozzi, F. Longo, A. Puliafito, O. Rana, Fog computing for the internet of things: a survey. ACM Trans. Internet Technol. **19**(2) (2019). https://doi.org/10.1145/3301443
16. H. Ali, G. Alsalman, A framework for optimization of location of fog servers and fog network formation to minimize latency (2019)
17. F. Bonomi, R. Milito, J. Zhu, S. Addepalli, Fog computing and its role in the internet of things, in *MCC'12—Proceedings of the 1st ACM Mobile Cloud Computing Workshop* (2012), pp. 13–15. https://doi.org/10.1145/2342509.2342513
18. N. Hassan, S. Gillani, E. Ahmed, I. Yaqoob, M. Imran, The role of edge computing in internet of things. IEEE Commun. Mag. **56**(11), 110–115 (2018). https://doi.org/10.1109/MCOM.2018.1700906
19. P.P. Ray, D. Dash, D. De, Edge computing for internet of things: a survey, e-healthcare case study and future direction. J. Netw. Comput. Appl. **140**, 1–22 (2019). https://doi.org/10.1016/j.jnca.2019.05.005
20. B. Al-Athwari, J. Altmann, Utility-based smartphone energy consumption optimization for cloud-based and on-device application uses, *Lecture Notes in Computer Science (including Subseries Lecture Notes in Artificial Intelligence, Lecture Notes in Bioinformatics)*, vol. 9512 (2016), pp. 164–175. https://doi.org/10.1007/978-3-319-43177-2_11
21. Y.L.K. Kumar, Cloud computing for mobile users: can offloading computation save energy? Comput. (Long. Beach. Calif) **43**(4), 51–56 (2010)
22. Y. Zhang, J. He, S. Guo, Energy-efficient dynamic task offloading for energy harvesting mobile cloud computing, in *2018 IEEE International Conference on Networking, Architecture and Storage, NAS 2018—Proceedings* (2018), pp. 1–4. https://doi.org/10.1109/NAS.2018.8515736

23. H. Wu, Y. Sun, K. Wolter, Energy-efficient decision making for mobile cloud offloading. IEEE Trans. Cloud Comput. **6**(X), 1–15 (2018). https://doi.org/10.1109/TCC.2018.2789446
24. E. Cuervoy et al., MAUI: making smartphones last longer with code offload, in *MobiSys'10—Proceedings of 8th International Conference on Mobile Systems, Applications and Services* (2010), pp. 49–62. https://doi.org/10.1145/1814433.1814441
25. X. Meng, W. Wang, Z. Zhang, Delay-constrained hybrid computation offloading with cloud and fog computing. IEEE Access **5**, 21355–21367 (2017). https://doi.org/10.1109/ACCESS.2017.2748140
26. I. Giurgiu, O. Riva, D. Juric, I. Krivulev, G. Alonso, Calling the cloud: enabling mobile phones as interfaces to cloud applications, *Lecture Notes in Computer Science (including Subseries Lecture Notes in Artificial Intelligence, Lecture Notes in Bioinformatics)*, vol. 5896 LNCS (2009), pp. 83–102. https://doi.org/10.1007/978-3-642-10445-9_5
27. R. Wolski, S. Gurun, C. Krintz, D. Nurmi, Using bandwidth data to make computation offloading decisions, in *IPDPS Miami 2008—Proceedings of 22nd IEEE International Parallel and Distributed Processing Symposium, Program and CD-ROM* (2008). https://doi.org/10.1109/IPDPS.2008.4536215
28. J. Altmann, M.M. Kashef, Cost model based service placement in federated hybrid clouds. Future Gener. Comput. Syst. **41**, 79–90 (2014). https://doi.org/10.1016/j.future.2014.08.014
29. J. Altmann, B. Rupp, P. Varaiya, Effects of pricing on internet user behavior. Netnomics **3**(1), 67–84 (2001). https://doi.org/10.1023/A:1009944726255
30. Y. Wang, L. Wu, X. Yuan, X. Liu, X. Li, An energy-efficient and deadline-aware task offloading strategy based on channel constraint for mobile cloud workflows. IEEE Access **7**, 69858–69872 (2019). https://doi.org/10.1109/ACCESS.2019.2919319
31. S. Guo, J. Liu, Y. Yang, B. Xiao, Z. Li, Energy-efficient dynamic computation offloading and cooperative task scheduling in mobile cloud computing. IEEE Trans. Mob. Comput. **18**(2), 319–333 (2019). https://doi.org/10.1109/TMC.2018.2831230
32. F. Xia, F. Ding, J. Li, X. Kong, L.T. Yang, J. Ma, Phone2Cloud: exploiting computation offloading for energy saving on smartphones in mobile cloud computing. Inf. Syst. Front. **16**(1), 95–111 (2014). https://doi.org/10.1007/s10796-013-9458-1
33. S. Kosta, A. Aucinas, P. Hui, R. Mortier, X. Zhang, ThinkAir: dynamic resource allocation and parallel execution in the cloud for mobile code offloading, in *Proceedings of IEEE INFOCOM* (2012), pp. 945–953. https://doi.org/10.1109/INFCOM.2012.6195845
34. B.G. Chun, S. Ihm, P. Maniatis, M. Naik, A. Patti, CloneCloud: elastic execution between mobile device and cloud, in *EuroSys'11—Proceedings EuroSys 2011 Conference* (2011), pp. 301–314. https://doi.org/10.1145/1966445.1966473
35. M. Nir, A. Matrawy, M. St-Hilaire, Economic and energy considerations for resource augmentation in mobile cloud computing. IEEE Trans. Cloud Comput. **6**(1), 99–113 (2018). https://doi.org/10.1109/TCC.2015.2469665
36. H. Wu, Multi-objective decision-making for mobile cloud offloading: a survey. IEEE Access **6**, 3962–3976 (2018). https://doi.org/10.1109/ACCESS.2018.2791504
37. X. Zhang, A. Kunjithapatham, S. Jeong, S. Gibbs, Towards an elastic application model for augmenting the computing capabilities of mobile devices with cloud computing. Mob. Netw. Appl. **16**(3), 270–284 (2011). https://doi.org/10.1007/s11036-011-0305-7
38. H.F. Atlam, A. Alenezi, A. Alharthi, R.J. Walters, G.B. Wills, Integration of cloud computing with internet of things: challenges and open issues, in *Proceedings 2017, IEEE International Conference on Internet of Things, IEEE Green Computing and Communications, IEEE Cyber, Physical and Social Computing. IEEE Smart Data, iThings-GreenCom-CPSCom-SmartData 2017*, vol. 2018, Janua (2018), pp. 670–675. https://doi.org/10.1109/iThings-GreenCom-CPSCom-SmartData.2017.105
39. Y. Liu, K. Akram Hassan, M. Karlsson, Z. Pang, S. Gong, A data-centric internet of things framework based on azure cloud. IEEE Access **7**, 53839–53858 (2019). https://doi.org/10.1109/ACCESS.2019.2913224
40. A. Botta, W. De Donato, V. Persico, A. Pescapé, Integration of cloud computing and internet of things: a survey. Future Gener. Comput. Syst. **56**, 684–700 (2016). https://doi.org/10.1016/j.future.2015.09.021

41. S.H. Shah, I. Yaqoob, A survey: internet of things (IOT) technologies, applications and challenges, in *2016 4th IEEE International Conference on Smart Energy Grid Engineering, SEGE 2016*, vol. i (2016), pp. 381–385. https://doi.org/10.1109/SEGE.2016.7589556
42. L. Atzori, A. Iera, G. Morabito, The internet of things: a survey. Comput. Netw. **54**(15), 2787–2805 (2010). https://doi.org/10.1016/j.comnet.2010.05.010
43. J. Zhou et al., CloudThings: a common architecture for integrating the internet of things with cloud computing, in *Proceedings of 2013 IEEE 17th International Conference on Computer Supported Cooperative Work in Design, CSCWD 2013* (2013), pp. 651–657. https://doi.org/10.1109/CSCWD.2013.6581037
44. J. Osborne, Internet of things and cloud computing, in *Internet of Things and Data Analytics Handbook* (2017), pp. 683–698. https://doi.org/10.1002/9781119173601.ch42
45. H.P. Breivold, Internet-of-things and cloud computing for smart industry: a systematic mapping study, in *Proceedings—2017 5th International Conference on Enterprise Systems ES 2017* (2017), pp. 299–304. https://doi.org/10.1109/ES.2017.56
46. N. Alhakbani, M.M. Hassan, M.A. Hossain, M. Alnuem, A framework of adaptive interaction support in cloud-based internet of things (IoT) environment, in *Lecture Notes in Computer Science (including Subseries Lecture Notes in Artificial Intelligence, Lecture Notes in Bioinformatics)*, vol. 8729 (2014), pp. 136–146. https://doi.org/10.1007/978-3-319-11692-1_12
47. F. Li, M. Voegler, M. Claessens, S. Dustdar, Efficient and scalable IoT service delivery on cloud, in *IEEE International Conference on Cloud Computing* (2013), pp. 740–747. https://doi.org/10.1109/CLOUD.2013.64
48. Z. Ghanbari, N. Jafari Navimipour, M. Hosseinzadeh, A. Darwesh, Resource allocation mechanisms and approaches on the internet of things. Cluster Comput. **22**(4), 1253–1282 (2019). https://doi.org/10.1007/s10586-019-02910-8
49. A.A. Alsaffar, H.P. Pham, C.S. Hong, E.N. Huh, M. Aazam, An architecture of IoT service delegation and resource allocation based on collaboration between fog and cloud computing. Mob. Inf. Syst. **2016** (2016). https://doi.org/10.1155/2016/6123234
50. X. Xu et al., Dynamic resource allocation for load balancing in fog environment. Wirel. Commun. Mob. Comput. **2018** (2018). https://doi.org/10.1155/2018/6421607
51. B. Pourghebleh, V. Hayyolalam, A comprehensive and systematic review of the load balancing mechanisms in the internet of things. Cluster Comput. (2019). https://doi.org/10.1007/s10586-019-02950-0
52. Y. Choi, Y. Lim, Optimization approach for resource allocation on cloud computing for IoT. Int. J. Distrib. Sens. Netw. **2016** (2016). https://doi.org/10.1155/2016/3479247
53. J. Altmann, C. Courcoubetis, M. Risch, A marketplace and its market mechanism for trading commoditized computing resources. Ann. Telecommun. Telecommun. **65**(11–12), 653–667 (2010). https://doi.org/10.1007/s12243-010-0183-1
54. K.T.J. Violos, V.M. de Lira, P. Dazzi, J. Altmann, B. Al-Athwari, A. Schwichtenberg, Y. Jung, T. Varvarigou, User behavior and application modeling in decentralized edge cloud infrastructures, in *International Conference on the Economics of Grids, Clouds, Systems, and Services* (2017), pp. 193–203
55. N.H.J. Altmann, B. Al-Athwari, E. Carlini, M. Coppola, P. Dazzi, A.J. Ferrer, N. Haile, Y.W. Jung, J. Marshall, E. Pages, E. Psomakelis, BASMATI: an architecture for managing cloud and edge resources for mobile users, in *International Conference on the Economics of Grids, Clouds, Systems, and Services* (2017), pp. 56–66
56. R.G. Aryal, J. Altmann, Dynamic application deployment in federations of clouds and edge resources using a multiobjective optimization AI algorithm, in *2018 3rd International Conference on Fog and Mobile Edge Computing, FMEC 2018* (2018), pp. 147–154. https://doi.org/10.1109/FMEC.2018.8364057
57. A. Uzbekov, J. Altmann, Enabling business-preference-based scheduling of cloud computing resources, in *International Conference on the Economics of Grids, Clouds, Systems, and Services* (2016), pp. 225–236

58. J. Altmann et al., GridEcon: a market place for computing resources, in *Lecture Notes in Computer Science (including Subseries Lecture Notes in Artificial Intelligence, Lecture Notes in Bioinformatics)*, vol. 5206, LNCS (2008), pp. 185–196. https://doi.org/10.1007/978-3-540-85485-2_15
59. G.Z. Santoso, Y.W. Jung, S.W. Seok, E. Carlini, P. Dazzi, J. Altmann, et al., Dynamic resource selection in cloud service broker, in *International Conference on High Performance Computing & Simulation (HPCS), Genoa* (2017), pp. 233–235
60. J.A.R.G. Aryal, Fairness in revenue sharing for stable cloud federations, in *International Conference on the Economics of Grids, Clouds, Systems, and Services* (2017), pp. 219–232

The Application of Technology Acceptance Model to Assess the Role of Complexity Toward Customer Acceptance on Mobile Banking

Gustita Arnawati Putri, Ariyani Wahyu Wijayanti, and Kurnia Rina Ariani

Abstract This study wants to prove complexity's role as an external variable in accepting the use of mobile banking using the TAM model framework. The sample included BCA, BRI, Mandiri, and BNI mobile banking customers in Indonesia in the sum of 200 respondents. The hypothesis used in this study was tested using structural equation modeling (SEM) method. The results reveal that perceived ease of use has a positive effect on perceived usefulness and attitude toward using perceived usefulness has a positive impact on attitude toward using, behavioral intention to use, and perceived usage. Besides, attitude toward using influences behavioral intention to use and behavioral intention to use affects perceived usage. The external variable, complexity, affects perceived usefulness and perceived usage but negative uses.

Keywords Technology accepted model (TAM) · Complexity · Mobile banking

1 Introduction

Information technology is currently developing very fast, one of which is evidenced by the rapid use of cell phones, or commonly known as mobile phones or smartphones. The Indonesian people's positive response to this development is implicit in the results of the Fintech Financial Forum (2018) meeting, which explains that digital economic growth in Indonesia is five times faster than the global average: the most rapid. Besides, the results of the e-marketer research university in 2018 also support these findings by revealing the number of diligent smartphone users in Indonesia

G. A. Putri · A. W. Wijayanti (✉)
Veteran Bangun Nusantara Sukoharjo University, Kabupaten Sukoharjo, Indonesia
e-mail: aryn.3d071208@gmail.com

G. A. Putri
e-mail: gustita.ap@gmail.com

K. R. Ariani
Muhammadiyah Surakarta University, Surakarta, Indonesia
e-mail: kra123@ums.ac.id

P. K. Pattnaik et al. (eds.), *Proceedings of International Conference on Smart Computing and Cyber Security*, Lecture Notes in Networks and Systems 149,
https://doi.org/10.1007/978-981-15-7990-5_25

that are hundreds of millions. Based on data from the Ministry of Communication and Information Technology in 2018, Indonesia ranks fourth as the biggest diligent smartphone user in the world.

Nowadays, many aspects of life gain advantages by using the Internet, mobile media, and the banking industry. The positive community responses to information and technological development in the form of the extensive use of the smartphone provides an open opportunity for the banking sector to put mobile banking in hand as a part of people's daily life—advantages by Google Play Store (for the Android system) and Apple Store (for IOS system) users.

The banking service in hand, namely electronic Banking (e-banking) and mobile banking (m-banking), aims to increase client access to banking products and transactions in ease. M-banking serves as a facility available on mobile communication devices such as mobile phones provide benefits for several parties such as their customers and the bank itself. For customers, m-banking services conveniences to them at banking transactions, i.e., balance checking, money transfers, and so on, which previously delivered manually by personal attending in the bank. These cutoff activities consider as time and costs savers and likewise, the benefits for the bank as a stakeholder, i.e., new business models, business expansion, competitive advantage, customer loyalty, revenue and cost improvement, and fee-based income.

2 Literature Review and Hypothesis Development

There is an approach that determines the ease of acceptance of new technology—the technology acceptance model [1] TAM models are used most often in information systems research because of their ability to declare proven validity. Gardner and Amoroso developed TAM by adding four external variables to test customer acceptance of Internet technology use. These four external variables are complexity, gender, volunteerism, and experience [2] (Fig. 1).

Another study conducted in 2004, located in Finland [3] on the customers interests in using of mobile banking, provides us results which state that the perceived ease of use, perception of usability, perceived pleasure, digital banking information, privacy, and security significantly influenced the customer interest in using the M-banking [3]. The influencing factors to the customer's interest in using mobile banking have also investigated [4] in Sabah, Malaysia, inform us that credibility, pleasure, and self-efficacy serves as more critical factors than technology ease and technological usefulness [4].

Empirical studies conducted to examine the relationship between perceived usefulness, perceived ease of use, attitude towards the adoption of information systems technology provide us with mixed results. Such a study [1, 2, 5–12] investigations result in a positive and significant effect. However, research conducted [13] dan [14] show different results.

H_1: (PE) affects (PU) in implementing mobile banking.

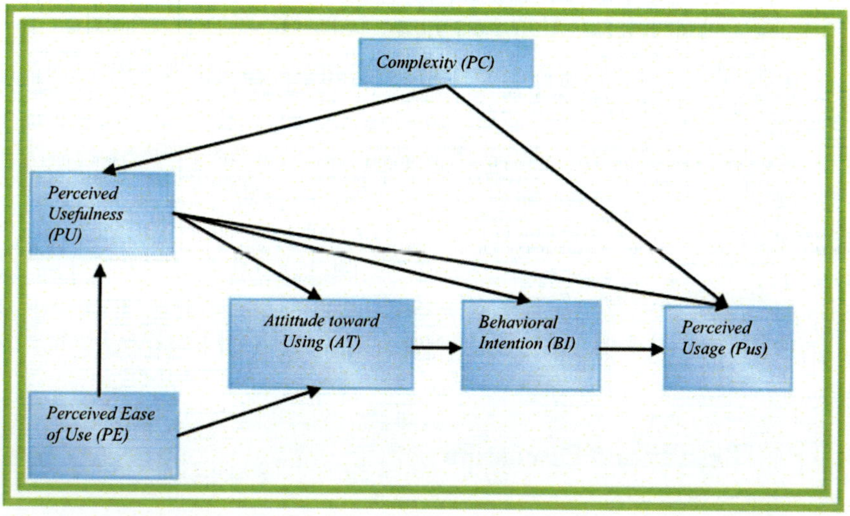

Fig. 1 Hypothesis framework

H_2: (PU) influences (AT) in implementing mobile banking.
H_3: (PE) affects (AT) in implementing mobile banking.
H_4: (AT) influences (BI) in implementing mobile banking.
H_5: (PU) influences (BI) in implementing mobile banking.
H_6: (BI) affects (PUs) in implementing mobile banking.
H_7: (PU) influences (PUs) in implementing mobile banking.
H_8: (PC) influences (PU) in implementing mobile banking.
H_9: (PC) influences (PUs) in implementing mobile banking.

3 Research Methods

3.1 Population and Sample

In this case, the intended research population is bank customers in Indonesia who stands as mobile banking services user in Indonesia and determined samples were customers of Bank Central Asia (BCA), Bank Rakyat Indonesia (BRI), Bank Mandiri, and Bank Negara Indonesia (BNI). The accidental sampling method was chosen as the non-probability sampling technique apply by meeting respondents by coincidence converge to the researcher [15] during data gathering activities.

3.2 *Operational Definition and Variable Measurement*

Exogenous Construct

Exogenous construct is known as the source variables or independent variables, which are not recognized by other variables in the model. Within this paper, complexity is the exogenous constructs, which defined as the perceived difficulty level of computer technology to be understood and used by their users [16].

Endogenous Construct

In this research, endogenous constructs involved, i.e., (PE), (PU), (AT), (BI),(PUs).

4 Results and Discussion

Instrument Quality Test

Validity test on the six research variables shows that the items in the instrument meet the required validity criteria, classified in good validity as well as the reliability test score. The reliability test resulted is (PU) of 0.858; (PE) of 0.845; (AT) of 0.881; (BI) of 0.828; (PUs) of 0.838; and (PC) of 0.811.

Hypothesis Test

Hypothesis testing in this study uses the structural equation modeling (SEM) method. Structural equation modeling (SEM) applies to test the hypotheses in this study. Two reasons for the use of SEM for testing hypotheses are the analysis of the suitability of the model and the path coefficient's analysis. With 200 respondents and producing valid statement items 21, the CFI, TLI, and RMSEA values show ethical values, so that the proposed model as a whole is supported.

Path Coefficient Analysis

Path coefficient analysis indicates that the significance test of all (nine) hypotheses was proven as significantly supported since the overall probability value resulted is smaller than 0.05 at a significance level of 5% (Table 1).

Discussion

The results of the significance testing and path coefficient analysis in hypothesis 1 found a significant relationship and positive—in line with the results of previous studies [17] dan [18], (PE) has a positive impact on (Pus).

Resulted test of significance and path coefficient analysis on respondent's perceived usefulness (PU) and perceived ease of use (PE) shows that influences attitude toward using (AT) since it shows significant positive results. This indicates that the higher a person's level of trust in the use of mobile banking will be followed

Table 1 Regression Weights

			Estimate	SE	CR	P
PU	←	PEoU	0.255	0.092	2.788	0.005
PU	←	PC	−0.527	0.150	−3.502	0.000
AtU	←	PEoU	0.257	0.082	3.127	0.002
AtU	←	PU	0.206	0.071	2.913	0.004
BI	←	PU	0.176	0.055	3.193	0.001
BI	←	AtU	0.248	0.085	2.903	0.004
Pus	←	PU	0.214	0.087	2.459	0.014
Pus	←	BI	0.314	0.115	2.730	0.006
Pus	←	PC	−0.320	0.138	−2.323	0.020

by improved performance, the more active they will be to use mobile banking. These results support the previous research by [3, 5–9, 11, 13, 16, 19–22].

The next test for analysis on behavioral intention to use shows that (AT) and (PU) influences (BI) since the path coefficient (standardized regression weight estimate) shows in a positive result. This result refers to the situation that if someone feels they have to use mobile banking, then the more they are interested in using mobile banking. While perceived usefulness is the leading cause of behavioral intention to apply for inexperienced users, also perceived value is the most significant construct of behavioral plans to use. The results of previous studies indicate that attitude toward using it is positive influences behavioral intention to use as the earlier research shows [6, 8, 14, 15, 17, 23–26]. The test on Hypotheses 6 and 7 shows significantly positive results, so it is in conclusion, that (BI) and (PU) affect (PUs). These results support the previous study [27] that perceived use if you have a behavioral intention to use to do it. If someone perceives that mobile banking improves their performance, it will make them use mobile banking continuously. Similar to research finding previously [25], that (PU) is the most significant construct determines the (PUs).

Testing the external hypothesis, complexity as an independent variable to perceived usefulness (PU), and perceived usage (PUs) resulted in significant negative results, indicates that the complexity of a technology (mobile banking) influences by decreasing the use of mobile banking technology. There is a strong relationship between variable complexity and perceived usage [28] dan [1].

5 Conclusion

This study provides empirical justification for the acceptance, determination, and application of mobile banking. It offers a structural model that examines the role of various motivators in attracting mobile banking users to banking customers. The results then provide substantial support for the (PE), (PU), and (PUs) models of variations in the use of mobile banking. This finding confirms the critical role of perceived benefits felt in publicizing the use of m-banking and shows the basis of a strategic decision to use m-banking. This is an explanation that individuals are expected to have a positive attitude in using mobile banking if they believe that it will improve their reputation and productivity. However, the finding that complexity negatively impacts on the use of mobile banking in daily life shows that people are reluctant to it, especially for customers in developing countries. This form of hesitant to complexity on mobile banking applications regarding its complicated user interface assumes that there is a low community interest to learn a new thing even it is potentially will convenience them in the future when they already get familiar with the intended application.

References

1. M. Igbaria, S. Parasuraman, J.J. Baroudi, A motivational model of microcomputer usage. J. Manag. Inf. Syst. **13**(1), 127–143 (1996)
2. J.C.C. Lin, H. Lu, Towards an understanding of the behavioral intention to use a web site. Int. J. Inf. Manage. **20**(3), 197–208 (2000)
3. T. Pikkarainen, K. Pikkarainen, H. Karjaluoto, S. Pahnila, Consumer acceptance of online banking: an extension of the technology acceptance model. Internet Res. **14**(3), 224–235 (2004)
4. H. Amin, R. Supinah, M.M. Aris, R. Baba, The receptiveness of mobile banking by Malaysian local customers in Sabah: an empirical investigation. J. Internet Bank. Commer. 1–12 (1970)
5. R. Govindaraju, N. Indriany, E.J. Bruijn, Studi Mengenai Penerimaan Sistem ERP: enhancement terhadap model Penerimaan Sistem ERP Berbasis technology acceptance model, in *National Industrial Engineering Conference*, no. 4 (2007), pp. 1–11
6. S.Y. Hung, C.M. Chang, T.J. Yu, Determinants of user acceptance of the e-Government services: the case of online tax filing and payment system. Gov. Inf. Q. **23**(1), 97–122 (2006)
7. C.T. Lu, S.Y. Huang, P.Y. Lo, An empirical study of the online tax filing acceptance model: integrating TAM and TPB. Afr. J. Bus. Manag. **4**(5), 800–810 (2010)
8. Y. Malhotra, D.F. Galletta, Extending the technology acceptance model to account for social influence: theoretical bases and empirical validation, in *Proceedings of 32nd Annual Hawaii International Conference on Systems Science* (1999), p. 14
9. B. Szajna, Empirical evaluation of the revised technology acceptance model. Manage. Sci. **42**(1), 85–92 (1996)
10. R.L. Thompson, C.A. Higgins, J.M. Howell, Personal computing: toward a conceptual model of utilization. MIS Q. 125–143 (1991)
11. L.R. Vijayasarathy, Predicting consumer intentions to use online shopping: the case for an augmented technology acceptance model. Inf. Manag. **41**(6), 747–762 (2004)
12. M.A. Islam, K.A.K. Daud, Factors that influence customer's buying intention on shopping online. Int. J. Mark. Stud. **3**(1), 128–139 (2011)
13. H. Kusuma, D. Susilowati, Determinan Pengadopsian Layanan Internet Banking: Perspektif Konsumen Perbankan Daerah Istimewa Yogyakarta. JurnalAkuntansidan Audit. Indones. **11**(2) (2007)
14. S. Taylor, P.A. Todd, Understanding information technology usage: a test of competing models. Inf. Syst. Res. **6**(2), 144–176 (1995)
15. J. Lee, Factors affecting intention to use online financial servicer (2003)
16. E.W.T. Ngai et al., Empirical examination of the adoption of WebCT using TAM. Comput. Educ. **48**, 250–267 (2007)

17. P.Y. Chau, P.J. Hu, Investigating healthcare professional's decisions to accept telemedicine technology: an empirical test of competing theories. Elsevier Inf. Manag. **39**, 297–311 (2002)
18. F.D. Davis, R.P. Bagozzi, P.R. Warshaw, User acceptance of computer technology: a comparison of two theoretical models. Manag. Sci. **35**(8), 982–1003 (1989)
19. J. Bugembe, Perceived usefulness, perceived ease of use, attitude, and actual usage of new financial management system: a case study of Uganda National Examinations Board (2010)
20. S. Chauhan, Acceptance of mobile by poor citizens of India: integrating trust into the technology acceptance model. Emerald Insight **17**(3), 58–68 (2015)
21. R.G. Saade, Is usage predictable using belief-attitude-intention paradigm? Informing Sci. Inf. Technol. **5**, 591–599 (2008)
22. S. Sulistiyarini, Pengaruh Minat Individu Terhadap Penggunaan mobile banking: model Kombinasi technology acceptance model (TAM) dan theory of planned behaviour (TPB), Univ. Brawijaya
23. K. Mathieson, Predicting user intentions: comparing the TAM with the TPB. Inf. Syst. Res. **2**(3), 173–191 (1991)
24. Y.Y. Shih, K. Fang, The use of a decomposed theory of planned behavior to study internet banking in Taiwan. Internet Res. **14**(3), 213–223 (2004)
25. H. Sun, P. Zhang, A methodological analysis of user technology acceptance, in *37th Hawaii International Conference on System Science, Big Island, Hawaii* (2004)
26. M. Tan, T. S. Teo, Factors influencing the adoption of internet banking. J. Assoc. Inf. Syst. **1**(5), 1–42 (2000)
27. Jogiyanto, *Sistem Informasi Keperilakuan*. Andi Offset (2007)
28. C. Gardner, D.L. Amoroso, Development of an instrument to measure the acceptance of internet technology by consumers, in *37th Annual Hawaii International Conference on System Sciences* (2004), p. 10

Exploring the Volatility of Large-Scale Shared Distributed Computing Resources

Md Azam Hossain, Baseem Al-athwari, Jik-soo Kim, and Soonwook Hwang

Abstract Scientific applications often require colossal amount of computing resources for running user's tasks. Grid computing has been proved to be powerful research testbed for accessing massive amount of computing resources at almost zero cost across various autonomous administrative institutes. It can seamlessly integrate hundreds of thousands of geographically distributed heterogeneous computing resources from multiple domains organized into virtual organization (VO). Unfortunately, existing Grid Information Service (GIS) suffers from providing exact dynamic resource information due to its scale and autonomous resource management policies. In this paper, we present a comprehensive volatility study of shared computing resources by VO in terms of characterizing resource performance related features of each computing element (CE) in computational Grids such as number of available CPU cores, average response time. We also performed experiments based on a large number of micro-benchmark tasks on real Grid environment to analyze implication of resources fluctuation. Evaluation results reveal that resource volatility studies tremendously help to decrease user response time and job completion rate.

Keywords Distributed computing · Computing element · Scientific computing · Virtual organization · Resource volatility

M. A. Hossain (✉) · B. Al-athwari
Department of Computer Engineering, Kyungdong University, Global Campus (Goseong), Goseong, Gangwon-do, Republic of Korea
e-mail: azam1708@kduniv.ac.kr

B. Al-athwari
e-mail: baseem_cs@kduniv.ac.kr

J. Kim
Department of Computer Engineering, Myongji University, Yongin, Republic of Korea
e-mail: jiksoo@mju.ac.kr

S. Hwang
Korea Institute of Science and Technology Information (KISTI), Daejeon, Republic of Korea
e-mail: hwang@kisti.re.kr

P. K. Pattnaik et al. (eds.), *Proceedings of International Conference on Smart Computing and Cyber Security*, Lecture Notes in Networks and Systems 149,
https://doi.org/10.1007/978-981-15-7990-5_26

1 Introduction

Distributed computing system facilitates to integrate seamlessly hundreds of thousands of geographically distributed heterogeneous computing resources from multiple domains organized into virtual organization (VO). During the past decade, Grid computing platform has emerged as one of the major distributed computing paradigms. It is intended to offer scalable, secure, coordinated resource sharing among dynamic collections of individuals, institutions to execute a large computational task [2] with much lower cost than cloud or supercomputers.

Moreover, Grid computing paradigm offers the opportunities for the research community with vast amounts of computational power for modeling and simulating complex scientific and engineering problems such as discovering new drug, diagnosing medical conditions, forecasting weathers, managing stock portfolios, simulating earthquake data. It offers computing resources beyond the capacity of even the largest parallel computer system and integrates extremely heterogeneous physical resources into a single virtual resource [11]. Examples of some large-scale Grids are enabling Grid for E-science (EGEE) [6], TeraGrid [1], Open Science Grid (OSG) [7]. Therefore, it is a viable choice to utilize the Grid computing resources as much as we can because they provide aggregated computation power on the order of PetaFLOPS and storage on the order of PetaByte at almost zero cost [5]. In order to effectively support these applications on Grids, it is crucial to locate the number of available resources (e.g., CPU cores) at each Grid site as well as their performance (e.g., response time).

However, discovering the number of free CPU cores as well as response time of each computing element (CE) of a virtual organization (VO) is very challenging and complex process because of its scale and the fact that resources are widely distributed on different locations under the control of different administrations [13]. In addition, resources in the VO are joining and leaving dynamically that make the entire Grid system vulnerable due to the fluctuation of participated resources in a VO. As a result, overall systems throughput and the makespan of user applications are degraded tremendously. Unfortunately, current Grid Information Service (GIS) suffers from providing these dynamic resources information accurately which can be crucial for an effective and efficient job scheduling in the Grid environment [3, 4].

To address the above issues, we need a mechanism that can find the number of available CPU cores and associated performance of Grid computing elements. We have developed [3, 4] which characterized each CE participating in the virtual organization (VO) periodically. This system prompts us to study the volatility of Grid computing resources so that we can learn about resource characteristics, and based on this information, user can schedule tasks efficiently across the Grid sites with potentially higher probability of allocating resources and executing the tasks in the shortest time.

This paper presents a comprehensive study to find and allocate the resources from each CE on Grid systems. The rest of this paper is structured as follows. In Sect. 2, we briefly describe the motivation and background study, and Sect. 3 presents results and discussion. Finally, Sect. 4 highlights the conclusion and our future work direction.

2 Motivation and Background Study

Grid computing leverages resources from multiple administrative domains, individuals and institutions to form virtual organization (VO) [10]. VO enables users to access the large amount of computing resources as a single virtualized computing platform scattered on all participating Grid sites. Figure 1 illustrates the relationship between users, VOs and participating Grid sites. A user may be a member of one or more VOs at the same time, and sometimes, same resources may be shared among multiples VOs. Figure 1 also illustrates that a VO consists a number of computing elements (CEs). A CE is a set of computing resources localized at a site (i.e., a cluster, a computing farm), and the cluster is a collection of worker nodes (WNs), where the jobs are actually run.

Grid sites are highly dynamic in nature, which means that a CE that was available at certain time or day might not be available in future for some period of time or a few days. Moreover, the capacity of each CE which means the number of available CPU cores is very different among the participating institutions, and also, it varies over time. In addition, the performance (i.e. response time) of each Grid site is also extremely variable possibly due to end-to-end network bandwidth, current system load, local scheduling policy and so on.

The Grid Information Service (GIS) is responsible for providing the information about the Grid resources and status. This information can be very important

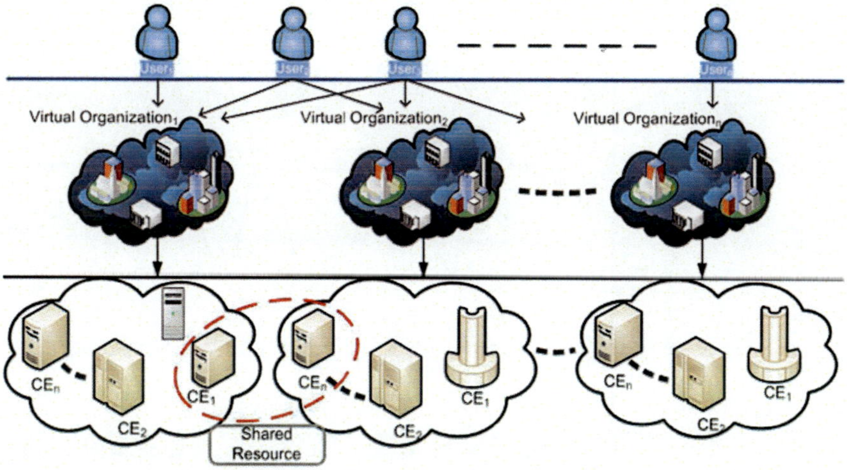

Fig. 1 Grid computing infrastructure

for discovering and monitoring the Grid resources and scheduling user tasks on the appropriate CEs. Unfortunately, GIS suffers from providing the exact dynamic resource information (as seen from our previous study [3, 4] where ***lcg-infosites*** shows completely misleading information) because Grid resources are managed by multiple independent autonomous administrators participating in the virtual organization (VO).

To address these obstacles and challenges, we are interested in understanding the sporadic nature of computing elements of Grids over some period of time. Our goal is to analyze the behaviors of CEs and use this dynamic information to discover a CE that can guarantee a given number of free CPU cores and acceptable response time to get resources allocated under the volatility nature of Grid environment.

3 Result and Discussion

In this section, first we will explore and study the volatility of computing resources in a real Grid platform and then present microbenchmark experimental results conducted on BIOMED virtual organization [12].

3.1 Volatility Study of Grid Systems

Computing elements of Grids is highly sporadic in nature, which means participated sites in VO can leave and join randomly. As a result, a CE that was available at certain time or day might not be available in future for some period of time or a few days. Figure 2 basically shows resent two-month history of how many CEs are usually

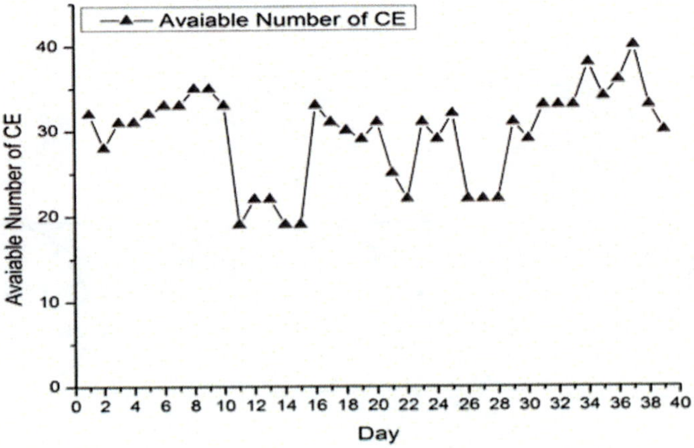

Fig. 2 Number of CEs available during two month of BIOMED VO

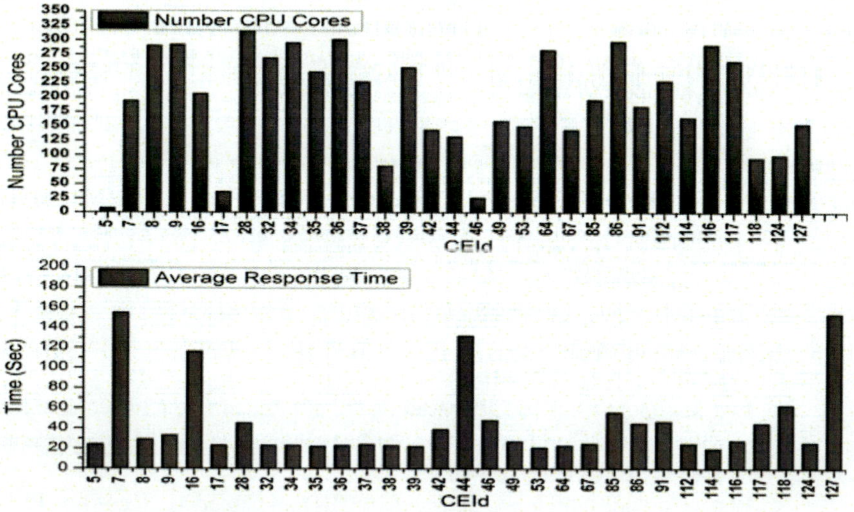

Fig. 3 Resource capacity and response time of each CE

available in each day at BIOMED Grid site [12]. This snapshot demonstrates typical CE fluctuation behavior observed in a volunteer Grids and also illustrates that it is very difficult to predict the availability of CEs. It can also be observed from Fig. 2 that average 30 or more CEs usually remain available from 5 to 10 days consecutively in BIOMED VO, and after this period of time, the number of available CEs is hovered around 20 for a few days possibly due to hardware/software maintenance or any other reasons.

We also studied the capacity of each CE which means the number of free CPU cores among the participating institutions over time. In addition, the performance (i.e., response time) of each Grid site is also extremely variable possibly due to end-to-end network bandwidth, current system load, local scheduling policy and so on. Figure 3a illustrates a typical one-day available CEs and their associated number of available free CPU cores at the BIOMED VO, and Fig. 3b depicts the average response time of each CE. As we can see from Fig. 3a, b, not only the number of CPU cores available in each CE but also the response times can be very different from each other which makes it very challenging to predict accurately the available resources and associate performance of each CE in a virtual organization.

3.2 *Microbenchmark Experimental Result*

In this section, we examine the effectiveness and importance of resources volatility study by carrying out benchmark experiments extensively on real Grid environments.

We evaluated the microbenchmark simulations by using short BoT (Bag of Tasks) jobs from the perspective of many-task computing [8, 9]. Each BoT is composed of different numbers of tasks ranging from 250 to 5000.

- Short BoT Jobs (Sleep Task 5 s)

To perform microbenchmark experiments and compare the results, total four lists of computing elements are selected each of which contains 10 CEs as follows:

- **List 1**: includes top 10 CEs based on maximum free CPU cores from our volatility study and labeled as "**Profiled Top 10 CE with Core**".
- **List 2:** contains top 10 CEs based on minimum response time and labeled as "**Profiled Top 10 CE with Time**".
- **List 3:** contains top 10 CEs based on the number of free CPU cores information provided by the Grid Information Service running the ***lcg-infosites*** command and labeled as "**Grid Info Top 10 CE**".
- **List 4:** includes randomly selected 10 CEs and named as "**Random 10 CE**".

Regardless of top 10 lists of CEs, we evenly distributed the load which means that the same number of tasks is submitted on each CE and the maximum response time is set to 30 min assuming that if the CEs have enough free resources then within this time limit tasks will be executed.

Figure 4 shows the makespans over various sizes of short running BoTs by comparing results with four lists of CEs as mentioned above. CEs with minimum response time ("**Profiled Top 10 CE with Time**") indeed exhibits the lowest makespan in comparison to the other three lists of CEs. As the number of tasks

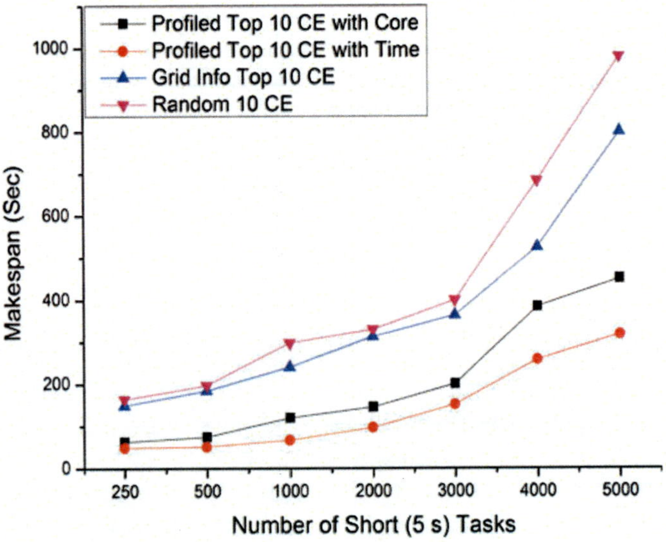

Fig. 4 Makespan comparison of short BoTs

increases, makespans of "**Grid Info Top 10 CE**" and "**Random 10 CE**" increase rapidly. From these results, we can observe that for short running tasks, minimum response time CEs perform better even the "**Profiled Top 10 CE with Core**" have potentially more free CPU cores.

4 Conclusion and Future Work

Grid computing elements have varying computing capacity and unreliable performance because of autonomous sharing and control from multiple resource providers. Grid Information Service (GIS) suffers from providing exact dynamic-state information that can be usable by Grid users or system (schedulers, resource brokers) to map their applications on the Grid sites.

To address this issue, we have studied the volatility of Grid computing resources so that users can schedule their tasks efficiently across the Grid sites with potentially higher probability of allocating resources and executing the tasks in the shortest time. Comprehensive evaluation results are also presented by experimenting the large number of micro-benchmark tasks. The experimental results demonstrate that if a user knows where free resources are available, then it will be very helpful to schedule tasks on Grids which will increase the system throughput by reducing response time and improving the job completion rate.

Our future work includes developing new prediction models and applying machine learning techniques to select the best resources from various virtual organizations.

References

1. C. Catlett, The philosophy of teragrid: building an open, extensible, distributed terascale facility, in *2nd IEEE/ACM International Symposium on Cluster Computing and the Grid, 2002* (IEEE, 2002), p. 8
2. I. Foster, C. Kesselman, S. Tuecke, The anatomy of the grid. Berman et al. [2] pp. 171–197 (2003)
3. M.A. Hossain, C.N. Nguyen, J.S. Kim, S. Hwang, Exploiting resource profiling mechanism for large-scale scientific computing on grids. Cluster Comput. **19**(3), 1527–1539 (2016)
4. M.A. Hossain, H.T. Vu, J.S. Kim, M. Lee, S. Hwang, Scout: a monitor and profiler of grid resources for large-scale scientific computing, in *2015 International Conference on Cloud and Autonomic Computing (ICCAC)* (IEEE, 2015), pp. 260–267
5. B. Javadi, K. Matawie, D.P. Anderson, Modeling and analysis of resources availability in volunteer computing systems, in *2013 IEEE 32nd International Performance Computing and Communications Conference (IPCCC)* (IEEE, 2013), pp. 1–9
6. E. Laure, B. Jones, Enabling grids for e-science: the egee project. Grid Comput. Infrastruct. Serv. Appl., 55 (2009)
7. R. Pordes, D. Petravick, B. Kramer, D. Olson, M. Livny, A. Roy, P. Avery, K. Blackburn, T. Wenaus, F. Wurthwein, et al., The open science grid. J. Phys. Conf. Ser. **78**, 012057 (2007)

8. I. Raicu, I. Foster, M. Wilde, Z. Zhang, K. Iskra, P. Beckman, Y. Zhao, A. Szalay, A. Choudhary, P. Little et al., Middleware support for many-task computing. Cluster Comput. **13**(3), 291–314 (2010)
9. I. Raicu, I. Foster, Y. Zhao, Many-task computing for grids and supercomputers, in *Proceedings of the Workshop on Many-Task Computing on Grids and Supercomputers (MTAGS'08)* (Nov 2008)
10. I. Rodero, D. Villegas, N. Bobro, Y. Liu, L. Fong, S.M. Sadjadi, Enabling interoperability among grid meta-schedulers. J. Grid Comput. **11**(2), 311–336 (2013)
11. H. Sanjay, S. Vadhiyar, Performance modeling of parallel applications for grid scheduling. J. Parallel Distrib. Comput. **68**(8), 1135–1145 (2008)
12. The biomed Virtual Organization: Available at https://lsgc.org/en/Biomed:home
13. Y. Wu, K. Hwang, Y. Yuan, W. Zheng, Adaptive workload prediction of grid performance in condence windows. IEEE Trans. Parallel Distrib. Syst. **21**(7), 925–938 (2010)

Business Transformations Within Intelligent Eco-Systems

Ana Hol

Abstract Each era of civilization brings new challenges and innovations to work and life. These revolutions require individuals and businesses to gain new skills, build knowledge and transform their society. Through this process, workers' interactions, work-hours, machines used, work locations and the nature of work itself have developed. This paper draws upon the past industrial changes, with specific focus on developments occurring during Industry 4.0 and provides transformative framework for business toward the Industry 5.0 and the phenomenon of intelligent eco-systems. Work and life in the intelligent eco-systems framework are seen to depend upon the successful partnerships between humans and machines. Humankind brings innovation, critical knowledge and strategy to a business, while technology provides a business infrastructure backbone, which enable processes to be executed through algorithms, human–machine co-creation and partnership.

Keywords Business transformation · Industry 5.0 · Artificial intelligence · Human–machine co-existence · Intelligent eco-system

1 Introduction

It is well known that humanity is continuously changing and transforming. Birth of the industrial era largely moved work from farms to factories. Down to dusk working hours of the agricultural era were transformed to regular shift work of the plants. The first industrial revolution, Industry 1.0, was characterized by work which was powered by machines which utilized steam or water power. Mechanical loom supported the production of the time. Industry 2.0 operations were introduced once it was possible to power the machines by electricity [1]. This was the time when some

A. Hol (✉)
School of Computer, Data and Mathematical Sciences, Western Sydney University, Crn. James Ruse Drive and Victoria Av., Rydalmere, NSW 2116, Australia
e-mail: a.hol@westernsydney.edu.au

P. K. Pattnaik et al. (eds.), *Proceedings of International Conference on Smart Computing and Cyber Security*, Lecture Notes in Networks and Systems 149,
https://doi.org/10.1007/978-981-15-7990-5_27

of the first assembly lines and mass productions were introduced. The next development was Industry 3.0 which saw first programable controllers and the emergence of the information technology systems [1]. In contemporary times, Industry 4.0 is well thriving. Its operations are characterized by ubiquitous production, connectivity, smart and integrated factory floors and well interconnected machines which allow both mass production and customizations.

Each business transformation brought new technological advancements and has consequently required more complex knowledge and skills of the employees. Today's business operations are largely characterized by the uses of the information technology and require a comprehensive systems architecture. Technology has not only transformed the nature of the work activities, but also changed business processes across the organizations [2]. It has influenced transformations of traditional businesses to online store and has also enabled the establishments of the innovative business models such as those of Amazon, Uber, Airbnb and Airtasker. Industry 4.0 has given power to the consumer, has allowed mass customization of products and has provided platforms for customers to share their skills, offer their goods and provide their services.

It is expected that technology will continue to shape future of business advancements. To investigate this phenomenon, review of Industry 4.0 and the current and upcoming business technological advancements have been studied with an aim to develop a model which could guide business transformations toward the Industry 5.0.

2 Business Transformation from Industry 4.0 to Industry 5.0

eTransformation model [2], which guides organizational transformations from the traditional 'brick and mortar' business to an online business, as a part of the transformation process, requires the transforming organizations to assesses their Strategy, Structure, Task and Processes and Tools and Systems. As a part of this journey, ***Strategy*** reviews are done first, so companies can define their business goals, business vision and the objectives. Following this, business ***Structure*** which in turn determines organizational decision-making hierarchy, governance and departmental divisions is assessed. Once Structure is defined, ***Tasks and Business Processes*** need to be identified and converted into the set operations. This analysis allows business to identify processes which can be streamlined and automated. To identify how operations are to be completed, ***Tools and Technologies*** need to be identified and selected, so hardware and software can be implemented.

The first online store transformations commenced with Amazon's inventions which allowed books to be sold online. Not long after software functions have rapidly started to transform the traditional shopping aisles into the online catalog, carts and the electronic point of sale systems. The next innovation which followed was the emergence of the new business models where companies like Dell have started to modify the traditional supply chain systems [3] which allowed manufacturers to sell goods directly to the customers and in addition also allow customers to purchase customized products.

Major business model innovations were later seen once companies as Uber and Airbnb were founded. The biggest difference to the previous changes was that none of these companies own the goods and services they provide. They own the platforms, effectively the technology business backbone. They do not own the places (houses, apartments) nor the transportation vehicles which are effectively main resources of the businesses.

Some other more recent business models have been seen with the innovations of the new banks, insurance companies and transaction providers. These companies often do not have physical branches such as ING in Germany [4], Xinja and Judo Capital [5]. Companies like these tend to operate without owning the goods, nor the business resources. Such business model was even further extended by companies such as Ant Financial [6], which uses the artificial intelligence to provide the critical company operations. To further understand the impacts of the powerplays which guide business to transform and change, factors guiding the Industry 1.0; 2.0; 3.0 and 4.0 are reviewed in detail.

2.1 Business Transformation and Technology Inventions

Business transformation journeys today seem to be much faster than the transformation journeys of businesses a century ago. It should be noted that journey from the first assembly line (Industry 2.0) to the first programable controller (Industry 3.0) took almost a century [1]. Transformation and development toward the current Industry 4.0, which is seen to have started around 2011, is depicted by devices that are interconnected and have brought smart production to manufacturing [7]. The journey from Industry 3.0 to 4.0 is seen to have taken around 42 years [1].

Considering the fast rate in which current technological innovations are introduced, it is apparent that it will not take too long for the Industry 5.0 to emerge. This transformational change [8, 9] is likely to be characterized by the emergence of the human–robot co-working and will require both humans and machines to be in harmony and well-integrated into the operational environments.

Industry 5.0 is likely to digitalize businesses and further transform a wide array of processes. Furthermore, digital change is going to require continuous and progressive [10] reviews of the technical systems as they are organic [11] and need to be updated, maintained and regulated, so they can meet the operations requirements.

Machine–human interaction is slowly allowing machine–human integrations where machines and humans co-work in teams to achieve common goals. This is, for example, visible during the customized car assemblies [12]. Such processes are also evident in medicine, transport, rescue operations and in other industries. For example, pilot training or rescue operations is quite risky and demanding. However, with the utilization of technologies in the mixed and virtual reality scenarios, staff are able to practice and learn and become more prepared to deal with hazardous or risky situations.

Artificial reality implementations also allow humans to extend their current vision and help them complete the required tasks through training, sophisticated work engagements and the enhancements of humans' capabilities which bridge physical and digital worlds [13]. Artificial reality implementations are also seen to support maintenance of machinery (Xerox) and training of staff (DHL, Boeing) as it creates extended and more holistic experiences, so users can gain the required information [13].

In medicine, data analytics is imbedded into the results analysis and robotics is utilized for sample analysis, data gathering and robotic surgeries. Auto-drivable cars are also examples of the machine–human co-existence where there is an awareness of process completion orders. Uses of drones in delivery, surveillance and remote operations are also now being integrated into organizational business process. Technologies such as 3D printing allow for the fast prototyping as well as well as the rapidly produced emergency equipment. Tools used today are increasingly 'Internet of Things' devices with sensors which share data and make decisions not just in manufacturing but also at homes and within communities. Smart fridges, TVs, washing machines, lights and wearables are all interconnected and are a part of the large eco-system which further links to intelligent housing, grids, CCTV security, healthcare, education and transport [14]. Consequently, we are increasingly seeing the emergence of the intelligent eco-system, which can collect data, learn, make decisions and complete tasks. So that such systems can allow humans and machines to co-exist and co-create a number of regulatory, legal and ethical perspectives needs to be reviewed as they develop further.

2.2 Legal Regulations

New business models require detailed legal considerations. For example, Uber, as a pioneer, started off without the full legal regulations in place, and consequently, those regulations need to be introduced at the later stages [15] in some countries even post the inititial introductions. Consequently, it can be concluded that new technologically enhanced business models need legal regulations well established, so that the future of machines–human teams can co-exist. Taking an example of the automated vehicle, it can be seen that multiple laws would need to be changed for the vehicles without the wheel and a driver in charge to be allowed on the Australian roads [16]. Among the questions which would need to be addressed is: Who is at fault if auto-driven car has an accident? At present, all responsibilities are in hands of the driver. If, however, an automated car is allowed to drive fully on its own, there would need to be strict legal standards and regulations. Some other concerns which may need to be addressed are: Who will be allowed to update the software which runs the car? Who could overwrite the systems?

Therefore, it can be seen that a clear definition of what robots are, what they do, what requirements they are going to have and what procedures they are going to follow would need to be identified [7]. If we are to allow machines/robots and humans to co-exist, they would need to be aware of one another's presence within the system. They would also need to know the order of precedence. Robots operate based on the expert knowledge, neural network learning [17], exposure and case scenario learning [18]. Based on this, they adapt, incorporate what they have learned and make future decisions. Important questions to address would be: At which point of time should the robot be able to overwrite the human decision? Or should it? Legal implementations are also closely aligned to ethical standards and regulations which should be considered in detail.

2.3 Ethical Regulations

Ethics plays a key role in how business operations are viewed and perceived. It is critical that new artificially intelligent systems are developed through considerations of their effectiveness within the eco-system they operate in. It is expected that in Industry 5.0, strict standards should be followed to ensure autonomous systems display ethical behaviors and abide by legal standards and regulations [8]. However, ethical considerations are much harder to convert into software code than the conversions of the legal clauses. It is also evident that machines do not have feelings and empathy which is also one of the key factors required when interpreting ethics.

It is important to note that new developments across the intelligent eco-systems will not only require comprehensive reviews of system security measures and implementations of specific legal regulations but may likely also have effects on possible social turbulences, global inequalities and consequently may place significant strains on regulatory agencies [6], which poses a question: Will machines be free of the artificial bias? Furthermore, it would be critical to review how specific financial or health decisions may impact minorities or people of a particular dissent when decision making is in question. Ethical considerations also raise a question of who is in power, who is leading and who is entitled to make a final decision [19].

2.4 Job Allocation and Decision Making

New business models will allow both machines and humans to engage in business operations. Therefore, in the future, jobs are likely to be distribute in a structured way. Jobs better done by machines and algorithms will be given to machines, while jobs requiring innovation, formations of the new business structures and complex tasks would be given to humans. This may mean that in some instances, companies or some department operations once established may become fully or partially run by the machines. This is already happening. For example, simple selections of the closest Uber car available for traveler pick up, to dynamic pricing on Amazon, to the selection of best deals are all computed by the artificially intelligently algorithms that play a role in data selection, computation, system learning and decision making [6]. The questions however which may need to be answered here are: Who should be in command? Should a robot be allowed to compete with humans? And, who will be responsible to oversee the achievements?

2.5 Characteristics Guiding Industry 5.0 Development

Industry 5.0 will bring further changes to how businesses operate. Based on the analysis of the industrial transformations, it is seen that each era's developments were guided by the sources which produced a driving power, the resources and the technological advancements of the time. With contemporary emphasis on the sustainable developments [20], it is expected that future business transformations will be guided by the sustainable energy resources. It is also expected that business operations which can be converted into algorithms will be converted as such. New business processes will require the interplay between the humans and machines particularly when considering skills required for operations, tools used, decision-making strategies and order of precedence. All operations, both those undertaken by humans and machines, will need to be safeguarded, so appropriate security standards, laws and ethics within the cyber-physical systems can be assured. Table 1 shows characteristics businesses operating in Industry 5.0 are likely to observe.

Table 1 Industry 5.0—business characteristics

Characteristics	Industry 5.0—business characteristics
Power	Electricity, renewable energy, natural sources wind, Sun
Operations	VR-virtual reality, AR-augmented reality, MR-mixed reality, actual scenarios
Machines/tools	Robots, drones, smart devices, 3D printing, AI-artificially intelligent devices
Machine skills	Monitoring, assessing, maintaining, teaching, sustaining
Employee skill	Specialized knowledge and skills, multidisciplinary analytical and innovative knowledge and understanding
Safety	Co-existence—robots and humans operate safely together and possess awareness of the environment and each other
Security	Physical systems, simulated and virtual systems all require security and adequate monitoring and safeguarding
Running of the operation	Autonomous production enhanced and guided by human knowledge and innovation

Based on the above data, it is apparent that intelligent eco-systems will be composed of humans and smart machines which will co-exist, co-work and co-invent, and thus require strict guiding principles to be in place.

3 Intelligent Eco-System Business Transformation

It is expected that Industry 5.0 will bring humans and machines closer together and will allow innovation and creativity. Humans will work alongside co-bots which will be able to do repetitive and largely predictive tasks yet will simultaneously be able to learn and adapt [8]. Such structures will require disruptive transformative innovations which will provide value differentiation [4]. Overtime machines will become smarter and will gain further capacities to learn, make more complex decisions and have abilities to even help amplify and extend the human skills [11].

It is evident that changes which companies are undertaking these days seem complex and require transformations to be guided by the digital advancements throughout the systems lifecycle as this is essential to ensure intelligent eco-system business transformations are successful. Figure 1 depicts intelligent eco-system transformation.

Based on the analysis, it is expected that Industry 5.0 ***Business Strategy*** will have its foundations on data, software and analytics and will therefore often require

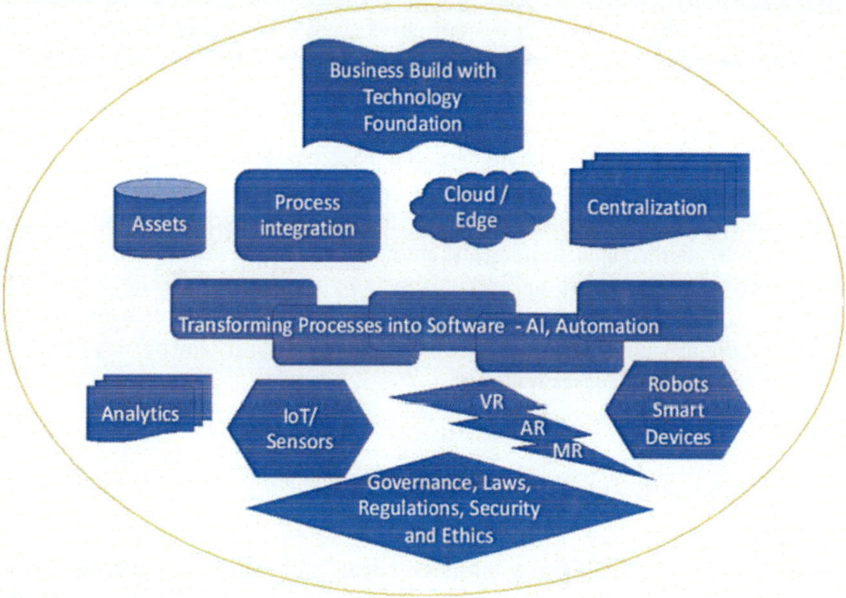

Fig. 1 Intelligent eco-system transformation

complete business re-design. Such design will need human expertise, so that systems can be well implemented.

The ***Structure*** of the business will be defined in the systems architecture which will ensure data asset integration and infrastructure accessibility such as access to cloud, edge computing, blockchain, data, software and service centralization architectures and other service technologies can be implemented.

Company's ***Business process*** operations will require processes that are built with technology in mind. Such developments will heavily rely on multidisciplinary teams that will allow for as many processes as possible to be translated into software parameters and functions, so that they can be understood and completed by the machines whenever possible and human's intelligence and innovations utilized to attend to the complex tasks.

Technologies to be implemented will require an understanding of the Eco-system in which business is to operate. It will be expected that smooth system operation will require machine–human co-existence. Such implementations will assume the use of data analytics, software functions and integrations of machine and human operations. Machines will have skills to recognize, compute, execute, learn and undertake a wide range of operations. Tools implemented may also include robots, 3D printers and smart devices, which can be utilized and applied in real and virtual worlds and can additionally be used for the completions of tasks, executions of skill, maintenance, security, risk prevention and other service operations.

Systems operating within the intelligent eco-system are going to be quite complex. Many will have the capabilities to deep learn, make decisions and even in some cases fully auto-run. Considering this, it is absolutely critical for such systems to have detailed and regular system checks, monitoring and maintenance. It is also important to note that machines should behave in the way humans intend them to, and consequently, their learning journey and training need to be guided and observed. It is also essential to have policies, rules and regulations clearly noted, so they are observed by both humans and machines. Humans need to be able to follow strict regulations when planning, developing, training and maintaining the systems. Consequently, their jobs are expected to be in the areas of innovation, programming, system maintenance, training and entrepreneurship [8]. Machines, on the other hand, need to be able to learn and execute functions based on the set rules, learned scenarios, standards, ethics and legal requirements. This would also assume that cyber-security paradigms would need to be established, so data security and privacy can be observed and algorithmic biases removed.

4 Conclusion

The future will bring new business models and with it new and exciting business challenges and opportunities. It is therefore essential that humans are mindful of how technologies are being sourced, how they are developed and integrated and how they are immersed into the intelligent eco-system environments. Technologies will change how humans perform their tasks, how they interact with the devices and how they allow decisions to be made, product to be developed and services to be offered. New innovations and business transformations will open avenues for future developments. Skills in demand will also be shaped by quickly arising new digital advancements. Intelligent eco-system transformation thus requires humans to be vigilant and mindful of how human–machine interactions are to be guided, led and developed. Furthermore, it is important for the human–machine teams working in the intelligent eco-systems to be supported through well developed legal, ethical and governance frameworks, so that they can work together in order to achieve sustainable developments which are expected to lead businesses toward emerging Industry 5.0 operations.

References

1. R. Waslo, T. lewis, R. Hajj, R. Carton, *Industry 4.0 and Cybersecurity: Managing Risk in an Age of connected production.* Deloitte Insights, 2017. March, 2017
2. A. Hol, *eTransformation Guide: An Online System to Guide eTransforming SMEs* (VDM Verlag Dr. Müller, Saarbrucken, Germany, 2009)
3. G. Fields, Innovation, time, and territory: space and the business organization of dell computer. Econ. Geogr. **82**(2), 119–146 (2006)
4. B. Grab, C. Ilie, *Innovation Management in the Context of Smart Cities Digital Transformation* (Varazdin Development and Entrepreneurship Agency (VADEA), Varazdin), pp. 165–174
5. Zahos, E. *The banks with no branches taking on Australia's Big Four*. My Money 2019 [cited 2020 20 July, 2020]; Available from: https://www.moneymag.com.au/neobanks-no-branches
6. M. Iansiti, K. Lakhani, Competing in the age of AI. Harvard Bus. Rev., 3 (2020)
7. K.A. Demir, G. Döven, B. Sezen, Industry 5.0 and human-robot co-working. Procedia Comput. Sci. **158**, 688–695 (2019)
8. N. Saeid, Industry 5.0—a human-centric solution. Sustainability **11**(16), 4371 (2019)
9. V. Özdemir, N. Hekim, Birth of Industry 5.0: making sense of big data with artificial intelligence, "The Internet of Things" and next-generation technology policy. Omics J. Integr. Biol. **22**(1), 65–76 (2018)
10. J. Lozic, *Core Concept of Business Transformation: From Business Digitization to Business Digital Transformation* (Varazdin Development and Entrepreneurship Agency (VADEA), Varazdin), pp. 159–167
11. H. Wilson, P. Daugherty, Collaborative intelligence: humans and AI are joining forces. Harvard Bus. Rev. (2018)
12. D. Alford, P. Sackett, G. Nelder, Mass customisation—an automotive perspective. Int. J. Prod. Econ. **65**(1), 99–110 (2000)
13. M. Porter, J. Heppelmann, Why every organization needs an augmented reality strategy. Harvard Bus. Rev. (2017)
14. R. Khanna, H. Kaur, Smart cities—the future of world. Int. J. Adv. Res. Comput. Sci. **8**(4) (2017)
15. A. Haylen, Uber and Airbnb: the legal and policy debate in NSW. e-Brief 2015 [cited 2020 June, 15]; Available from: https://www.parliament.nsw.gov.au/researchpapers/Documents/uber-and-airbnb-the-legal-and-policy-debate-in-n/Regulation%20of%20airbnb%20and%20uber%20in%20NSW.pdf
16. *Over 50 Laws Identified to Be Amended for Autonomous Vehicles.* Washington, D.C. (2017)
17. G. Park, J. Tani, Development of compositional and contextual communicable congruence in robots by using dynamic neural network models. Neur. Netw. **72**, 109–122 (2015)

18. G. Sun et al., Knowledge-intensive teaching assistance system for industrial robots using case-based reasoning and explanation-based learning. IFAC Proc. Vol. **47**(3), 4535–4540 (2014)
19. N. Radziwill, Quality considerations for ethical design of virtual and augmented reality. Softw. Qual. Prof. **21**(4), 34–47 (2019)
20. J. Bebbington, J. Unerman, Achieving the United Nations sustainable development goals. Account. Audit. Account. J. **31**(1), 2–24 (2018)

Detection of Network Intrusion and Classification of Cyberattack Using Machine Learning Algorithms: A Multistage Classifier Approach

Jay Sarraf, Vaibhaw, Sabyasachi Chakraborty, and Prasant Kumar Pattnaik

Abstract Rapid growth in the field of computer networks and applications has led to an extensive increase in the number of attacks, which raised many challenges for cybersecurity research. The primary concept of this study is to detect and classify the type of intrusion in the system using machine learning algorithms. In this paper, we use k-nearest neighbourhood, artificial neural network, random forest and XGBoost to classify the type of intrusion and compare for the better accuracy achievement. We perform binary classification for detecting whether there is an intrusion in the network traffic or not. If an intrusion is detected, the system aims to classify the type of attack which intrudes the network. The result clearly states that XGBoost outperformed the other three algorithms for intrusion classification with 83.92% accuracy achievement.

Keywords Network intrusion detection systems (NIDS) · Artificial neural network (ANN) · K-nearest neighbour (KNN) · Random forest · Decision trees · XGBoost · Denial of service (DoS) · Cyberattack · Machine learning

---

J. Sarraf (✉) · Vaibhaw · P. K. Pattnaik
School of Computer Engineering, KIIT University, Bhubaneswar, India
e-mail: jaysarraf596@gmail.com

Vaibhaw
e-mail: thevaibhaw@gmail.com

P. K. Pattnaik
e-mail: patnaikprasant@gmail.com

S. Chakraborty
Department of Computer Engineering, Inje University, Gimhae, South Korea
e-mail: c.sabyasachi99@gmail.com

P. K. Pattnaik et al. (eds.), *Proceedings of International Conference on Smart Computing and Cyber Security*, Lecture Notes in Networks and Systems 149,
https://doi.org/10.1007/978-981-15-7990-5_28

1 Introduction

Rapid growth in the field of computer networks and applications has led to an extensive increase in the number of attacks, which raised many challenges for cybersecurity research. Authentication, firewall, encryption and decryption, etc., are known as the first line of defence in the field of computer security which was used to detect intrusion traditionally [1]. However, the first line of defence is not so flexible and is unable to cover all region of network security and can be easily bypassed by different types of attacks [2]. The second line of defence may be an antivirus software which is limited to detect only those attacks whose signatures are already known and may not be able to detect any new type of attacks.

Network intrusion detection systems (NIDS) are responsible for monitoring the network traffic for possible threats of attacks. It gathers information from the network system and analyses it for any violation of security policies [3, 4]. NIDS can be divided into two categories: signature-based NIDS and anomaly-based NIDS. Signature-based NIDS monitor the traffic by matching the information with known attack to detect intrusion, whereas in case of anomaly-based NIDS, a normal profile is maintained by considering the normal behaviour of the network, and if there will be any deviation from this normal profile, then it will be considered as an attack [5]. Anomaly-based NIDS are mostly preferred as signature-based NIDS do not protect from an unknown attack [6, 7]. The effectiveness of a NIDS model is evaluated based on how accurately it classifies network traffic as an intrusion or not.

2 Literature Survey

Akashdeep et al. [8], presented work on feature-reduced intrusion detection system by utilizing the artificial neural network (ANN) classifier. Feature reduction was done by performing a ranking of the feature based on correlation and information gain. A novel approach was made to distinguish between useful and useless feature by combining the feature's rank obtained based on correlation and information gain. The feature-reduced dataset was then fed to the feed-forward neural network.

Ambusaidi et al. [9], proposed a mutual information-based feature selection algorithm which efficiently handles linear and nonlinear data features and selects the optimal feature which is utilized for classification. The extracted feature was then used to build the Least square support vector machine-based intrusion detection system (LSSVM-IDS). The model was evaluated using NSL-KDD, Kyoto 2006+ and KDDCUP datasets. The proposed feature selection algorithm contributed crucial features, also with lower computational cost as compared to state-of-the-art methods.

Farnaaz et al. [10], presented a random forest classifier-based network intrusion detection system. The NSL-KDD dataset was pre-processed, and the corresponding features were classified based on different categories of attacks, feature subset selection followed by selecting crucial feature subset. Extracted features are then used to train the random forest classifier.

Song et al. [11] proposed an intrusion detection system based on the time interval analysis of controller area network (CAN) message for the in-vehicle network. It has been found that time interval plays an important role in detecting network attacks on CAN.

Ugochukwu ct al. [12], compared various machine learning algorithms including Bayes net, J48, random forest, and random trees for classifying normal and abnormal network traffics on KDDCUP 99 dataset. The study showed that a random tree and random forest approach are most efficient in classifying the network traffic as compared to other algorithms.

3 Materials & Methods

In the study, primarily, we perform binary classification for detecting whether there is an intrusion in the network traffic or not. If an intrusion is detected, then our approach is to identify the type of attack which intrude the network. We classify the type of intrusion such as 'generic', 'exploits', 'fuzzers', 'DoS', 'reconnaissance'.

3.1 Dataset Description

Unavailability of the comprehensive network-based dataset which can reflect modern network scenarios along with low footprint attack environment is a major problem in the evaluation of network intrusion detection systems (NIDS) [13]. There exist many datasets such as KDDCUP 99 and NSL-KDD, but they are unable to reflect modern low footprint attack environment, and also, some of them contain redundant and missing records which are a major factor in changing the nature of the data [14].

Moustafa et al. [15], presented a UNSW-NB15 dataset for the evaluation of NIDS. The data was acquired by considering three virtual servers one out of which is used for introducing malicious/abnormal activities in the network traffic, and the other two are for normal network traffic. The presented dataset represents nine major families of network attacks with the help of IXIA PerfectStorm tool where we focus on only five major hits from the dataset. While comparing with the existing datasets, the UNSW-NB15 dataset shows promising results for NIDS research communities.

3.2 Feature Extraction

Instruction detection system dataset usually suffers from the curse of dimensionality which is responsible for increasing the time complexity [16]. Redundant and irrelevant features not only increase time complexity but also affect the classification accuracy by preventing the classifier to make accurate predictions, especially while dealing with large data [17]. The dataset contains 49 class labelled features which are used in the study for classification purposes.

3.2.1 Feature Classification Using Artificial Neural Network

An artificial neural network is a brain-inspired system which is intended to allow the computers to learn similarly as humans. ANN is modelled after the biological neural network in an attempt to narrow the gap between the learning process of a human brain and the computer.

The human brain has an interconnected network of neurons with dendrites for receiving inputs, and based on those, it produces output in the form of an electrical signal. Similarly, the simplest neural network, the perceptron, consists of one or more inputs and a processor for processing the input and providing output [18]. Generally, a neural network consists of an input layer, output layer and hidden layers that are responsible for processing the inputs. To model a complicated task, nonlinearity is introduced into the output of the neuron. This is done by the activating function, which decides whether a neuron should be activated or not by calculating weighted sum and adding bias to it [19].

3.2.2 Feature Classification Using *k*-Nearest Neighbour

k-nearest neighbour classifier is a supervised machine learning approach which is based on the fact that entities belonging to different classes will form different clusters in the feature space. It considers *k*-nearest neighbours to classify an entity. In this model, the distance between the feature of a test vector and the feature of a class in the feature vector space is considered as the measurement of similarity [20]. This can be represented mathematically as:

$$d(x, x') = \sqrt{(x_1 - x_1')^2 + \cdots + (x_i - x_i')^2} \tag{1}$$

where d is the similarity metric which can be obtained by considering the distance between the test vector x_i and the x_i' neighbours.

3.2.3 Feature Classification Using Random Forest

A decision tree is a tree-like model of decisions that continuously splits according to certain parameters of interest and leads to the predicted class at the end of the splitting. Random forest is a collection of such trees in which every tree has its prediction based on some parameters. These parameters are close randomly for every single tree at every split. The output of multiple decorrelated trees is combined to get the results for the model [21, 22].

Random samples are drawn from the training set with replacement during the training phase by using bootstrap aggregating (bagging). Prediction by random forest model can be obtained by combining the predictions for test data x' by individual trees.

$$\hat{f} = \frac{1}{B}\sum_{b=1}^{B} f_b(x') \tag{2}$$

3.2.4 Classification Using Extreme Gradient Boosting (XGBoost)

Boosting is a method in which a weak learner (mostly a decision tree) is modified to become a better learner by fitting the tree on the modified version of the original dataset. A loss function is defined depending upon the problem (generally logarithmic loss function is used for classification). Concept of gradient boosting was proposed by Friedman [23]. In boosting, the predictions are made sequentially, and not independently. The function $F(x)$ is learned by the model by executing the boosting algorithm s times, with the main goal to reduce the loss function. This is done by adding function $f(x)$ at every instance to make the model more accurate.

$$F_{s-1}(x) = F_s(x) + f(x) = y \tag{3}$$

3.3 Intrusion Detection

In intrusion detection phase, we perform binary classification for detecting whether there is an intrusion in the network traffic or not (0 for no intrusion and 1 for intrusion) using four different classification approaches: ANN, KNN, random forest classifier and XGBoost.

Figure 1 graphically describes the proportion of detected intrusions in the network traffic. We detect and classify 93,000 non-intrusion samples and 164,673 intrusion samples among which are generic (58,871 samples), exploits (44,525 samples), fuzzers (24,246 samples), DoS (16,353 samples), reconnaissance (13,987 samples),

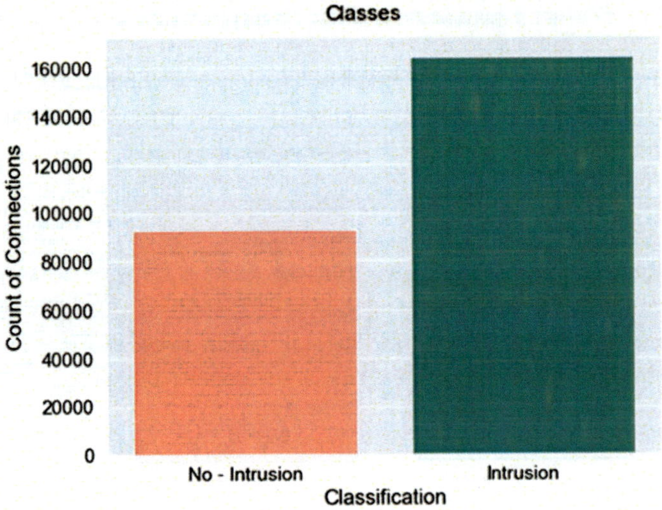

Fig. 1 Classifying intrusion and non-intrusion

analysis (2677 samples), backdoor (2329 samples), shellcode (1511 samples) and wroms (174 samples). XGBoost outperformed the other machine learning models by acquiring a classification accuracy of 95% as mentioned in Table 1.

Better the effectiveness of the model, better will be its performance in classifying the network traffics. To provide the performance measurements for the classification problem, the confusion matrix comes into the picture. Table 2 provides the confusion matrix for intrusion detection phase to evaluate different classification models based

Table 1 Classification accuracy for detecting intrusion/non-intrusion

Classifier	Accuracy %
KNN	85.64
ANN	93.72
Random Forest	94.67
XGBoost	95.00

Table 2 Performance analysis of classification algorithms for intrusion detection

Classification algorithm	Precision		Recall		F1 score	
	Non-intrusion	Intrusion	Non-intrusion	Intrusion	Non-intrusion	Intrusion
KNN	0.85	0.86	0.73	0.93	0.79	0.89
ANN	0.91	0.96	0.92	0.95	0.91	0.95
Random forest	0.93	096	0.93	0.96	0.93	0.96
XGBoost	0.93	0.96	0.94	0.96	0.93	0.96

on recall (out of all positive classes how much prediction is right), precision (out of correctly predicted positive classes, how many are positive) and F1 score (helpful in measuring both precision and recall at the same time).

3.4 Intrusion Classification

If an intrusion is detected in the intrusion detection phase, then it is further classified in intrusion classification stage to categorize it into a subcategory of attacks. For our study, we considered five different attacks including generic, exploits, fuzzers, denial of service (DoS) and reconnaissance. Classification is performed using the same four mentioned approaches.

Graphical representation of intrusion classification on detected intrusions is shown in Fig. 2; most of the detected intrusion were generic followed by exploits, fuzzers, DoS and lastly reconnaissance. As mentioned in Table 3 the highest accuracy of 83.92% in classifying the type of attack was achieved using the XGBoost classifier.

Table 4 shows the performance evaluation for different classification approaches in classifying the detected intrusions based on which family of attack they belong to. Different classifiers are evaluated based on recall, precision and F score for different classes of attacks. From the confusion matrices for the detection of the type of intrusion, it can be observed that a maximum of attacks which are from the DoS class

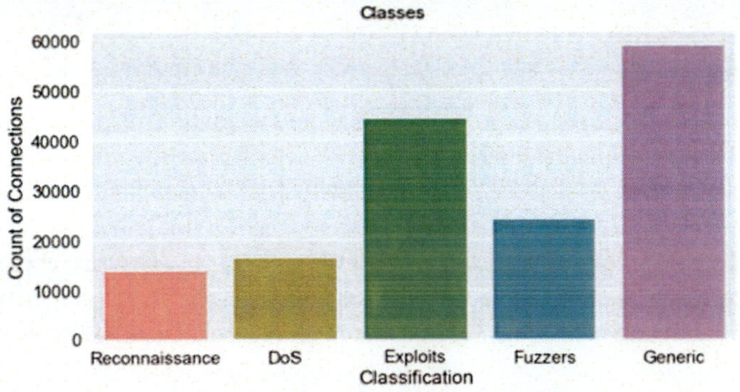

Fig. 2 Five types of intrusion classification

Table 3 Intrusion classification accuracy

Classifier	Accuracy %
KNN	68.85
ANN	79.25
Random forest	83.6
XGBoost	83.92

Table 4 Performance analysis of multiple algorithms for intrusion classification

Classification algorithm	Types of attacks	Precision	Recall	F1 score
KNN	Generic	1.0	0.97	0.98
	Exploits	0.49	0.81	0.61
	Fuzzers	0.57	0.32	0.41
	DoS	0.33	0.16	0.22
	Reconnaissance	0.92	0.34	0.49
ANN	Generic	1.0	0.97	0.98
	Exploits	0.64	0.90	0.74
	Fuzzers	0.79	0.79	0.79
	DoS	0.30	0.05	0.08
	Reconnaissance	0.74	0.58	0.65
Random forest	Generic	1.0	0.98	0.99
	Exploits	0.66	0.92	0.77
	Fuzzers	0.92	0.87	0.92
	DoS	0.44	0.12	0.18
	Reconnaissance	0.93	0.76	0.84
XGBoost	Generic	1.0	0.98	0.99
	Exploits	0.67	0.90	0.77
	Fuzzers	0.93	0.88	0.91
	DoS	0.44	0.16	0.24
	Reconnaissance	0.93	0.76	0.84

has been classified to exploits class. The reason for this mispredictions was analysed, and it was found that the feature 'sbytes' which corresponds to source to destination transaction bytes and the feature 'Spkts' which means Source to destination packet count are highly important features. Moreover, the two features have high correlation and are in a common range for the records pertaining to DoS and exploits class; hence, such mispredictions have taken place. Figure 3 shows the graphical description of the confusion matrix for ANN, random forest, KNN and XGBoost classifier.

4 Conclusion

Rapid development in the field of computer networks and applications has led to an extensive increase in the number of attacks. Despite the increase in network security awareness, the existing first and second line of defence is not capable of fully defending the network from modern cyberthreats. This is where the idea of network intrusion detection systems comes into the picture, which gathers information from the network system and analyses it for any violation of security policies. To protest

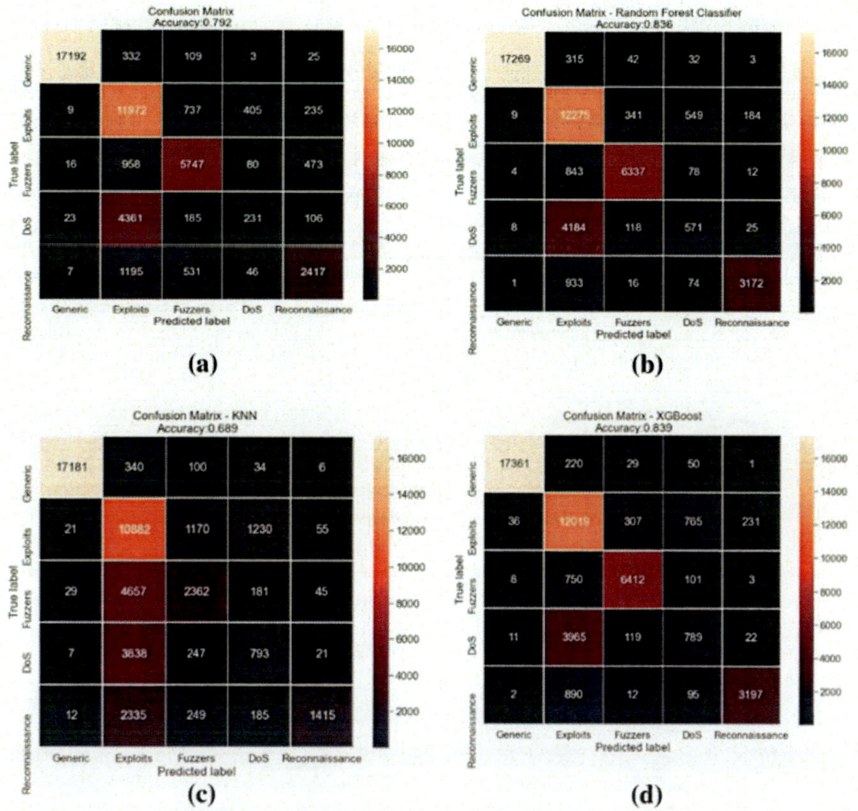

Fig. 3 Confusion matrix for intrusion classification for different classification algorithms. **a** Artificial neural network, **b** random forest classifier, **c** *k*-nearest neighbours and **d** XGBoost

the network system from unknown attacks, anomaly-based NIDS are used which maintains a normal profile based on normal network behaviour and deviation from such behaviour is considered as an attack. Recent advancements in intrusion detection systems (ISD) suggest that there are two main components essential for an efficient ISD. Efficient feature extraction and robust classification are the basic pillars of an efficient IDS. The primary concept of our study focused on detecting an intrusion in the network traffic and then classifying that intrusion in the system using a multistage classifier approach to distinguish it from other subsets of attacks.

References

1. N. Chaabouni, M. Mosbah, A. Zemmari, C. Sauvignac, P. Faruki, Network intrusion detection for IoT security based on learning techniques. IEEE Commun. Surveys Tutor. **21**(3), 2671–2701 (2019)
2. Z. Lu, G. Qu, Z. Liu, A survey on recent advances in vehicular network security, trust, and privacy. IEEE Trans. Intell. Transp. Syst. **20**(2), 760–776 (2018)
3. E. Hodo, X. Bellekens, A. Hamilton, P.L. Dubouilh, E. Iorkyase, C. Tachtatzis, R. Atkinson, Threat analysis of IoT networks using artificial neural network intrusion detection system, in *2016 International Symposium on Networks, Computers and Communications (ISNCC)* (IEEE, 2016), pp. 1–6
4. N. Sultana, N. Chilamkurti, W. Peng, R. Alhadad, Survey on SDN based network intrusion detection system using machine learning approaches. Peer-To-Peer Netw. Appl. **12**(2), 493–501 (2019)
5. C. Dartigue, H.I. Jang, W. Zeng, A new data-mining based approach for network intrusion detection, in *2009 Seventh Annual Communication Networks and Services Research Conference* (IEEE, 2009), pp. 372–377
6. N.T. Van, T.N. Thinh, An anomaly-based network intrusion detection system using deep learning, in *2017 International Conference on System Science and Engineering (ICSSE)* (IEEE, 2017), pp. 210–214
7. H. Bostani, M. Sheikhan, Hybrid of anomaly-based and specification-based IDS for Internet of Things using unsupervised OPF based on MapReduce approach. Comput. Commun. **98**, 52–71 (2017)
8. I. Manzoor, N. Kumar, A feature reduced intrusion detection system using ANN classifier. Expert Syst. Appl. **88**, 249–257 (2017)
9. M.A. Ambusaidi, X. He, P. Nanda, Z. Tan, Building an intrusion detection system using a filter-based feature selection algorithm. IEEE Trans. Comput. **65**(10), 2986–2998 (2016)
10. N. Farnaaz, M.A. Jabbar, Random forest modeling for network intrusion detection system. Procedia Comput. Sci. **89**(1), 213–217 (2016)
11. H.M. Song, H.R. Kim, H.K. Kim, Intrusion detection system based on the analysis of time intervals of CAN messages for in-vehicle network, in *2016 international conference on information networking (ICOIN)* (IEEE, 2016), pp. 63–68
12. C.J. Ugochukwu, E.O. Bennett, An intrusion detection system using machine learning algorithm. Int. J. Comput. Sci. Math. Theory **4**(1), 39–47 (2018)
13. R. Sharma, R.K. Singla, A. Guleria, A new labeled flow-based dns dataset for anomaly detection: PUF dataset. Procedia Comput. Sci. **132**, 1458–1466 (2018)
14. A. Javaid, Q. Niyaz, W. Sun, M. Alam, A deep learning approach for network intrusion detection system, in *Proceedings of the 9th EAI International Conference on Bio-inspired Information and Communications Technologies (formerly BIONETICS)*, pp. 21–26 (2016)
15. N. Moustafa, J. Slay, The evaluation of Network Anomaly Detection Systems: statistical analysis of the UNSW-NB15 data set and the comparison with the KDD99 data set. Inf. Secur. J. Glob. Perspect. **25**(1–3), 18–31 (2016)
16. T. Janarthanan, S. Zargari, Feature selection in UNSW-NB15 and KDDCUP'99 datasets, in *2017 IEEE 26th international symposium on industrial electronics (ISIE)* (IEEE, 2017), pp. 1881–1886
17. S.M. Othman, F.M. Ba-Alwi, N.T. Alsohybe, A.Y. Al-Hashida, Intrusion detection model using machine learning algorithm on Big Data environment. J. Big Data **5**(1), 34
18. F.M. Bayat, M. Prezioso, B. Chakrabarti, H. Nili, I. Kataeva, D. Strukov, Implementation of multilayer perceptron network with highly uniform passive memristive crossbar circuits. Nat. Commun. **9**(1), 1–7 (2018)
19. S. Sharma, Activation functions in neural networks. Towards Data Sci. 6
20. J.N. Myhre, K.Ø Mikalsen, S. Løkse, R. Jenssen, Robust clustering using a kNN mode seeking ensemble. Pattern Recogn. **76**, 491–505 (2018)

21. P. Dollár, C.L. Zitnick, Structured forests for fast edge detection, in *Proceedings of the IEEE International Conference on Computer Vision*, pp. 1841–1848
22. L.F. Nicolas-Alonso, J. Gomez-Gil, Brain computer interfaces, a review. Sensors **12**(2), 1211–1279
23. J.H. Friedman, Greedy function approximation: a gradient boosting machine. Ann. Stat., 1189–1232

Robotic Process Automation Implementation Challenges

Daehyoun Choi, Hind R'bigui, and Chiwoon Cho

Abstract Robotic process automation (RPA) is a new technology that enables the automation of high-volume, manual, repeatable, routine, rule-based and unmotivating human tasks. The main goal of RPA is to replace tedious human tasks with a virtual workforce or a digital worker performing the same work as the human worker was doing. This allows human workers to focus on difficult tasks and problem solving. RPA tools are considered simple and powerful for specific business process automation. However, since the technology is still emerging, RPA faces some challenges during the implementation. This paper provides an overview of RPA and its challenges.

Keywords Robotic process automation · RPA · Software robot · Digital workforce

1 Introduction

Robotics process automation (RPA) system enables organizations to automate high-volume, repeatable and unmotivated mass of tasks just like the human user performing across systems and applications, through the usage of graphical user interfaces. The main objective of robotic process automation is to substitute with a virtual workforce the repetitive and routine tasks handled by humans allowing them to focus more on valuable tasks and problem solving. RPA is defined as a category of tools that enable users to specify deterministic routines involving structured data, rules (if then-else), user interface interactions and operations accessible via

D. Choi · H. R'bigui · C. Cho (✉)
School of Industrial Engineering, University of Ulsan, Ulsan 680749, Republic of Korea
e-mail: chiwoon6@mail.ulsan.ac.kr

D. Choi
e-mail: dhchoi@nsoft.co.kr

H. R'bigui
e-mail: hind.rbigui@gmail.com

P. K. Pattnaik et al. (eds.), *Proceedings of International Conference on Smart Computing and Cyber Security*, Lecture Notes in Networks and Systems 149,
https://doi.org/10.1007/978-981-15-7990-5_29

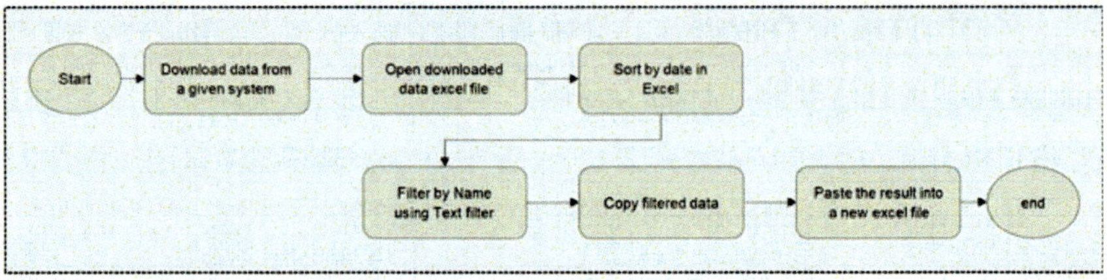

Fig. 1 Example of clerical tasks performed by a human

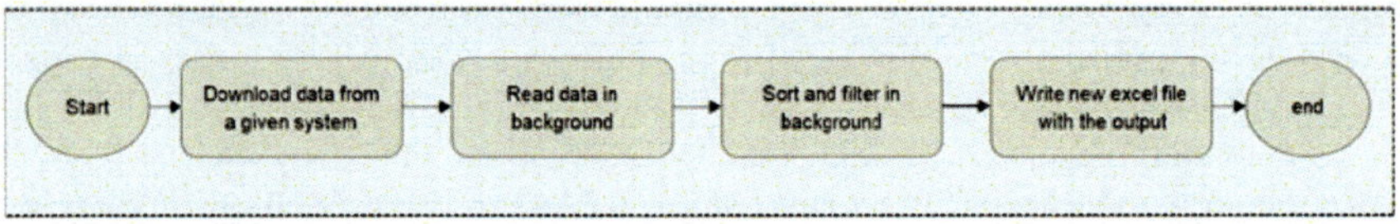

Fig. 2 Example of clerical tasks performed by a robot

APIs [1]. RPA reduces labor-intensive processes through simulating human effort to perform tasks, speeding up the execution of high-volume transactional process [2]. An example of clerical tasks, *as-is* process, performed by a human is depicted in Fig. 1. The to-be process performed by a robot is illustrated in Fig. 2. The *to-be* workflow looks like the *as-is* workflow, but it is optimized by using background tasks for sorting and filtering.

This work presents the components of robotic process automation, its advantages and some application area, as well as the challenges encountered during the implementation of this later.

2 RPA Components

Robotic process automation technology consists of three main elements: (1) RPA robots, virtual workforce, for executing repetitive human tasks, (2) RPA orchestrator for monitoring and managing the virtual workforce and (3) RPA studio for designing the workflow to be executed by the virtual workforce. The main components are illustrated in Fig. 3.

2.1 RPA Robots

RPA robots are virtual workers which perform the repetitive tasks of an employee. RPA robots are dedicated to handle unmotivated mass of tasks so that employees can

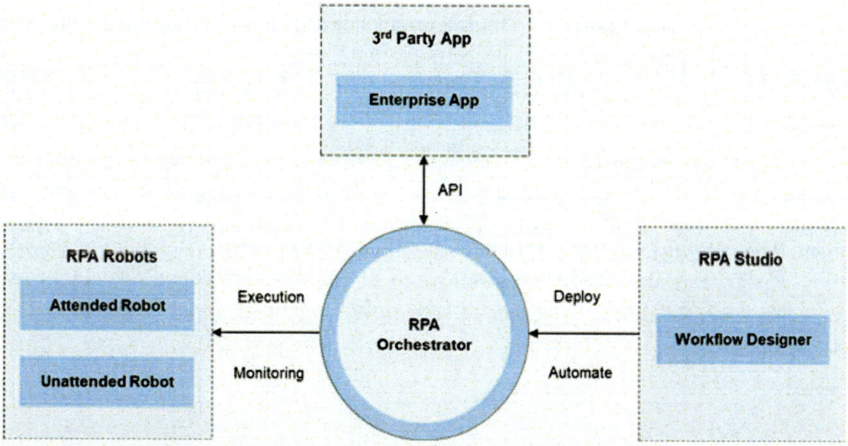

Fig. 3 RPA components

engage in valuable jobs and problem solving. There are two types of RPA robots, attended robots and unattended robots.

2.1.1 Attended Robots

Attended robots are robots designed to work side by side with a human user for speeding up repetitive tasks where the tasks can be triggered by the human user. It can be used in repetitive, manual and highly rule-based tasks which require human intervention for decision points.

2.1.2 Unattended Robots

Unattended robots are robots designed to work fully unattended in the back office. This type of robots operates on organization's server without the intervention of a human user and can be scheduled to be started automatically. The robots can be triggered by a satisfied rule or condition or by a business event. It can be used in repetitive, manual and highly rule-based tasks which do not require any human intervention.

2.2 RPA Orchestrator

RPA orchestrator is responsible for scheduling, monitoring, managing and auditing robots. It is considered the highly scalable control and management server platform.

The orchestrator sits in the middle connecting studio with the robots and exposes a REST API for third-party applications to be used.

2.3 *RPA Studio*

RPA studio is the designer tool used for development. It enables users to create, design and automate the workflow to be executed by robots. Business users can program the robots by record & screenplay capability and intuitive scenario design interface.

3 RPA Advantages and Application Area

Robotic automation interacts with the existing IT architecture without the need of complex system integration. RPA aims to transfer the process execution from humans to bots. An average person can work 8 h a day, while the robot can work 24 h without fatigue. The average human productivity is 60%, and the error is small, while the robot's productivity is 100%, without any error. Moreover, compared with humans, robots can handle multiple tasks. By assigning repetitive, routine and high-volume task to robots, human worker becomes able to perform extra tasks that before was not able to find the time to perform it, thus increasing the productivity. To benefit from its advantages, RPA technology can be applied to areas [4, 6, 7] where there are processes containing tedious and high-volume tasks to be accomplished by the employee. Some of the application area of RPA is depicted in Table 1.

Table 1 Application area of RPA

Industry	Usage
Healthcare	• Billing • Patient registration
Human resources	• New employee joining formalities • Payroll process • Hiring shortlisted candidates
Insurance	• Clearance & claims processing • Premium information
Manufacturing & retail	• Calculation of sales • Bills of material
Banking and financial services	• Discovery • Frauds claims • Cards activation

4 RPA Challenges

Robotic process automation (RPA) recently gained a lot of attention in industries and academia [4] as it speeds up business growth by reducing a lot of manual and repetitive-based work [5]. However, at present, the implementation of RPA still faces many challenges. According to the report of Global RPA Survey 2019 [3] depicted in Fig. 4, challenges at the organizational structure level include the inability to assess process priorities (40%), lack of risk management tools (28%), insufficient internal staff skills (24%) and the lack of sense of urgency (23%). At the technical risk level, it is information and data security (40%), difficulty in achieving scale (37%) and selection of a suitable development platform (30%). The financial and regulatory aspects include higher implementation costs (37%), inappropriate application scenarios (32%) and external legal regulatory requirements (30%). A further discussion on these challenges is presented in Table 2.

Prioritizing potential RPA initiatives is currently the starting point of robotic process automation (RPA) implementation. Therefore, any poor choice of processes for initial pilot will result in the failure of RPA implementation. In other words, establishing what is in and out of the scope for RPA, which processes should be automated and which routines should be automated in the first place since RPA can automate a wide range of routines can be seen as the main challenge for RPA implementation.

It is not difficult to see that although RPA can quickly achieve process automation functions with a low and lightweight code, due to its deep business integration and its direct impact on achieving business goals and processing business data, etc., RPA should take over companies great interest in business changes, management design, control, security, operational stability and mechanisms for dealing with exceptions.

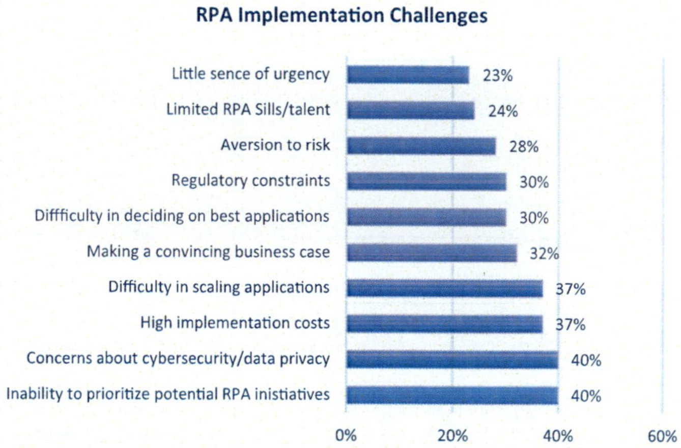

Fig. 4 RPA implementation challenges

Table 2 RPA implementation challenges by perspective

Perspective	Challenges	Comments
Organizational perspective	Prioritizing potential RPA initiatives	Identifying where RPA is highly likely to provide significant value is challenging. The huge effort will be in this stage when implementing RPA. A poor choice of processes to be automated may result in implementation failure. Approaches for identifying the suitable processes to be automated is strongly required
	Aversion to risk	Most of organizations prefer not to take the risk in adopting a technology emerging. This can be overcome by applying RPA to many areas as case studies
	Limited RPA Sills/talent	RPA is still an emerging technology. Therefore, there is a lack in skilled people in RPA
	Little sense of urgency	It might take a long time to decide implementing RPA for organizations having a little sense of urgency. Many use cases are important to serve for the growth of RPA implementation
Technical Perspective	Cybersecurity/data privacy	Cybersecurity and data privacy were always considered crucial. RPA is based on mining user interface data that may include private information. Thus, there is a need for a secure RPA development
	Scaling applications	When automating a core business process with RPA and finding that the business is growing rapidly, if that automated process cannot scale as required, the RPA technology can become an obstacle for growth. Therefore, techniques to make the scalability easy is needed
	Deciding on best applications	Ensuring that you are using the right application can be very challenging. There is a need for benchmarks on how to decide the best application

(continued)

Table 2 (continued)

Perspective	Challenges	Comments
Financial and regulatory	Implementation costs	Process analysis phase takes a long time in the implementation process. By speeding up this phase, one can reduce implementation costs. So, approaches for accelerating the process analysis phase are needed
	Convincing business case	A considerable number of use cases is needed to convince businesses
	Regulatory constraints	New technologies are required to meet regulatory constraints

5 Conclusion

Robotic process automation is a simple and powerful technology for a specific business process automation. With robotic process automation, you can use tools to create your own software robots that automatically execute defined business processes. Your "robot" is a configurable software that is used to perform tasks you assign and control. RPA robots can learn and can also be cloned. It has no code, is not interrupted and intrusive and is easy to use. However, there are some challenges that one can face during the implementation of RPA. In this paper, we provided an overview on robotic process automation, RPA components and the challenges faced during RPA implementation. Prioritizing potential RPA initiatives or identifying the processes to be automated with RPA is outlined to be the main challenges of RPA implementation. Since determining what is in and out of the scope for RPA is the starting point of RPA implementation, a poor choice of processes candidates for automation can widely results in RPA implementation failure. Therefore, a considerable number of research works in this area is required.

Acknowledgements This work was supported by the University of Ulsan in 2020.

Reference

1. V. Leno, A. Polyvyanyy, M. Dumas et al., Robotic process mining: vision and challenges. Bus Inf Syst Eng (2020). https://doi.org/10.1007/s12599-020-00641-4
2. S. Gupta, S. Rani, A. Dixit, Recent trends in automation—a study of RPA development tools, in *IEEE 3rd International Conference on Recent Developments in Control, Automation & Power Engineering (RDCAPE)*, 10–11 Oct 2019, NOIDA, India
3. https://econsultsolutions.com/wp-content/uploads/2019/05/2019-global-rpa-survey-protiviti.pdf

4. J. Gao, S.J. van Zelst, Lu X., W.M.P. van der Aalst, Automated robotic process automation: a self-learning approach, in *On the Move to Meaningful Internet Systems: OTM 2019 Conferences. OTM 2019. Lecture Notes in Computer Science*, vol. 11877, ed. by H. Panetto, C. Debruyne, M. Hepp, D. Lewis, C. Ardagna, R. Meersman (Springer, Cham, 2019)
5. N. Nawaz, Robotic process automation for recruitment process. Int. J. Adv. Res. Eng. Technol. (IJARET) **10**(2), 608–611 (2019)
6. M., Ratia, J. Myllärniemi, N. Helander, Robotic process automation—creating value by digitalizing work in the private healthcare, in *ACM International Conference Proceeding Series: International Academic Mindtrek Conference* (2018)
7. L. Ivančić L., D. Suša Vugec, V. Bosilj Vukšić, Robotic process automation: systematic literature review, in *Business Process Management: Blockchain and Central and Eastern Europe Forum. BPM 2019. Lecture Notes in Business Information Processing*, vol. 361, ed. by C. Di Ciccio C. et al. (Springer, Cham, 2019)

Blockchain Technology to Support Employee Recruitment and Selection in Industrial Revolution 4.0

Happy Rhemananda, Dima Roulina Simbolon, and Olivia Fachrunnisa

Abstract Blockchain will come to pervade the business environment, bringing huge implications for HR and the workforce. There was also a sense that technology has eroded trust in recent years, especially with the growth of cyber risks, blockchain presents a way to use technology to win back that lost trust. HR functions need to start to include blockchain alongside other emerging technologies in formulating their digital strategy. Assessing the potential blockchain to enhance efficiency and effectiveness should be considered alongside the broader implications for the future of work. Having identified and unpacked the problem to be solved, a good next step is to start to create prototypes that can develop into proofs of concept (POCs) that will target the most valuable case usage. This article develops the concept of blockchain technology application in the field of human resource management, especially in the employee recruitment and selection process. The benefit from HR blockchain is targeting productivity gains. The enhanced ability to match people's skills and performance to jobs would provide an uplift to productivity. Small and medium-sized enterprises (SMEs) may benefit particularly. The burden of finding and recruiting the right talent is especially difficult for smaller businesses, and anything that can help them do this more effectively and efficiently will boost their productivity. Hence, blockchain will simplify the process of employee selection and recruitment by helping to identify skills, knowledge, and experiences of potential candidates that are validated accurately. In the e-HRM context, the implementation of the blockchain system has the potential to improve the process of human resource management (HRM), specifically in the selection and recruitment process through screening of best candidates.

H. Rhemananda · D. R. Simbolon · O. Fachrunnisa (✉)
Department of Management, Faculty of Economics, Universitas Islam Sultan Agung, Jl. Kaligawe Raya Km. 4, Semarang 50111, Indonesia
e-mail: olivia.fachrunnisa@unissula.ac.id

H. Rhemananda
e-mail: happyrhemananda@std.unissula.ac.id

D. R. Simbolon
e-mail: dimasimbolon@std.unissula.ac.id

P. K. Pattnaik et al. (eds.), *Proceedings of International Conference on Smart Computing and Cyber Security*, Lecture Notes in Networks and Systems 149,
https://doi.org/10.1007/978-981-15-7990-5_30

Keywords Recruitment and selection · Blockchain · e-human resource management

1 Introduction

The entry of the Industrial Revolution 4.0 era, where the digital era continues to develop, is no exception in the system of employee recruitment at companies. The old manual recruitment method has begun to be abandoned, replaced with a database-based digital system. It is not impossible in the next few years, the limited manual recruitment method has begun to be abandoned and shifted to a more efficient digital-based recruitment trend, and everything will change into an integrated system called blockchain. Blockchain is a distributed ledger (open ledger) that is open and can record transactions between two parties efficiently and in a way that can be verified and is permanent. As quoted from the book Blockchain Revolution [1], blockchain is a digital ledger of uninterrupted financial transactions that can be programmed to record not only financial transactions but also everything of value. Blockchain is a technological breakthrough that creates trust through consensus by ensuring that all parties authorized to access the blockchain agree with any additions they make so that the data is guaranteed to be valid.

In the digital or e-HRM context, there is potential to improve the process of human resource management (HRM) using the blockchain for job screening, credentials and verification of education, contracts, and worker payments [2, 3]. That is, individual data such as education and work history can be stored and interact safely in digital ledgers in real time. This is an innovation that can raise high expectations for HRM so that every organization successfully adopts and applies it in the Industrial Revolution 4.0.

Interestingly, in the blockchain technology system, every slightest change must be verified in advance by the network. This means that all data entered is trusted and passes the authentication process. If there is one change, then the change will be updated to everyone who has access to the database. Conversely, old data storage systems tend to be centralized on one computer and run the risk of being lost or damaged. Databases stored in blockchain are error-free, transparent, and safe because every data is verified. That means, companies can get comprehensive detailed information.

During this time, one of the most popular blockchain technologies is bitcoin where the application of this technology is in the financial aspects of payment and recording. But actually, this blockchain technology has begun to develop and researchers have begun to explore the possibility of this technology being applied to the human resource function. This article focuses on the implementation or application of blockchain technology in its function in the field of human resource management, especially in the process of recruitment and selection of employees in the company. One way in which blockchain technology can simplify the selection and recruitment process is by helping to identify the experience and skills of potential candidates who are validated accurately.

2 Blockchain for Recruitment Tool

Recruitment and selection are one of the most important things in a company because this process involves the selection and placement of the most appropriate employees among the best for the continuity of the company's activities in creating company value. Recruitment is the process of attracting a group of candidates to a certain position, followed by a selection phase [4]. Recruitment involves actions and activities taken by an organization to identify and attract individuals who have the ability to help the organization realize its strategic goals. Such activities must produce the desired set of candidates, increase their interest and be interested in the organization as employers, and increase the likelihood that they will receive job offers.

The recruitment and manual selection process still have many shortcomings because the results of the process obtained are still not in accordance with reality and require a long time in the data verification process. Stages of recruitment that are too long also make the assessment process and decisions taken take a long time. In the manual recruitment process generally, it still requires a third party in making decisions and that too will also add to the company's costs. Employee data that has entered the company system is usually the previous data, and the latest data is only stored in the software. But the resilience of stored data is less effective because the software contains a lot of data which will result in software overload and data can be lost.

There are cases that some job applicants send fake applications with fake training and diplomas, references, awards, promotions, and so on to intentionally exaggerate their qualifications and abilities. For example, Amazon pays an employee $5000 bonus to cancel a work contract (Amazon 2014). As a result, organizations end up paying large amounts of fees just to get rid of bad recruitment. Another survey revealed that 74% of employers had the wrong recruitment. It also identified that the skills of 45% of workers did not match their claims and 33% had lied about their qualifications [5].

The emergence of blockchain technology in the digital era can minimize the disadvantages of using manual recruitment in order to produce quality prospective employees. Unlike the old recruitment system that relies on manual records, in the blockchain everything is digitally connected. No more decision making that takes too long because the information is more organized. Thus, companies can conclude more quickly or compile shortlist of candidates who meet the criteria.

The cost during the recruitment process is also much cheaper and efficient. For example, there are prospective employees who register and 5 years previously had worked in the same company. Through the manual method, it could be that the data has been lost. But with blockchain, everything can be traced easily. The data verification process is guaranteed, considering the authenticity of the prospective employee's profile is one of the most important things. The long process that has been done manually can be replaced with digital verification.

The phases or digital database-based recruitment processes will then become more concise, efficient, and effective. The data collection and verification process and the decision-making process are also expected to be faster and more precise.

Recruitment and selection stages using blockchain technology are as follows.

2.1 The Registration of Prospective Employees

In the first stage, prospective employees must register themselves by registering a curriculum vitae (CV) into the company's blockchain. Prospective employees are directed to fill in the registration form and input their data and CV on the company's blockchain Web site. After submitting, the prospective employee will automatically get a code from the system. Blockchain will check the validity of the inputted data by matching the code, and if a discrepancy is found between the data codes with one another, the blockchain will automatically declare that the data is invalid. Therefore, the data inputted by employees must be in accordance with the reality of the prospective employee's identity. This of course reduces the risk of data fraud and increases the time efficiency of data validation.

Furthermore, when the data has been submitted, prospective employees cannot replace the data unless there is a consensus between the prospective employee and the company. Illustration: A prospective employee applies for a job at company A and submits data and CVs on the company's blockchain Web site. Prospective employees no longer need to come to the company office and bring a hard copy of their data files. He only needs to submit data on the company's blockchain system to be further validated automatically by the system. When a prospective employee applies for a job at company A and submits data and CVs on the company's blockchain Web site, the data and CV he inputs state that he is a graduate of a well-known university and has participated in a certified apprenticeship program at a large company B. However, the company's blockchain system has detected invalid data which states that the university code mentioned is correct but the code for experience in a certified apprenticeship at company B is flawed or invalid. This means that the prospective employee is indeed registered with the university but he has never participated in a certified apprenticeship program organized by company B. Then this indicates an indication of data fraud committed by the prospective employee.

2.2 Administrative Selection of Prospective Employees

According to the HireRight employment screening benchmark report in 2018, 84% of applicants falsified data on their resumes. Blockchain-based credential verification systems can help reduce the time spent doing background checks, reduce fraud, and build more trust in the recruitment ecosystem. File or data of prospective employees who enter the system will be validated using blockchain technology between companies and agencies that issue files or certificates. The certificate issuer can verify it automatically without the company asking for it, for example, validation of prospective employee's college data, work experience, educational history, self-identity, etc., using the blockchain. Blockchain directly gets data that is very real time and accurate because in this system data cannot be replaced or falsified without the knowledge and agreement of the blockchain users involved. This means that the blockchain can minimize fraud or falsification of data by prospective employees even though they have collaborated with agencies, governments, and so on.

When a prospective employee includes data stating that he is a fresh graduate of university A who graduated at the end of the year, the blockchain can detect the recorded data code which means there is a mismatch between the data code inputted by the prospective employee and the data code that is in the university blockchain system A. Furthermore, the blockchain found that the prospective employee was still a student and was scheduled to graduate at the beginning of the following year. Although the prospective employee cooperates with the university, the code on the blockchain is confidential and cannot be known and changed without any consensus between all elements of the blockchain. Moreover, any changes made are also recorded in the system automatically and are permanent.

In the blockchain system, the quality of recruitment and HRM activities is ensured by machine. This system can handle, store, validate, and rank information with complete transparency and security. The security of a blockchain system cannot be easily violated by hacking or system managers. Because the blockchain system is protected by multiple layered keys and hash encryption where the risk of information leakage and data changes is very low. Because this system is distributed, so changes in information can be easily tracked and original data can be retrieved.

2.3 Faster Decision Making

Blockchain technology simplifies and accelerates the decision-making process in the selection and recruitment process. In the process of finding the truth of the prospective employee's data already, it uses sophisticated technology and does not need to use a third party to search for data to go to agencies, companies, and governments to find the truth. Blockchain can store employment contract information such as applicants' electronic signatures, payroll details, security access codes, performance reports, and even psychometrics. Blockchain technology automatically verifies the correctness of its data through blockchain agencies, companies, and governments. Blockchain working time in data search takes approximately seven seconds to find out the truth. In theory, a candidate can be hired immediately and even can get a contract and their payroll number is set at a glance. Thus, the company can immediately see the results of real-time data obtained and can immediately make decisions as to who is determined as an employee who passes.

The application of blockchain technology enables companies and prospective employees to obtain and find out the results of the selection and recruitment process faster. With a short data input and validation stage, the company can quickly abort prospective employees who have data defects. Furthermore, companies can use the blockchain system to make decisions directly. With the validity of personal data and the competencies of prospective employees who have been confirmed, the system can directly communicate to the prospective employee whether he is accepted or rejected.

2.4 *Placing Employees in the Right Position and Field*

Reporting from a Career survey conducted in 2014, nearly 60% of job seekers misrepresent themselves on their resumes. This fact is certainly surprising to the company, but with blockchain technology, it allows the recruitment team to have instant, accurate, and complete access to all the potential and work history of employees. The truth of prospective employee data obtained from blockchain technology can be used as a reference for placing employees in their fields. For example, with the data of prospective employee A who is a graduate of a well-known university with a GPA of 3.90 and gets good grades in accounting courses, then prospective employee A can be employed or placed in corporate finance. Companies can trust data obtained by the blockchain system which can also be used as a reference for decision making and employee placement in accordance with their fields.

3 Conclusion

HR must begin to adopt this technology in line with the development of technologies in formulating their digital strategies. Assessing the potential of the blockchain in increasing efficiency and effectiveness must be considered along with the wider implications of future work. Eventually, employers will have a stronger and more trusted talent pool to recruit, and candidates will know that they do not waste time on reckless job searches. Through the blockchain, the HR department can introduce trust and transparency back to a broken system and meet the upcoming global talent shortages [6].

In the blockchain era, recruiters will be challenged to create more value than before. But they will also have better tools to help make it happen. The next right step is to start making prototypes that can develop into proof of concepts (POCs) and which will target high-value cases. It can be concluded that blockchain is a technology that has developed in the HR field. If the goal is a more complete and more cost-effective recruitment process, then it is time to use the blockchain. The race to seize competitive advantage through the blockchain has begun, and the HR function must participate in the race or the risk may be left behind.

Reference

1. D. Tapscott, A. Tapscott, Blockchain revolution (2018)
2. R. Maurer, Blockchain could phase out employment screening, secure candidate data. Society for Human Resource Management (SHRM) (2018). Retrieved from https://www.shrm.org/ResourcesAndTools/hr-topics/talent-acquisition/Pages/BlockchainEmployment-Screening-Secure-Candidate-Data.aspx. Accessed on April 6, 2018
3. D. Zielinski, Is HR ready for blockchain? Society for human resource management (SHRM) (2018). Retrieved from https://www.shrm.org/hr-today/news/hr-magazine/0318/pages/ishr-ready-for-blockchain.aspx. Accessed on April 9, 2018
4. A. M. Saks, in *The impracticality of Recruitment Research*, Eds. by A. Evers, N. Anderson, & O. Voskuijl. The Blackwell Handbook of Personnel Selection (Blackwell Publishing, Ltd, UK, 2005), pp. 47–72.
5. New Career Builder Survey, Employers share most memorable lies they discovered. https://www.careerbuilder.com/share/aboutus/pressreleasesdetail.aspx?sd=8%2F7%2F2014&id=pr837&ed=12%2F31%2F2014 (2014). Accessed on 9 Nov 2019
6. Francis Gregory, Utilising blockchain in recruitment. https://www.recruitment-international.co.uk/blog/2018/09/utilising-blockchain-inrecruitment (2018). Accessed on 20 Nov 2019
7. M. Olivas-Lujan, Blockchains 2019 in e-HRM: hit or hype?, in *HRM 4.0 For Human-Centered Organizations. Advanced Series in Management*, vol. 23 (Emerald Publishing Limited, 2019), pp. 117–139
8. M.M.H. Onik, M.H. Miraz, C.-S. Kim, A recruitment and human resource management technique using blockchain technology for Industry 4.0, in *Proceeding of Smart Cities Symposium (SCS-2018)*, Manama, Bahrain (IET, 2018), pp. 11–16
9. C. Brandão, R. Silva, J.V. dos Santos,Online recruitment in Portugal: theories and candidate profiles. Journal of Business Research (2018)
10. C. Murray,How will blockchain technology impact HR and the world of work? (2017)
11. St.Louis, Charles. 2019. *"Blockchain and HR: How They Intertwine"*.
12. M. Fincher, 5 Ways Blockchain Technology Can Revolutionize Human Resource Management (2019)
13. A.K. Moin, Z. Safdar, et al., Securing IoTs indistributed blockchain: analysis,requirements and openissues. Future Gener. Comput. Syst. **100**, 325–343 (2019)
14. Wikipedia, Blockchain. https://id.wikipedia.org/wiki/Blockchain (2019). Accessed on 10 Nov 2019
15. Talent Mind, Using Blockchain technology for HR recruitment, why not? https://talentmind.ai/id/blog/insight/915/ (2018). Accessed on 10 Nov 2019
16. R. Poojary, Here's what you should know about Blockchain in recruitment. https://www.peoplematters.in/article/hr-technology/heres-what-you-should-know-about-blockchain-in-recruitment-17869 (2018). Accessed on 9 Nov 2019
17. HireRight, *Employment Screening Benchmark Report* (2018)
18. J. Bezos, Letter for Amazon's shareholders. https://www.sec.gov/Archives/edgar/data/1018724/000119312514137753/d702518dex991.h (2014). Accessed on 20 Nov 2019

Android-Based Online Attendance Application

Panji Rachmat Setiawan, Abdul Syukur, Novendra Kurniadi, and Amrizal Amrizal

Abstract Nowadays, technology development can be found anywhere. Especially, at Informatics Technology development. It can be seen in our lives, everyday, we always involving technology in our activity like using mobile phone for paying bills, direction if we want to go somewhere, and anything else. But many jobs that we still not using technology to achieve the data that we want. Usually, a lot of papers involved when doing this job. For example, attendance. Attendance is an important task to determine something, either at office, or at school. School or university has an attendance to measure for their student, and it writes on attendance list. But using attendance list means using more papers. We know paper is expensive and using paper to much it can damage our nature. For school or university, attendance become an assessment for each of their students. Every student has their limit for absence. But students have their own way to manipulate their attendance, like ask for their friend to present student who does not come to class, or if teacher or lecturer does not aware, they sign their missing attendance. This can be harmful to the school or university, another student, and students who manipulate their attendance. Another problem, attendance list using for report can be damaged, because school or university is using papers that can easily damage.

Keywords Mobile phone · Attendance · Paper

P. R. Setiawan (✉) · A. Syukur · N. Kurniadi · A. Amrizal
Department of Informatics, Universitas Islam Riau, Pekanbaru, Indonesia
e-mail: panji.r.setiawan@eng.uir.ac.id

A. Syukur
e-mail: abdulsyukur@eng.uir.ac.id

N. Kurniadi
e-mail: novendratkj@gmail.com

A. Amrizal
e-mail: it.amrizal@gmail.com

P. K. Pattnaik et al. (eds.), *Proceedings of International Conference on Smart Computing and Cyber Security*, Lecture Notes in Networks and Systems 149,
https://doi.org/10.1007/978-981-15-7990-5_31

1 Introduction

Nowadays, the development of technology has an impact. Especially, for Informatics Technology. It can be seen everywhere, how people always use technology for doing jobs. But many people do not know what is the benefit to utilize technology. There are so many jobs that can be done using technology, but why we are always using the old-fashioned way to do our job or to get some data.

On this occasion, author has a vision how to do our jobs using technology, especially using mobile technology. But the question is, what job can be done using technology? The answer is attendance for college student. Attendance is a simple activity that has an extraordinary impact. For universities, attendance was done when the class started. For college student, attendance is a way to prove their activity for a semester, and for university, attendance is part of assessment.

Attendance activities always become part of assessment between lecturer and college student, which can affect student's final grade. And some of college students often do something wrong to achieve their presence, so they can fulfill minimum attendance for semester. Some of them asking their friends to leave absent for them, or if their lecturer is not paying any attention, they sign their presence for last week, or the day before current meeting. This can be detrimental to students and lecturers. Attendance sheet used for attendance often occurs damaged. It can make the process for assessment and the report to universities slower. And also, printing attendance sheets is expensive.

Looking for the problem, authors done the research to avoid attendance activity which is can be detrimental to students and lecturers, also to avoid from damage of attendance sheet, and how to reduce cost for print attendance sheet.

2 Research Methodology

Before doing this research, author has done previous research. There are many papers that become references for this research. Because this research especially in Android, author only looking for papers only to develop an application or system for Android.

First paper has been done by Husain and friends [1]. They have said, nowadays, almost every company needs a system for attendance. System that can provide efficiency, effective, and fast. Development is done by utilizing Android devices owned by each employee. Then, using local area network that is only within around the company. They are doing this because to avoid employee who is absent outside from company area.

Next research reference is conducted by Aini and friends [2]. They stated that application of attendance using QRCode can avoid fraud in the attendance process. Because the system can store data in the form of Student identity number, lecturer, class, and time schedule.

Further research references, according to Fitri Andini [3], in the research conducted has produced an Android-based online attendance system application. Students can do attendance by online, and it shows time when attendance is done, also lecturers can control students from their Android application.

Author also conducted literature review of research by Ronny et al. [4] in which they succeeded creating an attendance application system using GPS Lock from Android at PT. PLN APP Malang. With this application, employees can assist by doing attendance without going to attendance place that have considerable distance. Attendance can be made when they are in their work area. This system using GPS Lock Technology.

Taufik Ramadhan stated that using website and notice board for class schedule and information is less effective for students. Because this method not able to deliver news or information quickly and precisely to students. Therefore, researchers develop an application that is run on an Android device, with the aim of providing information about lectures that will be given directly to students through their respective Android devices [5].

In collecting data, the writer goes directly to the relevant agency and conducts an interview, and makes direct observations. The data will be used as a basis for developing this application. For example, author will see in a month, how many cases of errors, how big the impact, and what are losses to be received both students and lecturers.

3 Result and Discussion

In this section, author will explain the needs of the system designed. Based on the analysis produced, continued by design consisting of application design, database, and system interface design.

3.1 Application Design

The application design on Android-based online attendance system involves six features that will be used by both lecturers and students. The features involved are login, course, course schedule, lecturer course schedule, lecturer schedule details, and student attendance. Design of the application can be seen in Fig. 1.

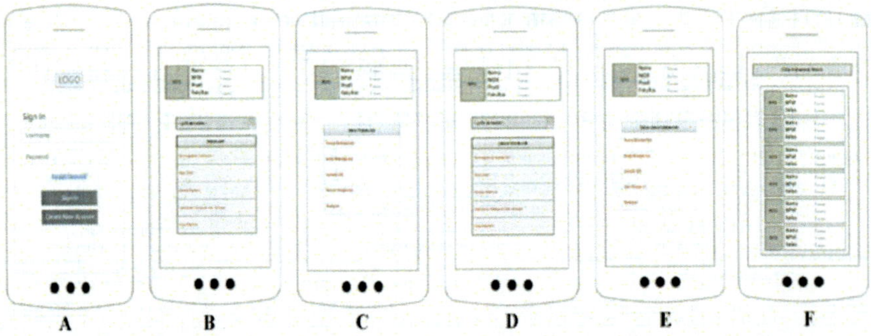

Fig 1. A. Login, B. Course, C. Course Schedule, D. Lecturer Course Schedule, E. Lecturer Schedule Details, F. Student Attendance

3.2 *Entity Relationship Diagram*

Database design is done according to the system requirements needed. It uses five tables, which consist of lecturer, course, schedule, classroom, and student. The design of database can be seen in Fig. 2.

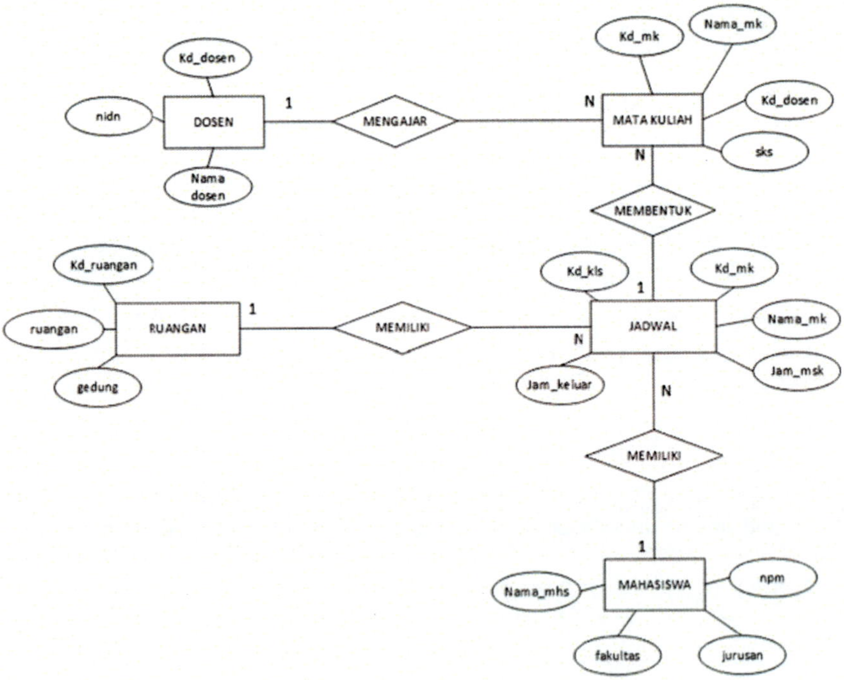

Fig 2. Entity relationship diagram for database

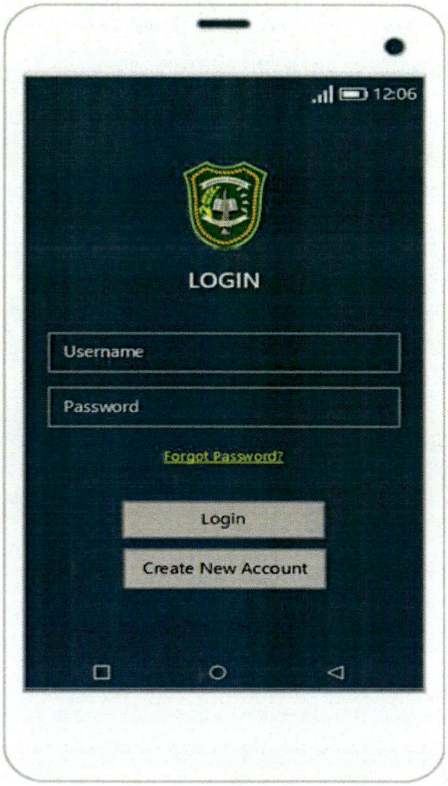

Fig 3. Login page

3.3 *Interface Design*

Next, the design of interface is done according to the design of the application. Application will be used on Android devices; therefore, design of interface will be used on Android device. Consist of login, course, course schedule, lecturer course schedule, lecturer schedule details, student attendance. It can be seen in Figs. 3, 4, 5, 6, 7, and 8.

3.4 *Application Test*

Testing is the most important thing that aims to find errors in the system before the system is implemented. The test intends to ascertain whether the system that has been created is running as expected or not. Testing technique used is black box testing

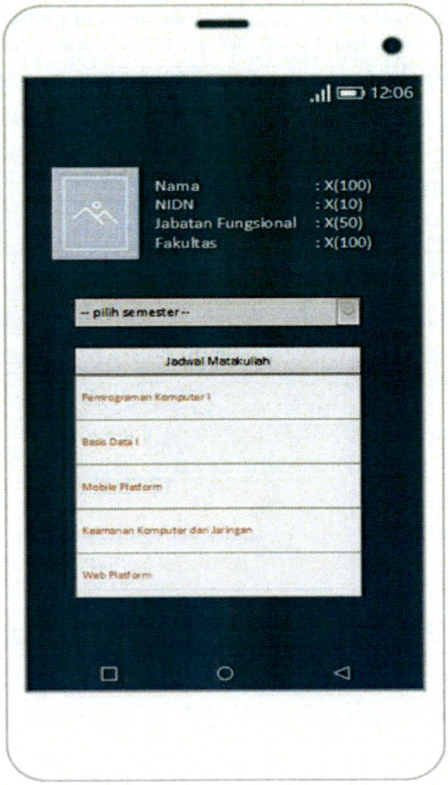

Fig 4. Course page

method, which is focused on the function of the application that has been made. The description of testing through black box can be seen in Table 1.

4 Conclusion

Based on research that has been done regarding this Android-based online attendance application, the following conclusions can be made:

- Android-based online attendance application was built to provide convenience in carrying attendance activities between students and lecturers in class according to predetermined schedule.
- This application can help lecturer in making student attendance reports and report to university.
- This application can reduce fraud committed during attendance activities.

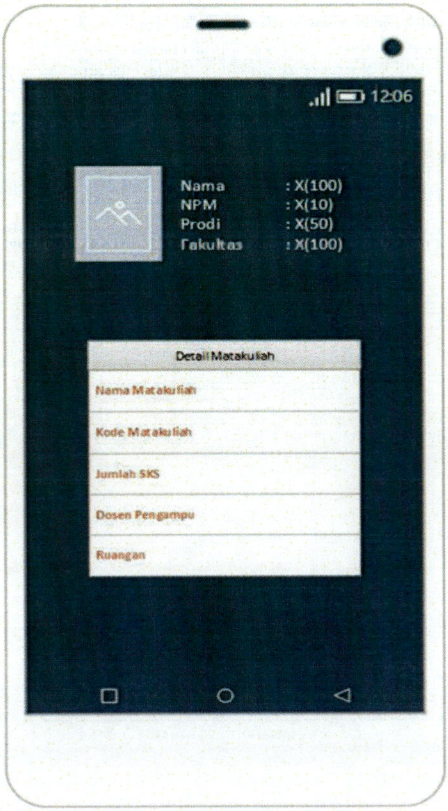

Fig 5. Course schedule

This research also not free from shortcomings. Therefore, for the development of a better system, it is necessary to pay attention for several things, including the following:

- It is expected that researchers will further add features for Android-based attendance application to further assist attendance activities between students and lecturers, such as by adding how far the lesson is explained.
- The appearance of the application can be developed further to get more attractive appearance.
- Can be used as a reference for comparison with other concepts in similar research or reference material as a research reference.
- It is expected for the development of further research, there will be a warning feature of the entry lecture time, and it will be developed in the exam activities that take place at the university, whether the Midterm Examination or the End of Semester Examination.

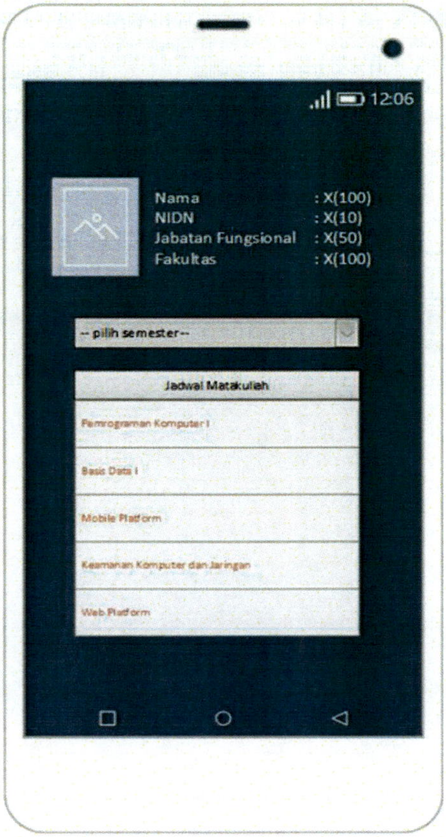

Fig 6. Lecturer schedule

Fig 7. Schedule details

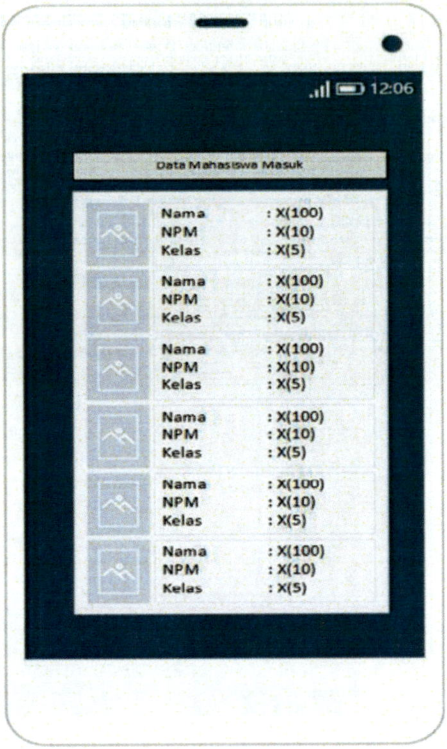

Fig 8. Student attendance

Table 1. Description of application test through black box

No.	Test class	Details	Result
1	Login	Verification of username and password	Success
2	Student course schedule	Verification of course data taken by students	Success
3	Student course detail	Verification of course data detail taken by students	Success
4	Lecturer course schedule	Verification of course data that is taught by lecturers	Success
5	Detailed lecturer course schedule	Verification course data detail that is taught by lecturers	Success
6	Student attendance	Verification of student attendance	Success

References

1. A. Husain, A.H.A. Prastian, A. Ramadhan, Perancangan Sistem Absensi Online Menggunakan Android Guna Mempercepat Proses Kehadiran Karyawan Pada PT. Sintech Berkah Abadi. Technomedia J. **2**(1), 105–116 (2017). https://doi.org/10.33050/tmj.v2i1.319
2. Q. Aini, Y.I. Graha, S.R. Zuliana, Penerapan Absensi QRCode Mahasiswa Bimbingan Belajar pada Website berbasis YII Framework. Sisfotenika **7**(2), 207 (2017). https://doi.org/10.30700/jst.v7i2.145
3. R.A. Fitri Andini, M. Irzal, Perancangan dan implementasi sistem absensi online berbasis android di lingkungan Universitas Negeri Jakarta Anantassa Fitri Andini, Med Irzal, Ria Arafiyah Program Studi Ilmu Komputer, FMIPA UNJ. Sist. Inf. **1**(1), 1–10 (2017)
4. R.A. Makhfuddin, N. Prabowo, Aplikasi Absensi Menggunakan Metode Lock GPS dengan Android di PLN APP Malang Basecamp Mojokerto. Issn **5**(2), 55–63 (2015)
5. T. Ramadhan, V.G. Utomo, Rancang Bangun Aplikasi Mobile Untuk. J. Teknol. Inf. dan Komunikasi **5**, 47–55 (2014). https://doi.org/10.1234/JTIK.V5I2.93

Customer Sentiment Analysis Using Cloud App and Machine Learning Model

P. Manjula, Neeraj Kumar, and Ahmed A. Al-Absi

Abstract The customer sentiments are very important to any business, as positive or negative feedback can affect the sales and adoption of the product in the market and subsequently define the product's success. The monthly active usage of major social media platform such as Facebook is 2.32 billion monthly active users (MAU) and of Twitter is 126 million; hence, the market for understanding the customer sentiment through social media can be a game changer for a company and can help define the success of the company in the future. If the sentiments of the users are not captured correctly, it could lead to catastrophic failure of the product and hamper company's reputation. Existing systems require a lot of manual tasks such as customer surveys, aggregating the sentiments then generating excel reports which are not very interactive and require a lot of time to gather results. These reports also do not show real-time data. People express their opinion on social media. Companies can use such platforms to capture honest and transparent opinions of the consumers. The cognitive service evaluates tweet texts and returns a sentiment score for each text, ranging from 0 (negative) to 1 (positive). This capability is useful for detecting positive and negative sentiments in social media such as Facebook, Twitter, customer reviews, and discussion forums. The machine learning model used by the cognitive service helps determine sentiment using data provided by the user. This feedback can allow the company to know the acceptance of the product in prototype stages and can use the same to modify the product as per the customer feedback before making the product generally available. The implementation environment uses Azure services

P. Manjula (✉)
SRM University, Sonepat, India
e-mail: gpmanjula28@gmail.com

N. Kumar
VIT University, Vellore, India
e-mail: neerajsinghofficial@gmail.com

A. A. Al-Absi
Global Academic Collaboration Centers, Kyungdong University-Global Campus, Wonju-si, South Korea
e-mail: absiahmed@kduniv.ac.kr

P. K. Pattnaik et al. (eds.), *Proceedings of International Conference on Smart Computing and Cyber Security*, Lecture Notes in Networks and Systems 149,
https://doi.org/10.1007/978-981-15-7990-5_32

(Logic apps, Cognitive Services, SQL Database, App Services) with Power BI used to generate real-time business intelligent reports to capture customer sentiment, and a Windows 10 workstation can be used to access all these services.

Keywords Customer sentiments · Machine learning model · Azure services · Business intelligent

1 Introduction

1.1 Background

The customer sentiments are very important to any business, as positive or negative feedback can affect the sales and adoption of the product in the market and subsequently define the product's success [1]. The need for sentiment analysis arises from the market research needed by the organization for launching the product at mass scale, if the product has a positive outlook at various platforms, the likelihood of products success is increased tenfold, likewise negative sentiment can help organization to either improve or discard product and focus on products that are more likely to succeed [2].

1.2 Problem Statement

The problem statement can be understood by looking at the monthly active usage of major social media platform such as Facebook is 2.32 billion monthly active users (MAU) and of Twitter is 126 million [3]; hence, the market for understanding the customer sentiment can be a game changer for a company and can help define the success of the company in the future. People express their opinion on social media. Companies can use these platforms to capture honest opinions of its customers [4, 5].

1.3 Motivation

For the project arises by looking at the everyday business problem of solving complex business intelligence tasks that shape the future of companies using top of the class cutting edge cloud technologies that analyze user sentiments and help business shape its future decisions.

1.4 Existing Work

The Study by Fang and Zhan [6] related to sentiment analysis using product review data of amazon to calculate the polarity categorization of products is a past study referenced to model the architecture of current application. It is great to capture the product reviews on Amazon but can fail if working with other platforms. Fake reviews and spammers can skew and affect the correct results. The system present before the emergence of sentiment analysis involves more manual work. This further leads to consumption of time and human stress. These drawbacks are eliminated in cloud sentiment analysis.

1.4.1 Drawbacks

- Maintenance costs high.
- Need more man power.
- Publicity cannot be assured.
- The existing infrastructure allows people to post their opinions and the honesty of opinion cannot be guaranteed [6].
- Fraudulent opinions with spams that do not relate to the product can be posted.
- Fake opinions might be posted which may lead to skew in the expected results.
- Sarcasm can lead to cognitive service not determine correct outcome of sentiment.

1.5 Challenges

The company requires adequate data from a diverse demographic to correctly capture the sentiment and this can be a challenge, as the sentiments will only be posted by the people if they have used the product or are aware about its functioning, marketing of the product can be a challenge for the company.

1.6 Essence of Approach

The product will use Logic app with Twitter connector to pull tweets from Twitter and fill the SQL DB table and the table will be used by Web App to transfer data to cognitive services text analysis API to evaluate text input and returns a sentiment score for each document, ranging from 0 (negative) to 1 (positive) which is useful for detecting positive and negative sentiments in social media such as Facebook, Twitter, customer reviews, and discussion forums. The machine learning model used by the cognitive service helps determine sentiment using data provided by the user.

1.7 Assumptions

The project assumes that the appropriate permissions of various cloud services is available and a working pay as you go account has been created. Also, adequate data (tweets from Twitter) is available to analyze the product sentiment. RDP requirements are also met to get remote of VM's required in the project. Also, a windows 10 client machine is available to access azure services.

1.8 Aim

The aim of the project is to analyze sentiments from Twitter and meet the objectives of classifying sentiments based on cognitive services API between 0 (Negative) to 1 (Positive) and help organization get business intelligence reports based on the sentiment.

2 Literature Review

Fang and Zhan [6] described that sentiment analysis or opinion mining is a field of study that analyzes people's sentiments, attitudes, or emotions toward certain entities. This paper tackles a fundamental problem of sentiment analysis, sentiment polarity categorization. Online product reviews from Amazon.com are selected as data used for this study. A sentiment polarity categorization process has been proposed along with detailed descriptions of each step. Experiments for both sentence-level categorization and review-level categorization have been performed.

Shayaa et al. [7] reviewed that the datasets used for OMSA, data from Twitter seemed to dominate the data sets. This is further aligned with the growth of the sentiment analysis in which most of the data is captured from social media more profoundly on Twitter as compared to the other data sources. Previously, Twitter has been used as a tool for disseminating and propagating information rather than simply a social networking site. Previous research shows that top users, as measured by the number of followers on Twitter, are mostly celebrities and those who attract the keen interest of the mass media. An electronic platform for word-of-mouth influence in marketing, Twitter also serves as a political sentiment analysis predictor of elections and as a stock market movement predictor. This has widened its exploitation to politicians and other pundits. It can be used to quickly share information with people, promote new products, and communicate with celebrities' fans or political supporters.er: IEEE.

Xu et al. [8] described that the era of rapid development of Internet technology and social networks, it is very meaningful to explore the emotional tendency of comments through artificial intelligence technology. In this paper, a sentiment analysis method of comments based on BiLSTM is proposed and applied to the comment sentiment analysis task. According to the deficiency of the word representation method in the current researches, the sentiment information contribution degree is integrated into the TF-IDF algorithm of the term weight computation, and a new representation method of word vector based on the improved term weight computation is proposed. In addition, BiLSTM model fully considers the context information and can better obtain the text representation of the comments. Finally, through the feedforward neural network and softmax mapping, the sentiment tendency of the text is obtained. The experiments of different word representation methods prove the validity of the proposed word representation method in this paper. Through the comparison experiments with other traditional sentiment analysis methods, the accuracy of the proposed comment sentiment analysis method is improved.

Zhou et al. [9] reviewed that sentiment analysis on Chinese microblogs has received extensive attention recently. Most previous studies focus on identifying sentiment orientation by encoding as many word properties as possible while they fail to consider contextual features (e.g., the long-range dependencies of words), which are, however, essentially important in the sentiment analysis. In this paper, we propose a Chinese sentiment analysis method by incorporating a word2vec model and a stacked bidirectional long short-term memory (Stacked Bi-LSTM) model. We first employ the word2vec model to capture semantic features of words and transfer words into high-dimensional word vectors. We evaluate the performance of two typical word2vec models: continuous bag-of-words (CBOW) and skip-gram. We then use the Stacked Bi-LSTM model to conduct the feature extraction of sequential word vectors. We next apply a binary softmax classifier to predict the sentiment orientation by using semantic and contextual features. Moreover, we also conduct extensive experiments on the real dataset collected from Weibo (i.e., one of the most popular Chinese microblogs). The experimental results show that our proposed approach achieves better performance than other machine learning models.

Cambria [10] explained that understanding emotions is an important aspect of personal development and growth, and as such it is a key tile for the emulation of human intelligence. Besides being important for the advancement of AI, emotion processing is also important for the closely related task of polarity detection. The opportunity to automatically capture the general public's sentiments about social events, political movements, marketing campaigns, and product preferences has raised interest in both the scientific community, for the exciting open challenges, and the business world, for the remarkable fallouts in marketing and financial market prediction. This has led to the emerging fields of affective computing and sentiment analysis, which leverage human–computer interaction, information retrieval, and multimodal signal processing for distilling people's sentiments from the ever-growing amount of online social data.

Park and Seo [11] providing an enhancing experience is one of the most significant current issues in the user's research. A process that improves user's experience should be required to evaluate the usability and emotion. Above all, sentiment analysis based on user's opinions can be used to understand user's tendency. This paper aims to make a criterion what artificial intelligence assistant is statistically better. User's opinions about three artificial intelligence assistants from Twitter were collected and classified into positive, negative, neutral opinions by a lexicon named Valence Aware Dictionary and Entiment Reasoner (VADER). Also, we analyzed tweets through independent samples T-test, Kruskal-Wallis test, and Mann-Whitney test to show the statistical significance among groups. The results suggested the highest rank of three artificial intelligence assistants by using statistical analysis.

2.1 Gaps in Available Research

- The existing infrastructure allows people to post their opinions and the honesty of opinion cannot be guaranteed [6].
- Fraudulent opinions with spams that do not relate to the product can be posted [12, 13].
- Fake opinions might be posted which may lead to skew in the expected results [14, 15].
- Sarcasm can lead to cognitive service not determine correct outcome of sentiment [16, 17].

3 Materials and Methods

In this section, we give the materials and methods used in customer sentiment analysis using cloud application. Architecture of the proposed method is illustrated in Fig. 1.

Tweets from different locations, posted by different users on Twitter are pulled by the Logic app using Twitter connector, and this connector performs the workflow integration of connection to SQL DB; the data is inserted into the table populating different columns and rows except the score column. The cognitive service uses text analytics API with natural language processing to process the sentiment from the tweets by using the connection from Web application that connects SQL DB and the cognitive service and populates the sentiment column with a score between 0 and 1, with a score toward 0 showing a negative sentiment and a score toward showing a positive sentiment. The Power BI Desktop client can now be used to import or direct query SQL DB to generate reports as per the customer requirements.

4 Implementation of the System/Methodology

1. Create a new Resource Group (RG).
2. Create a new SQL database in the resource group.
3. Click create to complete the SQL DB creation.
4. Run the query to create a table in SQL DB.
5. Create a new cognitive service.

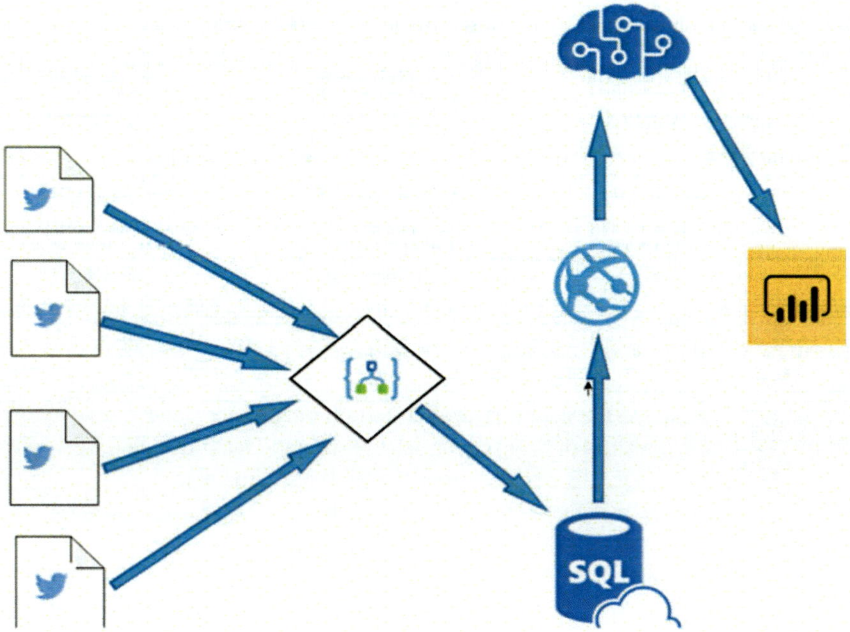

Fig. 1 System architecture

6. Create a new Logic app.
7. From Logic app designer, create a new Blank Logic app.
8. Search a Trigger "when a new tweet is posted".
9. Login to Twitter for the connector to work.
10. Select a search text and set the interval and frequency as per required.
11. Here, the search text chosen is "#iPhone" and interval (3) and frequency (Minutes) were chosen.
12. In Choose an action search "SQL server".
13. Select Insert row.
14. Choose connection name of row, select the SQL Database, username and password of SQL DB.
15. In Insert row, select appropriate row.
16. Search and select appropriate rows that need to be inserted in the table.
17. Save and Run the Logic app.
18. Post positive and negative tweets on Twitter, so that appropriate table rows can be populated.
19. Login to SQL DB created earlier
20. Run Select *From Tweet From Twitter table, we will see all the rows and columns populated as data is pulled from Twitter.
21. Run visual studio 2017 and create a new Project, .net core console app.
22. Configure the new project.
23. To connect SQL server with webapp and cognitive services together.

Table 1 Network capacity and utilization metric

Timestamp	Dimension "IP"	Dimension "Direction"	Metric value (Kbps)
1/3/2019 8:14	IP = "192.168.5.2"	Direction = "Send"	646.5
1/3/2019 8:14	IP = "192.168.5.2"	Direction = "Receive"	420.1

24. Run the query SELECT Top (1000) *from TweetFromTwitter, we can see the sentiment score given by the cognitive service.
25. Now, we can optionally create a Web Job to publish our web app we created.
26. Publish the application and select existing publishing target.
27. Select appropriate Web app created earlier.
28. Click Publish.
29. We can see the appropriate Web Job running in the cloud.
30. From power BI desktop client, select get data > Database > SQL server Database.
31. Import SQL DB or direct query to fetch real-time data.
32. Enter credentials to connect to SQL DB.
33. Choose appropriate Table, click Load.
34. Create appropriate reports as per the requirement.
35. Similarly, we can create more reports like AWS versus Azure Sentiment country wise.

5 Results and Discussions

5.1 Performance Metrics

The various metrics maintained at Azure Datacentre help us in keeping the application performance as expected

- Azure Availability and Response Rate—Most azure services are available 99.9% of the time and have a quick response rate.
- Network capacity and utilization metrics can be measured in the metrics section of each service and network utilization is measured as in Table 1.
- Storage capacity and scalability—Depends on the SKU chosen, higher configuration SKUs have better storage capacity and scalability.
- Azure recovery—The RTO and RPO depend on the design of the infrastructure, a more robust infrastructure will have low RTO (15 min) and low RPO (5 min).

5.2 Parameter Under Study

- The parameter under study is the sentiment of the people tweeting about the product or elaborating their reviews, cognitive service API evaluates text input

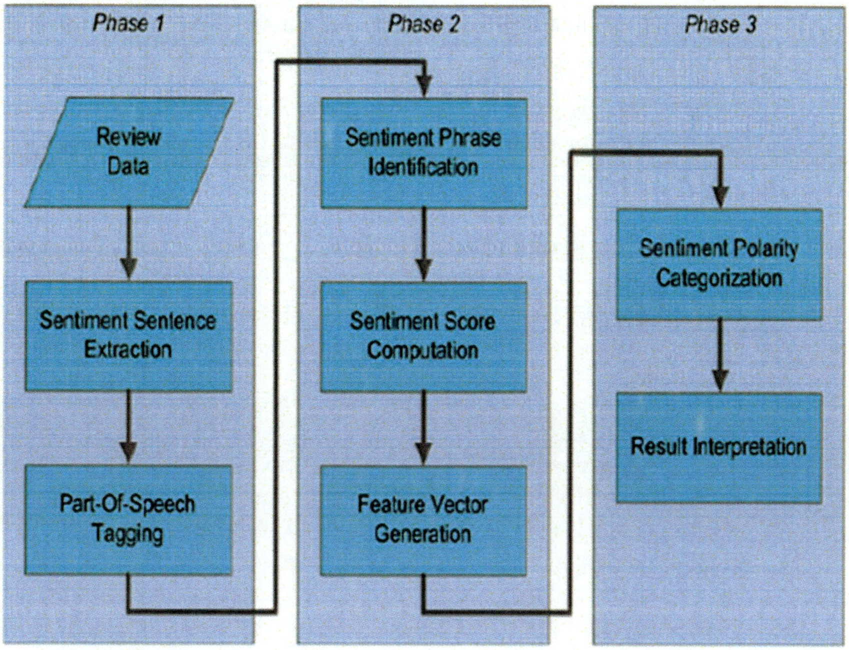

Fig. 2 Sentiment polarity categorization process [6]

and returns a sentiment score for each document, ranging from 0 to 1 which is useful for detecting positive and negative sentiment in social media such as Facebook, Twitter, customer reviews, and discussion forums.

- The machine learning model used by the cognitive service helps determine sentiment using data provided by the user.

5.3 Design Comparisons

The design used by Xing Fang and Justin Zhan, "Sentiment analysis using product review data [6], is divided into three phases, whereas the design used to detect sentiment using azure has various services integrated with each other as shown in Fig. 2.

5.4 Expected and Obtained Results

The expected results were to see the sentiment score for #iPhone product on the custom power BI Dashboard based on location to get the positive and negative reviews of the product and score them using cognitive api, and we were successfully able to capture the sentiment on the dashboard which is shown in Fig. 3.

- Similarly, we created more custom reports like AWS versus Azure sentiment country wise with a map depicting presence of CSP's around the world (Fig. 4).

6 Conclusion and Future Work

6.1 Conclusion

In this project, we used Logic app with Twitter connector to pull tweets from Twitter and fill the SQL DB table and the table was used by Web app to transfer data to cognitive services text analysis API to evaluate text input and return a sentiment score for each document, ranging from 0 (negative) to 1 (positive). In conclusion, it helped in detecting positive and negative sentiments on Twitter; the machine learning model used by the Cognitive service helps determine sentiment using data provided by the user. This provides the organizations huge datasets to use business intelligence and reduce cost on surveys and other traditional methods which are time consuming and are very costly with unreliable results.

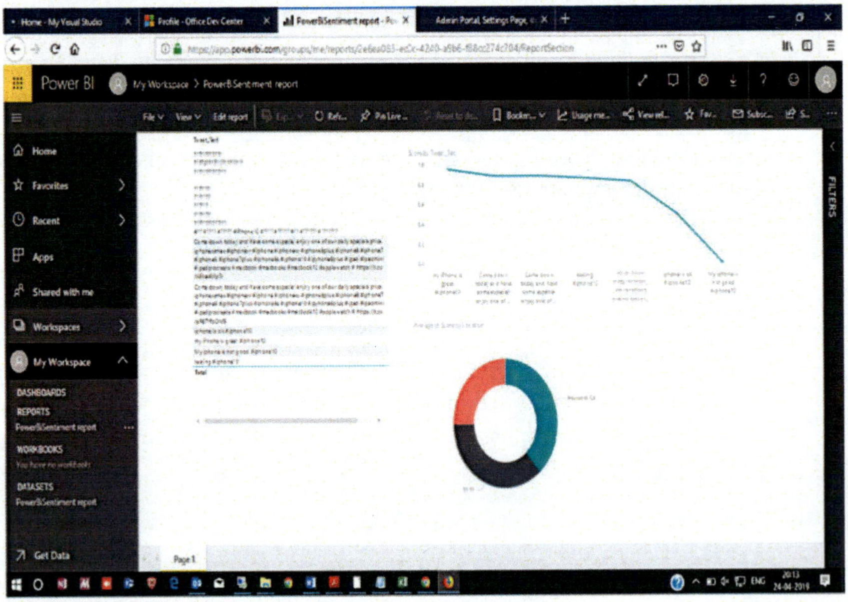

Fig. 3 Create appropriate reports as per the requirement

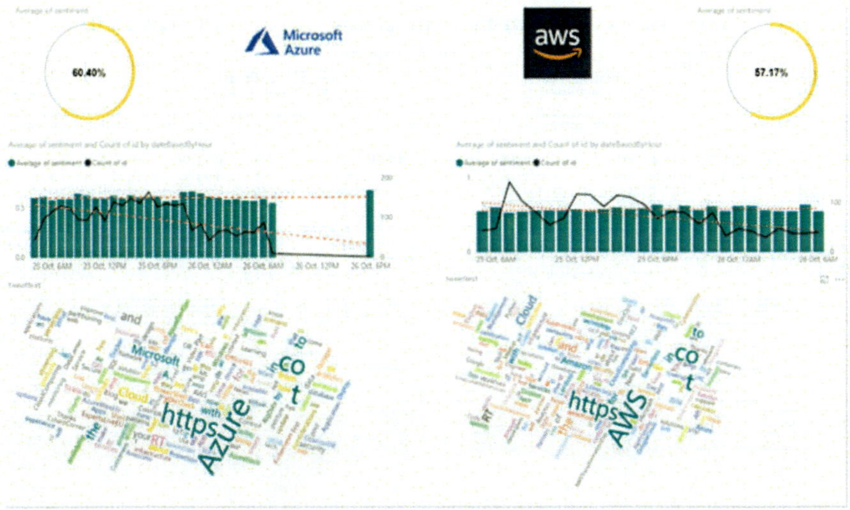

Fig. 4 Create more reports like AWS versus azure sentiment country wise

6.2 *Future Work*

- Tackling fake reviews and opinions by tagging each tweet or comment with correct training model will ensure only relevant training data is captured.
- This application can be used to capture sentiment on other social media as well such as Facebook, customer reviews, and discussion forums.
- The model might not understand sarcasm and people's emotions which might lead to incorrect capturing of sentiment.
- The sentiment analysis method of comments based on BiLSTM consumes a long time in the training model. In the future work, the method to effectively accelerate the training process of the model will be studied [8].
- Techniques need to be developed using ML studio to create robust models that can capture correct sentiment and robust processed data should help tackle fake reviews and opinions and prevent spams from skewing data.

References

1. N. Jindal, B. Liu, Opinion spam and analysis, in *Proceedings of the 2008 International Conference on, Web Search and Data Mining, WSDM '08* (ACM, New York, 2008), pp. 219–230
2. A. Mukherjee, B. Liu, N. Glance, Spotting fake reviewer groups in consumer reviews, in *Proceedings of the 21st, International Conference on World Wide Web, WWW '12* (ACM, New York, 2012), pp. 191–200
3. B. Pang, L. Lee, A sentimental education: sentiment analysis using subjectivity summarization based on minimum cuts, in *Proceedings of the 42nd Annual Meeting on Association for*

Computational Linguistics, ACL '04 (Association for Computational Linguistics, Stroudsburg, 2004)
4. P.D. Turney, Thumbs up or thumbs down? Semantic orientation applied to unsupervised classification of reviews, in *Proceedings of the 40th Annual Meeting on Association for Computational Linguistics, ACL '02* (Association for Computational Linguistics, Stroudsburg, 2002), pp. 417–424
5. C. Whitelaw, N. Garg, S. Argamon, Using appraisal groups for sentiment analysis, in *Proceedings of the 14th ACM International Conference on Information and Knowledge Management, CIKM '05* (ACM, New York, 2005), pp. 625–631
6. X. Fang, J. Zhan, Sentiment analysis using product review data. J. Big Data **2**, 5 (2015)
7. S. Shayaa, N.I. Jaafar, S. Bahri, A. Sulaiman, P.S. Wai, Y.W. Chung, A.Z. Piprani, M.A. Al-Garadi, Sentiment analysis of big data: methods, applications, and open challenges. IEEE Access **6** (2018)
8. G. Xu, Y. Meng, X. Qiu, Z. Yu, X. Wu, Sentiment analysis of comment texts based on BiLSTM. IEEE Access **7** (2019)
9. J. Zhou, Y. Lu, H.-N. Dai, H. Wang, H. Xiao, Sentiment analysis of Chinese microblog based on stacked bidirectional LSTM. IEEE Access **7** (2019)
10. E. Cambria, Affective computing and sentiment analysis. IEEE Intell. Syst. **31** (2016)
11. C.W. Park, D.R. Seo, Sentiment analysis of Twitter corpus related to artificial intelligence assistants, in *2018 5th International Conference on Industrial Engineering and Applications (ICIEA)*. IEEE
12. C. Clavel, Z. Callejas, Sentiment analysis: from opinion mining to human-agent interaction. IEEE Trans. Affect. Comput. **7**(1) (2016)
13. S.-M. Kim, E. Hovy, Determining the sentiment of opinions, in *Proceedings of the 20th International Conference on Computational Linguistics*. Association for Computational Linguistics, Stroudsburg, PA, USA (2004), p. 1367
14. B. Liu, Sentiment analysis and subjectivity, in *Handbook of Natural Language Processing*, 2nd edn. (Taylor and Francis Group, Boca Raton, 2010)
15. A. Pak, P. Paroubek, Twitter as a corpus for sentiment analysis and opinion mining, in *Proceedings of the Seventh Conference on International Language Resources and Evaluation* (European Languages Resources Association, Valletta, 2010)
16. A. Go, R. Bhayani, L. Huang, Twitter sentiment classification using distant supervision, 1–12. CS224N Project Report, Stanford (2009)
17. B. Liu, M. Hu, J. Cheng, Opinion observer: analyzing and comparing opinions on the web, in *Proceedings of the 14th International Conference on World Wide Web, WWW '05* (ACM, New York, 2005), pp. 342–351

Mood Enhancer Based on Facial Expression Using Machine Learning and Virtual Assistant Technology—An Android App

P. Manjula, Akshay Nagpal, and Ahmed A. Al-Absi

Abstract Ansel Adams once said, "When words become unclear, I shall focus on the photographs." The images hold more power of expression than words, and thus human face conveys more meaning with expression. Face expressions play an inevitable role in verbal as well as non-verbal communication. Based on same philosophy, the IT industry is also showing keen interest in developing the technology to better understand the human expressions or human emotions. Big IT giants, like Google Inc., are spending a huge portion of their budget and manpower to develop the face expression recognition (FER) systems. However, the idea of developing this area for behavioral studies and in medical rehabilitation is still a challenge. Since every face holds a different story, studying the same using existing technology and understanding toward machine learning is a vast area to develop. Thus, standardizing the facial expression understanding for the computers becomes the major issue in the current environment. 'AI-Based Mood Enhancer—An Android App' is an attempt to make use of machine learning and virtual assistant technology, and making the computer systems ready for better understandings on facial expressions. This would be one of its kind applications that will communicate with its users, based on the user's facial expressions. The application, which will run on the Android mobile devices, will capture the video stream of the user's facial expressions using the device front camera. The inputs will then be processed using the machine learning algorithms of the open-source libraries like Weka, OpenCV, Dlib, and TensorFlow Lite. The unique feature of this application is that, based on the evaluated expression classification,

P. Manjula
SRM University, Sonepat, India
e-mail: gpmanjula28@gmail.com

A. Nagpal
VIT University, Vellore, India
e-mail: akshaynagpal94@gmail.com

A. A. Al-Absi (✉)
Global Academic Collaboration Centers, Kyungdong University-Global Campus, Wonju-si, South Korea
e-mail: absiahmed@kduniv.ac.kr

P. K. Pattnaik et al. (eds.), *Proceedings of International Conference on Smart Computing and Cyber Security*, Lecture Notes in Networks and Systems 149,
https://doi.org/10.1007/978-981-15-7990-5_33

the application will attempt to share the dynamic response to better and enhance the user's mood, hence named as '**The Mood Enhancer**.'

Keywords Mood enhancer · Android application · Face expression recognition · Machine learning · Virtual assistant technology

1 Introduction

Facial expressions are important in facilitating human communication and interactions, as they help us understand the intentions of others. Also, they are used as a valuable tool in behavioral studies and in medical rehabilitation. In general, people infer the emotional states of other people, such as joy, sadness, and anger, using facial expressions and vocal tone. Facial image-based mood detection techniques may provide a fast and practical approach for non-invasive mood detection. According to different surveys, verbal components convey one-third of human communication, and non-verbal components convey two-thirds. Among several non-verbal components, by carrying emotional meaning, facial expressions are one of the main information channels in interpersonal communication. Therefore, it is natural that research of facial emotion has been gaining lot of attention over the past decades with applications not only in the perceptual and cognitive sciences, but also in affective computing and computer animations.

Facial expressions and related changes in facial patterns give us information about the emotional state of the person and help to regulate conversations with the person. Moreover, these expressions help in understanding the overall mood of the person in a better way. Facial expressions play a significant role in human interactions and non-verbal communication. Their analysis deals with visually recognizing and analyzing different facial motions and facial feature changes.

The purpose of the present project is to develop an intelligent system that address these domains for facial image-based expression classification using input as images or video stream, and upon understanding the nuances, share dynamic responses with intend to better the user's mood. This is one of its kind approaches to use machine learning and face recognition to understand the user's mood and based on the same, respond with feedback to better the user's present state of emotion.

An average human today spends more time with the technology than any other physical being, where mobile devices account for the largest share, with Android being the major market holder in mobile OS. Being engaged so much into the virtual environment, one tends to experience gaps with the social bonds. Thus, keeping these environment factors in mind, this application is an attempt to bring the users one step closer to positivity and happy life, in the same virtual world. We hold the vision to create a self-learning AI program that will learn the user's behavior with time, to adapt in changing environments.

In the domain of artificial intelligence, face recognition started to gain interest of many engineers back in 1960s. One of the first researches on this subject was

Woodrow W. Bledsoe. In 1960, Bledsoe, along other researches, started Panoramic Research, Inc., in Palo Alto, California. The majority of the work done by this company involved AI-related contracts from the US Department of Defense and various intelligence agencies [1]. Today, developing the correct and accurate techniques for facial expression recognition using these methods is gaining pace. A lot of work is ongoing in this direction. There are many projects and researches in the market on this domain that uses deep learning methods with Python language to formulate a computer application for face expression recognition (FER). On portable platforms like, Android OS, the industry is blooming to establish the capabilities of user expression recognition that can contribute to enhance user experiences on various industry applications like sales, stocks, gaming, e-learning, etc.

1.1 How Emotion Recognition APIs Work?

Emotive analytics is an interesting blend of psychology and technology. Though arguably reductive, many facial expression detection tools lump human emotion into seven main categories: joy, sadness, anger, fear, surprise, contempt, and disgust. With facial emotion detection, algorithms detect faces within a photograph or video and sense microexpressions by analyzing the relationship between points on the face, based on curated databases compiled in academic environments.

1.2 Basic Terminologies

1.2.1 Face Detection

Face detection is to determine that a certain picture contains a face we need to be able to define the general structure of face. Luckily human faces do not greatly differ from each other; we all have noses, eyes, foreheads, chins and mouths; these compose the general structure of a face.

1.2.2 Face Identification

In this, the system compares the given individual to all the other individuals in the database and gives a ranked list of matches.

1.2.3 Facial Expression

Facial expression is one or more motions or positions of the muscles beneath the skin of the face. These movements express the emotional state of the person to

Fig. 1 Steps for recognizing emotions

observers. It is a form of non-verbal communication. It plays a communicative role in interpersonal relations. The common ones are in Fig. 1.

Further variability is introduced because the expression databases are classified only into seven basic facial expression types (contempt, angry, disgust, fear, happy, sad, and surprised). In reality, an expression is often a combination of two or more of the prototypic expressions, also, expressions are assumed to be singular and to begin and end with a neutral position. In reality, facial expressions are much more complex and occur in different combinations and intensities. Therefore, an identified expression could be a combination of two different expressions with one of them being more dominant in intensity. The classifier, therefore, should be smart enough to correctly identify the combination of expressions and each expression's individual intensity.

The idea of this project comes from the busy routine of a common man, which in the modern world have lost the true essence of happiness and personal being. Thus, keeping the same in mind, this project is an attempt to collaborate the domains of artificial intelligence and virtual assistant, to give the user a personal experience of entertainment. On the same idea, we call this application as 'AI-Based Mood Enhancer—Android App,' a virtual friend.

2 Literature Survey

Over the last ten years or so, face recognition has become a popular area of research in computer vision and one of the most successful applications of image analysis and understanding. Because of the nature of the problem, not only computer science researchers are interested in it, but neuroscientists and psychologists also. It is the general opinion that advances in computer vision research will provide useful insights to neuroscientists and psychologists into how human brain works, and vice versa.

Face recognition systems use computer algorithms to pick out specific, distinctive details about a person's face. These details, such as distance between the eyes or shape of the chin, are then converted into a mathematical representation and compared to data on other faces collected in a face recognition database. The data about a particular face is often called a face template and is distinct from a photograph because it is designed to only include certain details that can be used to distinguish one face from another.

2.1 Image Processing and Feature Extraction

In [2], two types of parameters are extracted from the facial images of 97 subjects: (1) real valued parameters and (2) binary parameters.

2.1.1 Real Valued Parameters

- Eyebrow raise distance—The distance between the junction point of the upper and the lower eyelid and the lower central tip of the eyebrow.
- Upper eyelid to eyebrow distance—The distance between the upper eyelid and eyebrow surface.
- Inter-eyebrow distance—The distance between the lower central tips of both the eyebrows.
- Upper eyelid–lower eyelid distance—The distance between the upper eyelid and lower eyelid.
- Top lip thickness—The measure of the thickness of the top lip.
- Lower lip thickness—The measure of the thickness of the lower lip.
- Mouth width—The distance between the tips of the lip corner.
- Mouth opening—The distance between the lower surface of top lip and upper surface of lower lip.

2.1.2 Binary Parameters

- Upper teeth visible—Presence or absence of visibility of upper teeth.
- Lower teeth visible—Presence or absence of visibility of lower teeth.
- Forehead lines—Presence or absence of wrinkles in the upper part of the forehead.
- Eyebrow lines—Presence or absence of wrinkles in the region above the eyebrows.
- Nose lines—Presence or absence of wrinkles in the region between the eyebrows extending over the nose.
- Chin lines—Presence or absence of wrinkles or lines on the chin region just below the lower lip.
- Nasolabial lines—Presence or absence of thick lines on both sides of the nose extending down to the upper lip.

3 Materials and Methods

In this section, we give the materials and methods used in AI-Based Mood Enhancer—An Android App. Architecture of the proposed method is illustrated in Fig. 2. Now let us try to understand the highlights of our project through some simple DFDs.

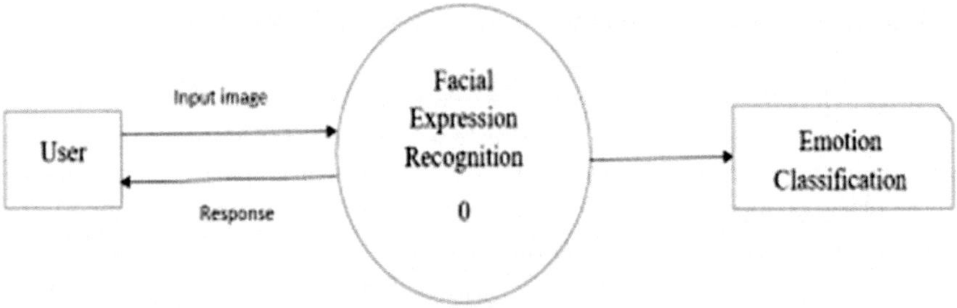

Fig. 2 Level 0 DFD: basic overview of emotion classification

The following steps were undertaken to develop this project:

- Analysis of the problem statement.
- Gathering of the requirement specification.
- Analyzation of the feasibility of the project.
- Development of a general layout.
- Going by the journals regarding the previous related works on this field.
- Choosing the method for developing the algorithm.
- Starting the development of the project.
- Installation of software like Android Studio.
- Incorporating the open-source libraries for OpenCV, Dlib, and TensorFlow Lite.
- Testing the output for bugs in Device Emulator.
- Testing final working of the application on the physical device using USB debugging.
- Repeating the above steps to solve the bugs and changes in the final product.

Thus, our idea is to make use of both of these technologies and incorporate them on a single platform and as a result produce an application that can provide a custom experience to the user and act as a virtual friend to enlighten his mood. Since, Android holds the largest share to the OS available to the portable hand-held devices; we have chosen to develop this application specifically for Android devices. This way, this application can reach out to maximum users. Let us try to better understand the salient features and specs of the application and proposed system. The 'AI-Based Mood Enhancer App' is purely developed in Android Studio IDE, using other open-source libraries, such as OpenCV, Dlib, and TensorFlow. The application can run on any android device having inbuilt front camera with Android OS version 5 (Android Lollipop) and above.

4 Implementation of the System/Methodology

Now that we have a foundation level understanding of in and out of the entire application, start from its functional working till its actual working experience; we are in good shape to begin out understanding toward the more specific and details aspects of the layouts as well as the background codes.

Please note, that in Android, each screen layout is designed and coded in XML format. So, we will first begin to understand the visuals followed by code snippets for the layout and backend functional tagging.

4.1 The Splash Screen

The app visuals begin with a short-lived welcome screen, named as the splash screen. This is the first and foremost screen visible to the user. It presents our brand logo as well as the graphics depicting the main idea of the application, the moods. We have given this layout a distinct black background to enhance its visual attraction. The layout was made in a full screen activity, to keep the distractions from the device notifications area away from the sight and allowing the user to focus completely on the screen design and animations which is shown in Figs. 3 and 4.

Fig. 3 Splash screen

Fig. 4 Home screen

4.2 The Home Screen

The next screen where the application lands is the main home screen, or named as the second activity in the project code. This will be the layout which holds all the available functionalities of the application. We can also call this pane as the deciding pane, which upon the user's click further decides which option to preview to the user. In other words, this pane is a connecting link for all the other layouts and panes of the application. Upon every other activity, the application returns to this very page. Thus, it was highly important to develop this layout with perfection and upmost attention. Based on the same idea, we have given this page a clean look without overloading it with unnecessary option or animations. All the quick access items were placed on the central screen, rest other options were masked into the menu options present on the toolbar (Fig. 5).

Fig. 5 Toolbar and its menu options

4.3 *The Toolbar and Its Menu Options*

The toolbar is the topmost rectangular pane, which hosts the application logo, application title along with the menu options. In order to give it a cleaner look, the option for 'How it Works?' is given a prominent look with an elegant icon, composed of multiple 'gear' embedded in the two-faced symbol, giving an overall look of the settings option. Since this can be handy every time a new user encounters the application, we have given this option to be always available attribute, no matter which device the application will run on (Fig. 6).

4.4 *The Camera Screen*

As described in the above section for the home screen, there is a floating button with an elegant camera icon, which tells the user to launch the camera screen and start checking the mood through a live camera stream. We have designed this layout very strategically, to place each option for the camera preview as well as the response panes

Fig. 6 Camera screen

in such a manner that the entire work space can be utilized without compromising with the minimum space required by each section. The camera preview is given the dimensions that irrespective of the target device screen size, the preview pane will not lose its aspect ratio. On every screen size, it should give the maximum layout size in the middle of the screen, leaving the top and the bottom of the screen available for the dynamic responses that the app will present post-predicting the user's emotion. For the camera live stream activity, we have given it a dark color layout, since this preview will be dynamic from the camera input. Thus, we have chosen the background of this layout and 'Jet Black.' The above pane will share a dynamic response to confirm the user the emotion that the application has predicted. And based on this predicted emotion, the below pane will present a dynamic response to the user. For example, for any given emotion among the seven classes, the app can share back either of the minimum of two dynamic response each. We have given both the dynamic response text fields a unique view in 'Silk White' color with the top pane having the 'Red' shadow glow option. The entire layout remains for every session; however, the responses are all dynamic and chosen directly at the runtime at random by the system.

4.5 The Gallery Selection Screen

The second button, with an image like icon, present on the bottom right of the screen, allows user to select a static photograph from the device default gallery and run the emotion detection algorithms over it. The app analyzes the image and share back the predicted emotion as well as the response to the user. Adjacent is a sample output screenshot which depicts the actual layout in the application. Contrastingly to the camera preview layout, this design was intended to be presented like a new nature. We tried to give this layout a different look, while maintaining the background feel same as the home screen. However, to make it different from the camera preview, we inverted the colors, by placing the chosen image in the center of the pane and allowing it to cover the maximum possible space without losing its aspect ratio. Moving further, the predicted emotion will be presented over the top of the screen and the dynamic response based on the predicted emotion will be shown at the bottom of the page (Fig. 7).

Fig. 7 Gallery section screen

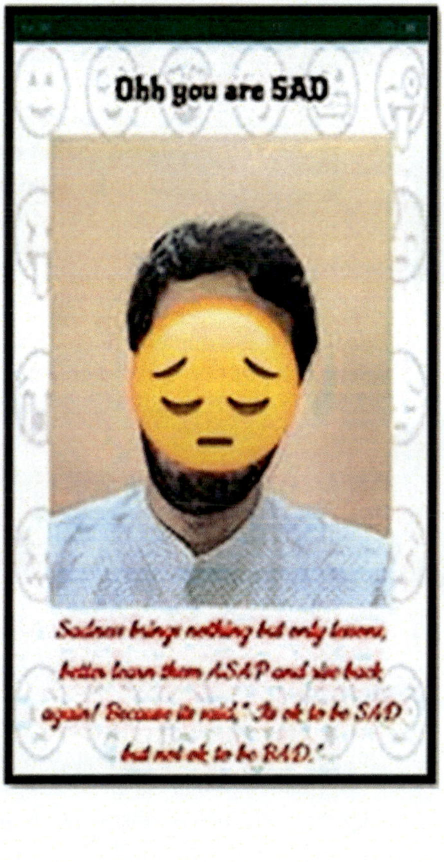

5 Results and Discussions

The application interface woks efficiently within the standard limits on different android devices. The tests were performed on the MI devices of screen size 4.0″ and 5.0″ and each gave the same experience.

Also, the performance testing was conducted, to check the consistent running of the application and it passed successfully. The application ran to and fro without crashing a single time in its test, however sometimes due to memory constraints it does land into 'out of memory' error. Under this situation, the user is expected to close the application and run it again.

There are various approaches that are undertaken to perform the task of emotion prediction, however the results vary drastically upon the size of the dataset over which the respective model is trained on. Under the scope of this project, we could not equip the application with sufficient time in training the model, which could further enhance the ability of the application to present the correct analysis. Still, to enable the application performs well in the standard conditions, we equipped the application with the pretrained models and libraries using which it shares the responses that are close to the expected results. There is also an achievement in developing this application model, that upon right prediction of the emotion we were successfully able to share back the correct responses that were chosen randomly and presented to the user. And this was possible in every single run without any failure.

The correctness of the output response varies on various factors. Let us try to have a quick look on some of the factors affecting the results. The first factor that we could determine was the amount of light the image carries or the amount of light that falls on the user in the live camera preview. Since our model relies on the facial landmarks to calculate the emotions probability and comparison with the test images, the amount of sufficient lightening plays a very crucial role in the computations and analysis. If the amount of light falling on the subject is not sufficient, it was observed that the result becomes fuzzy in majority of the outputs. Thus, the proper lightening conditions become the primary factor driving the effectiveness of the application's outcomes.

The second factor is of the orientation of the device placed in the live preview or the orientation of the clicked image in the gallery preview. For more accurate results while testing the camera preview, it was observed that the app requires the mobile device perpendicular to the subject's face. If the application catches the tilted sections of the images, the results vary drastically and hamper the overall experience of the user. Thus, it is always advisable to use the application or mainly the camera preview in upright conditions for more accurate results. Further, the third factor that controls the output of the application is the distance between the mobile device and the subject face. It was observed that as the subject face covers exactly the entire frame of the camera preview, the probability of predicting the correct emotion increases manifolds. In contrast, if the subject was quite at a distance from the camera, the app presents the responses that are sometimes incorrect and vary in every run. Based on these observations, we can say that the combinations of all these factors play a crucial role in determining the accuracy of the output. Thus, in order to obtain the right results, the user expected to take these prerequisites in consideration always.

Using the current project plan, we can equip the application with endless dynamic responses which allows the customization options never ending. We have made the application ready with minimum of two responses per emotion which directs the user every time toward an amusing mood, thereby enlightening the thoughts. All the responses were developed readily and can further be enhanced and increased anytime based on the requirements. Even the responses in the present system are chosen in laymen's verbiage which gives the application more of a personal touch. Thus, using the application successfully fulfills our idea of actually enhancing the user's mood.

We also ran the application with some test cases to check its working efficiency as well as its performance under typical mobile conditions. Let us have a quick look over some of the test cases and their test results.

6 Conclusion and Future Work

The complete journey of developing the entire project had been fascinating. It provided new learning opportunities as well as tremendous brain-storming checkpoints. Based on the observations drawn from the test run of the application as well as from the results from the test cases, we can state that the idea of the project was successfully executed. We were successfully able to equip the application with the said functionalities and enable the same to work flawlessly in sharing the dynamic responses. However, there remains some of the factors, like adequate lightening conditions, mobile device, subject orientation, etc., that become the bottleneck for the accuracy of the application results.

This application holds endless areas of enhancements under the future scope. Let us have a glance on the possible future work under our project domain:

1. This application can be added with multiple additional features like '*The Happiness Meter*.' Using the same algorithms, which are present in the system, we attempt to create a submodule in the current application which will take human emotions input from the camera preview and calculate the probability percentage of Happiness emotion and will present to the user.
2. To make the application more appealing and interactive, we can add a functionality of text to speech into the responses of the application.
3. Further, the future scope of our project domain of human emotion detection is very vast.

References

1. M. Ballantyne, R.S. Boyer, L. Hines, Woody bledsoe: his life and legacy. AI Magazine **17**(1), 7–20 (1996)
2. Y. Biao, C. Jinmeng, N. Rongrong, Z. Yuyu, Facial expression recognition using weighted mixture deep neural network based on double-channel facial images. IEEE Access **6**, 4630–4640 (2018)
3. B.Y.L. Li, A.S. Mian, W. Liu, A. Krishna, Using Kinect for face recognition under varying poses, expressions, illumination and disguise. IEEE Workshop on Applications of Computer Vision (WACV) **2013**, 186–192 (2013)
4. M.A. Ponti, L.S.F. Ribeiro, T.S. Nazare, T. Bui, J. Collomosse, Everything you wanted to know about Deep Learning for Computer Vision but were afraid to ask, in *30th SIBGRAPI Conference on Graphics Patterns and Images Tutorials (SIBGRAPI-T)*, 2017, pp. 17–41
5. M. Viraj, R. Shanmuganathan, K.P. Miyapuram, Facial expression recognition using visual saliency and deep learning, in *IEEE International Conference on Computer Vision Workshops (ICCVW)*, 2017
6. S. Yang, P. Luo, C.C. Loy, X. Tang, Faceness-net: face detection through deep facial part responses. IEEE Trans. Pattern Anal. Mach. Intell. **40**(8), 1845–1859 (2017)
7. P. Hu, D. Cai, S. Wang, A. Yao, A. Yao, Learning supervised scoring ensemble for emotion recognition in the Wild, in ICMI '17, 2017, pp. 553–560
8. R. Kumar, R. Kant, G. Sanyal, Facial emotion analysis using deep convolution neural network, in *International Conference on Signal Processing and Communication (ICSPC)*, 2017, pp. 369–374
9. K. Zhang, Z. Zhang, Z. Li, Y. Qiao, Joint face detection and alignment using multitask cascaded convolutional networks. IEEE Signal Process. Lett. **23**(10) (2016)
10. M. Shin, M. Kim, D.-S. Kwon, Baseline CNN structure analysis for facial expression recognition, in *25th IEEE International Symposium on Robot and Human Interactive Communication (ROMAN)*, 2016.
11. K. Shan, J. Guo, W. You, D. Lu, R. Bie, Automatic facial expression recognition based on a deep convolutional-neural-network structure, in *IEEE 15th International Conference on Software Engineering Research Management and Applications (SERA)*, 2017, pp. 123–128
12. A.T. Lopes, E. de Aguiar, A.F. De Souza, T. Oliveira-Santos, Facial expression recognition with convolutional neural networks: coping with few data and the training sample order. Pattern Recognit. **61**, 610–628 (2017)
13. B. Yang, J. Cao, R. Ni, Y. Zhang, Facial expression recognition using weighted mixture deep neural network based on double-channel facial images. IEEE Access **6**, 4630–4640 (2018)
14. S. Kasim, R. Hassan, N.H. Zaini, A.S. Ahmad, A.A. Ramli, A study on facial expression recognition using local binary pattern. Int. J. Adv. Sci. Eng. Inf. Technol. **7**(5), 1621–1626 (2017)
15. A. Eleyan, Comparative study on facial expression recognition using gabor and dual-tree complex wavelet transforms. Int. J. Eng. Appl. Sci. **9**(1), 1–13 (2017)
16. Y. Liu, Y. Li, X. Ma, R. Song, Facial expression recognition with fusion features. Sensor **712**(17), 1–18 (2017)
17. A. Jan, *Deep Learning Based Facial Expression Recognition and Its Application* (Brunel University London, 2017)
18. A. Uçar, Deep convolutional neural networks for facial expression recognition, in *IEEE International Conference on INnovations in Intelligent SysTems and Applications (INISTA)*, 2017, pp. 371–375
19. R. Ginne, K. Jariwala, Facial expression recognition using CNN: a survey. Int. J. Adv. Electron. Comput. Sci. **5**(3) (2018)
20. C. Garcia, M. Delakis, Convolutional face finder: a neural architecture for fast and robust face detection. IEEE Trans. Pattern Anal. Mach. Intell. **26**(11) (2004)
21. S. Li, W. Deng, Deep facial expression recognition: a survey. Comput. Vis. Pattern Recogn. (2018)

22. https://opencv.org/about/, 7 June 2019
23. https://dlib.net/, 7 June 2019
24. https://www.ncbi.nlm.nih.gov/pmc/articles/PMC2731770/, 10 June 2019
25. https://en.wikipedia.org/wiki/Convolutional_neural_network, 2 Aug 2019
26. https://www.quora.com/What-are-the-advantages-of-a-convolutional-neural-network-CNN-compared-to-a-simple-neural-network-from-the-theoretical-and-practical-perspective, 2 Aug 2019
27. https://en.wikipedia.org/wiki/Logistic_regression, 2 Aug 2019
28. https://machinelearning-blog.com/2018/04/23/logistic-regression-101/, 4 Aug 2019
29. https://en.wikipedia.org/wiki/Naive_Bayes_classifier, 5 Aug 2019
30. https://www.analyticsvidhya.com/blog/2017/09/naive-bayes-explained/, 5 Aug 2019

Integrating Complete Locomotive Assistance and IoT-Based Health Care for the Disabled

S. Aravinthkumar, Ajayveer Singh Chandel, and Ahmed Abdulhakim Al-Absi

Abstract Employing the growing IoT technologies with efficient biometric and motion sensors, the prototype is developed to address one of the most critical issues that humanity faces today. It is not only a locomotive aid that enables the disabled and the elderly to commute with ease, but also a safe way of health monitoring, and a gateway to the digital world of endless possibilities. The plight of people with disabilities and the elderly has continued to evoke serious concern as a disabled life in a different world of helplessness and deprivation. While many hi-tech products available in the market aim at addressing different issues of the disabled, due to the sheer cost and inefficiency of such solutions, they have failed to reach the masses. This product is being developed with an intention to provide complete self-dependence to disabled people and the elderly in performing everyday work at extremely low cost. Locomotive assistance is provided by a customizable-accelerometer/voice command/IOT/touchscreen-enabled wheelchair, which can either be controlled by the disabled person himself, or by remote assistance (using IoT), supported by live camera feed. The personal assistant robot shall help in performing tasks like cleaning & mopping, moving objects like medicines, parcels, food, water, etc. This robot can be controlled by Voice control, RF module, or by Bluetooth interface (customizable) and shall optionally provide live camera feed too. The digital home aspect of the project shall automate the lighting & cooling. The door access, light, fan, entertainment system, power sockets, and other household appliances can also be controlled wirelessly using an IR remote. The emergency alert system shall work by manual control. This alert system shall notify the support teams in case of an emergency.

S. Aravinthkumar (✉)
SRM University, Sonepat, India
e-mail: tsaravintkumar@gmail.com

A. S. Chandel
VIT University, Vellore, India
e-mail: pkschandel07@gmail.com

A. A. Al-Absi
Kyungdong University, Global Campus, Wonju, South Korea
e-mail: absiahmed@kduniv.ac.kr

P. K. Pattnaik et al. (eds.), *Proceedings of International Conference on Smart Computing and Cyber Security*, Lecture Notes in Networks and Systems 149,
https://doi.org/10.1007/978-981-15-7990-5_35

Keywords Touchscreen-enabled wheelchair · Voice control · IOT · Locomotive assistance

1 Introduction

Old age and physical disability can be cruel companions [1]. The elderly and the disabled live in a hostile world of helplessness and deprivation [2–6]. They are unable to commute from one place to another with ease and often rely on the help of others to do so. Furthermore, being unaware of different modes of health monitoring, health issues invoke panic, and even the slightest of issues go on to become serious concerns due to lack of diagnosis and proper medical assistance [7–9]. For the old and disabled, getting the daily essential chores done is another matter of concern. While manual help from the acquaintances may not always be available, the paid assistance is mostly unaffordable or burns a hole in the pocket anyways [10–13]. Furthermore, the old and disabled have been stereotyped as outcast from the modern digital world, which keeps them deprived of the implementations of latest technologies—making the life easier [14–16].

This prototype being developed is to address the above-mentioned issues [17–21]. It is not only a locomotive aid that enables the disabled and the elderly to commute with ease, but also a safe way of health monitoring, and a gateway to the digital world of endless possibilities [22, 23]. While there are a few products available in the market that aim at addressing different issues of the disabled, due the sheer cost and inefficiency of such products, as the disability comes in many forms, they have failed to reach the masses [24, 25].

In existing system [26], there are very few products available in the market that aim at addressing different issues of the disabled, due the sheer cost and inefficiency of such products, as the disability comes in many forms, they have failed to reach the masses. Joystick-based electronic wheelchairs are the most common example for the said solutions [27–31]. These products do not provide assistance to those who cannot move their lower arm, or cannot move altogether, or the blind. The old and the physically disabled often rely on the help of others to commute from one place to another with ease [32].

For the senior citizens and the especially abled, doing the daily essential chores is troublesome, and manual help from the acquaintances is not always available [33]. The paid assistance, if available, is mostly unaffordable or expensive. Furthermore, being unaware of different modes of health monitoring, health issues invoke panic, and even the slightest of issues go on to become serious concerns due to lack of diagnosis and proper medical assistance [34]. The digital world solutions have complex UI, which keeps them deprived of the implementations of latest technologies that is making the life easier [35–39].

1.1 Objective

- To provide locomotive assistance by a customizable—accelerometer/voice command/IOT/touchscreen-enabled wheelchair, which can either be controlled by the disabled person himself, or by remote assistance (using IOT), supported by live camera feed.
- To support in performing daily chores—the personal assistant robot shall help in performing tasks like cleaning & mopping, moving objects like medicines, parcels, food, water, etc. This robot can be controlled by Voice control, RF modulev or by Bluetooth interface (customizable).
- Offering a gateway to the digital world—the digital home aspect of project shall automate the lighting & cooling. The door access, light, fan, entertainment system, power sockets, and other household appliances can also be controlled wirelessly using IR remote.
- Emergency alert system shall work by manual control. This alert system shall notify the support teams in case of an emergency.

2 Materials and Methods

In this section, we give the materials and methods used to integrate complete locomotive assistance and IoT-based health care for the disabled. Architecture of the proposed method (entity relationship diagram) is illustrated in Fig. 1.

Hardware requirements for implementing this project are Raspberry Pi 3, 5 MP Raspberry Pi 3 camera, Arduino Uno, L293d driver, 500 rpm geared motor with circuitry, three-axis accelerometer ADXl335 shown in Fig. 2, 7' LCD touchscreen, 433 MHz RF module, Bluetooth module—Hc05, Hc-Sr04 ultrasonic sensor, SIM900a GSM module, Arduino Ethernet shield 2, USB microphone with sound card, basic electronic components—capacitors and resistors, mechanical parts for robot gripper, and any Android smartphone. Software requirements are OS—Raspbian Stretch, Windows 10 for IOT, Visual Studio 2010, and Python 3.4.9 release, Arduino IDE, Android studio 3.4.1, or APP Inventor.

There was a need to build a platform to address the above-mentioned issues. It is not only a locomotive aid that enables the disabled and the elderly to commute with ease, but also a safe way of health monitoring, and a gateway to the digital world of endless possibilities. As an advantageous solution, the product shall provide locomotive assistance by a customizable—accelerometer/voice command/IOT/touchscreen-enabled wheelchair, which can either be controlled by the disabled person himself or by remote assistance (using IOT), supported by live camera feed. Furthermore, it will support in performing daily chores, as the personal assistant robot shall help in performing tasks like cleaning & mopping, moving objects like medicines, parcels, food, water, etc. In order to provide a window for the digital world—the digital home aspect of project shall automate the lighting & cooling. The door access, light, fan,

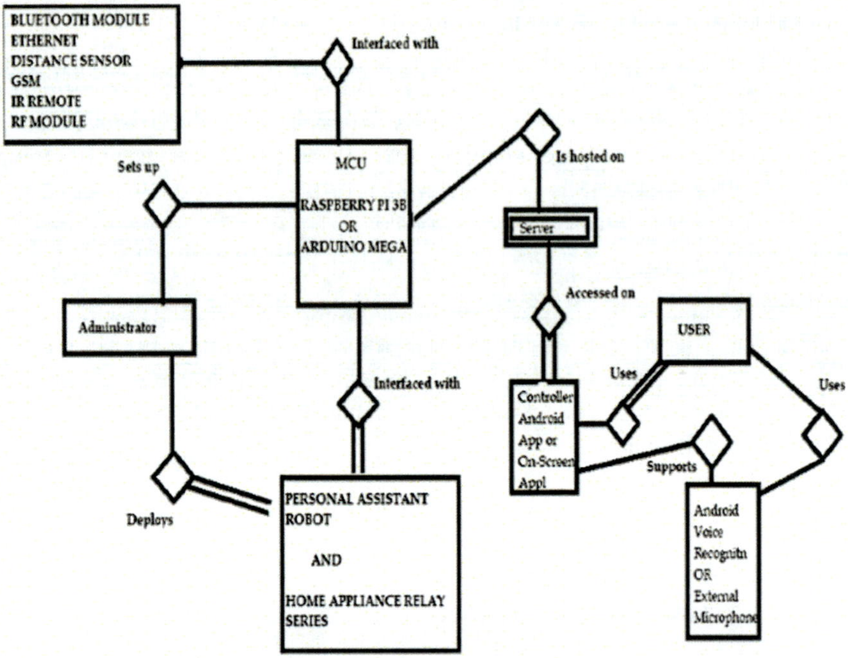

Fig. 1 Entity relationship diagram

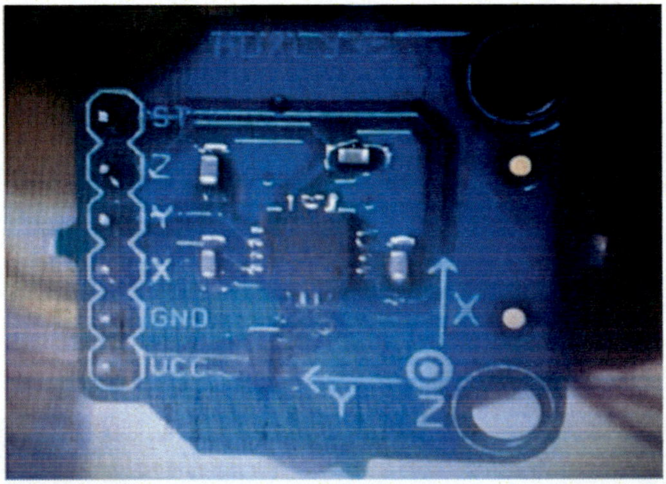

Fig. 2 Three-axis accelerometer ADXl335

Table 1 Control the wheelchair using touch buttons—User's case

Purpose	Enables user to control the wheelchair using touch buttons on the 7' screen or optionally on Android app
Description of field(s)	• User must first open the controller application on the 7' inch screen or optionally preinstalled controller Android app • User shall be directed to the activity page where the movement control touch buttons shall be accessible
Validation checks	• The movement control panel may take a while to adapt to • The loading of online controls shall be subjected to the Internet connectivity speed

Table 2 Control the wheelchair using hand gestures—User's case

Purpose	Enables user to control the wheelchair using hand gestures
Description of field(s)	• User must wear the module with the installed accelerometer • User must press the provided tactile button to activate hand gestures • User must now be able to control the wheelchair using hand gestures
Validation checks	• It is mandatory for all the hardware connections to be in place • The movement control panel may take a while to adapt to • The loading of online controls shall be subjected to the Internet connectivity speed

entertainment system, power sockets, and other household appliances can also be controlled wirelessly using IR remote.

On the top of above-mentioned functionality, emergency alert system shall work by manual control. This alert system shall notify the support teams in case of an emergency. We have emphasized on achieving a user friendly GUI in order to attain a broad user base.

3 Detailed Design of the System—Input Output Design

See Tables 1 and 2

4 Results and Discussion

The tests shall be implemented on individual units under unit testing. Each of functionalities inputs shall be divided as valid or invalid inputs, and then, a combination of these shall be tested upon the concerned functional modules. The results of these inputs shall be described in test case results area. Under system testing, the complete functionality of the system as per the requirements shall be tested. Every requirement was studied, and its implementation with the concerned functionality shall

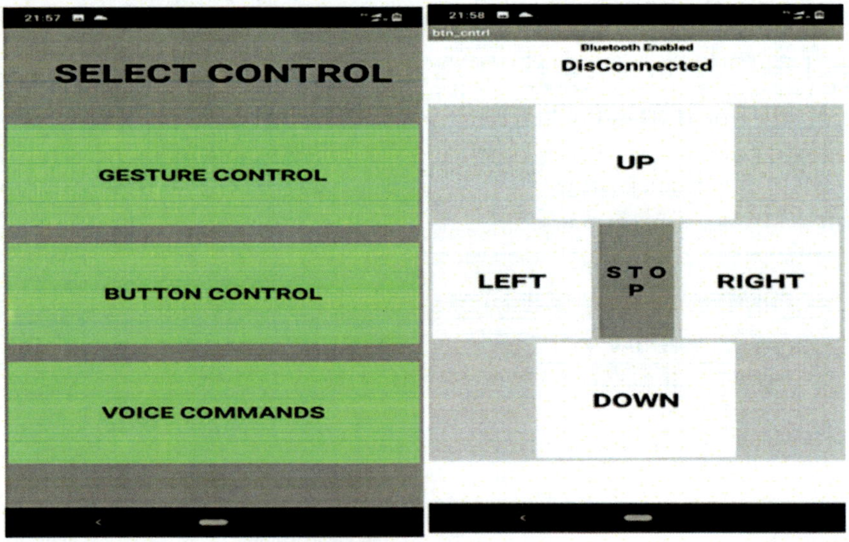

Fig. 3 Home page and touch screen-based control for wheelchair

be provided. The Android application and Web-based UI developed for wheelchair control are shown in Figs. 3, 4, 5, 6 and 7 and hardware components in Figs. 8 and 9.

5 Conclusion

Locomotive assistance by a customizable—accelerometer/voice command/IOT/touchscreen-enabled wheelchair, which can either be controlled by the disabled person himself or by remote assistance (using IOT), supported by live camera feed can be implemented successfully. The personal assistant robot helped in performing daily chores like cleaning & mopping, moving objects like medicines, parcels, food, water, etc. It can be controlled by Voice control, RF module, or by Bluetooth interface (customizable) and shall optionally provide live camera feed too. The digital home aspect of the project automated the lighting & cooling. The door access, light, fan, entertainment system, power sockets, and other household appliances can also be controlled wirelessly using an IR remote. The emergency alert system worked by manual control. This alert system notified the support teams in case of an emergency.

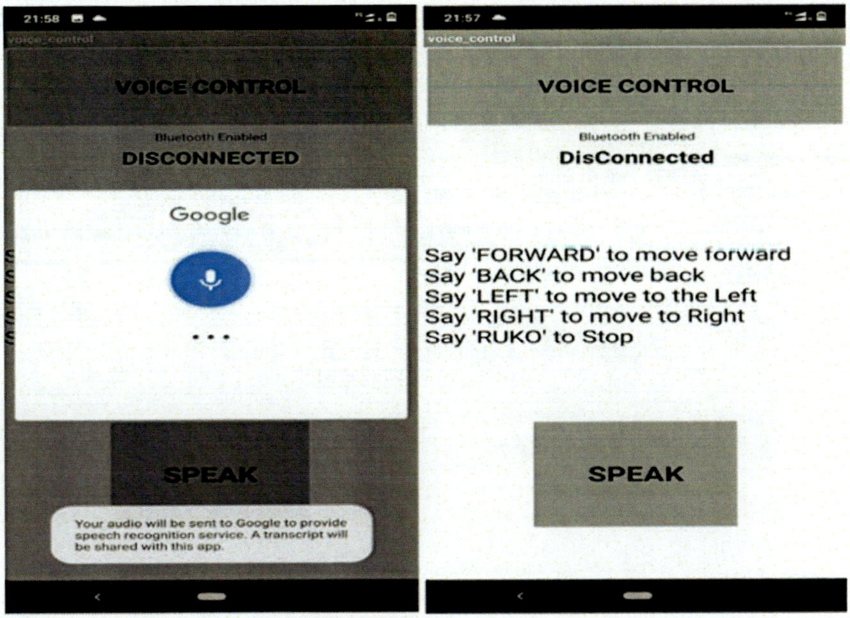

Fig. 4 Voice-based control or wheelchair

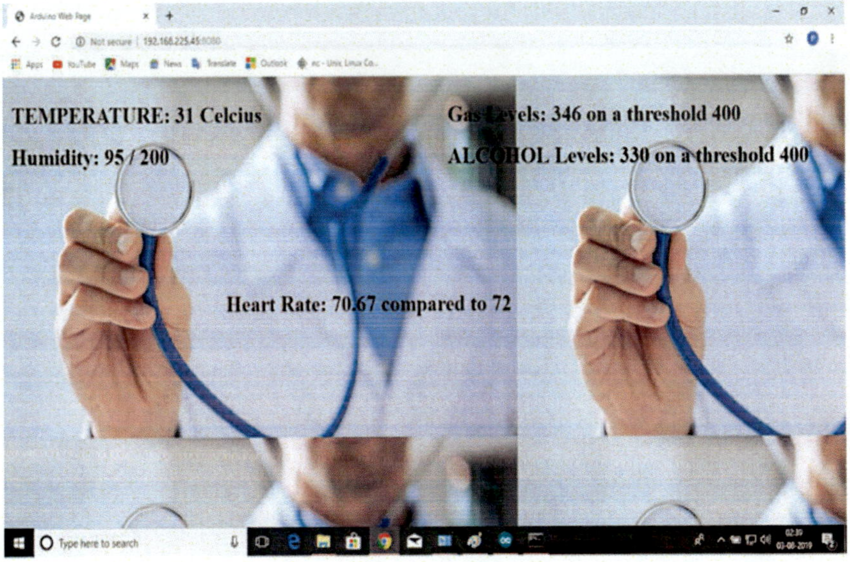

Fig. 5 Health care monitoring UI

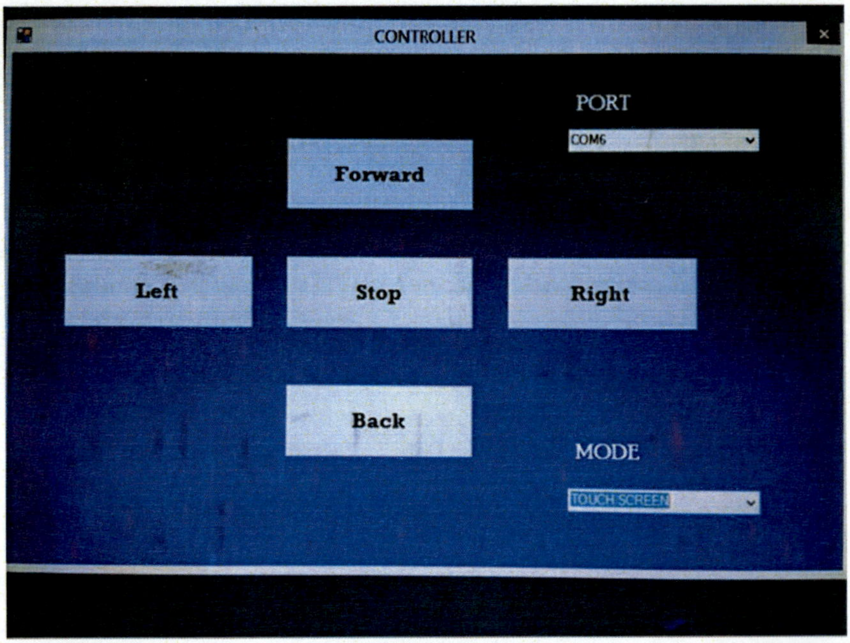

Fig. 6 UI for Windows-based controller for wheelchair

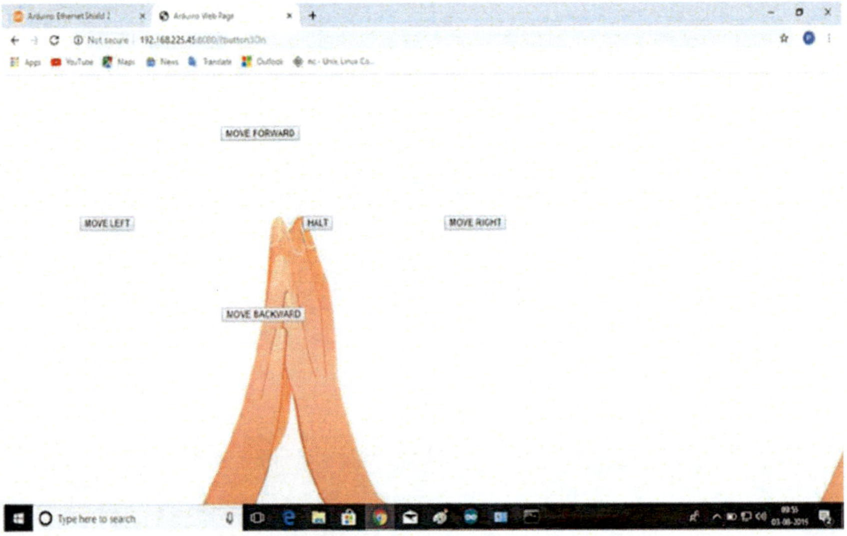

Fig. 7 IOT-based control for wheelchair—UI

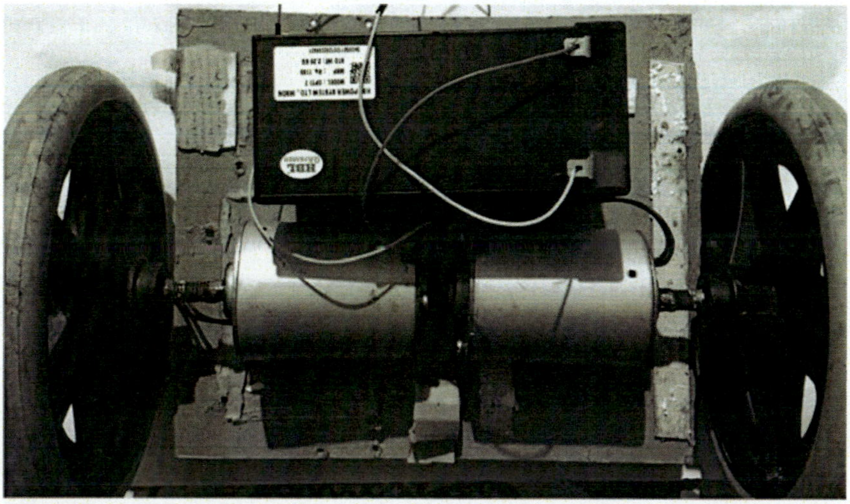

Fig. 8 Hardware module to be added to manual wheelchairs to add required functions

Fig. 9 Hardware testing modules for touch screen, voice-based, and gesture control

References

1. S. Tyagi, A. Agarwal, P. Maheshwari, A conceptual framework for IoT-based healthcare system using cloud computing, in *Proceedings of the 2016 6th International Conference-Cloud System and Big Data Engineering (Confluence)*, pp. 503–507, Noida, India (2016)

2. S.H. Almotiri, M.A. Khan, M.A. Alghamdi, Mobile health (m-health) system in the context of IoT, in *Proceedings of the 2016 IEEE 4th International Conference on Future Internet of Things and Cloud Workshops (FiCloudW)*, pp. 39–42, Vienna, Austria (2016)
3. M.M. Dhanvijay, S.C. Patil, Internet of things: a survey of enabling technologies in healthcare and its applications. Comput. Netw. **153**, 113–131 (2019)
4. Y. Karaca, M. Moonis, Y.-D. Zhang, C. Gezgez, Mobile cloud computing based stroke healthcare system. Int. J. Inf. Manage. **45**, 250–261 (2019)
5. A.H. Sodhro, Z. Luo, A.K. Sangaiah, S.W. Baik, Mobile edge computing based QoS optimization in medical healthcare applications. Int. J. Inf. Manage. **45**, 308–318 (2019)
6. B. Kitchenham, S. Charters, Guidelines for performing systematic literature reviews in software engineering. Software Engineering Group School of Computer Science and Mathematics, Keele University Keele, Staffs ST5 5BG, UK, Tech. Rep. EBSE-2007-01 (2007)
7. T. Saheb, L. Izadi, Paradigm of IoT big data analytics in healthcare industry: a review of scientific literature and mapping of research trends. Telematics Inform. **41**, 70–85 (2019)
8. S. Asadi, R. Abdullah, Y. Yah, S. Nazir, Understanding institutional repository in higher learning institutions: a systematic literature review and directions for future research. IEEE Access **7**, 35242–35263 (2019)
9. S. Pal, M. Hitchens, V. Varadharajan, T. Rabehaja, Policy-based access control for constrained healthcare resources, in *Proceedings of the 2018 IEEE 19th International Symposium on "A World of Wireless, Mobile and Multimedia Networks" (WoWMoM)*, pp. 588–599, Chania, Greece (2018)
10. S. Nazir, S. Shahzad, N. Mukhtar, Software birthmark design and estimation: a systematic literature review. Arab. J. Sci. Eng. **44**(4), 3905–3927 (2019)
11. A. Onasanya, M. Elshakankiri, Smart integrated IoT healthcare system for cancer care. Wirel. Netw. **25**(165), 1–16 (2019)
12. C. Wohlin, Guidelines for snowballing in systematic literature studies and a replication in software engineering, in *Proceedings of the 18th International Conference on Evaluation and Assessment in Software Engineering*, Berlin, Germany (2014)
13. M.-H. Maras, A.S. Wandt, Enabling mass surveillance: data aggregation in the age of big data and the Internet of Things. J. Cyber Policy **4**(2), 160–177 (2019)
14. A.A. Mutlag, M.K.A. Ghani, N. Arunkumar, M.A. Mohammed, O. Mohd, Enabling technologies for fog computing in healthcare IoT systems. Future Gener. Comput. Syst. **90**, 62–78 (2019)
15. M. Bhatia, S.K. Sood, Exploring temporal analytics in fog-cloud architecture for smart office healthcare. Mobile Netw. Appl. **24**(4), 1392–1410 (2019)
16. S.A. Khowaja, A.G. Prabono, F. Setiawan, B.N. Yahya, S.-L. Lee, Contextual activity based healthcare internet of things, services, and people (HIoTSP): an architectural framework for healthcare monitoring using wearable sensors. Comput. Netw. **145**, 190–206 (2018)
17. P. Kaur, R. Kumar, M. Kumar, A healthcare monitoring system using random forest and Internet of Things (IoT). Multi. Tools Appl. **78**(14), 19905–19916 (2019)
18. M.N.D. Tuan, N.N. Thanh, L.L. Tuan, Applying a mindfulness-based reliability strategy to the internet of things in healthcare–a business model in the vietnamese market. Technol. Forecast. Soc. Chang. **140**, 54–68 (2019)
19. S. Mohapatra, S. Mohanty, Smart healthcare: an approach for ubiquitous healthcare management using IoT, in *Big Data Analytics for Intelligent Healthcare Management*, pp. 175–196, Elsevier, Amsterdam, Netherlands (2019)
20. A.G. Zahid et al., A smart healthcare monitoring system using smartphone interface, in *Proceedings of the 2018 4th International Conference on Devices, Circuits and Systems (ICDCS)*, pp. 228–231, Coimbatore, India (2018)
21. A. Dumka, A. Sah, "Smart ambulance system using concept of big data and internet of things, in *Healthcare Data Analytics and Management*, pp. 155–176, Elsevier, Amsterdam, Netherlands (2019)
22. P. Verma, S.K. Sood, Fog assisted-IoT enabled patient health monitoring in smart homes. IEEE Internet Things J. **5**(3), 1789–1796 (2018)

23. A. Kumari, S. Tanwar, S. Tyagi, N. Kumar, Fog computing for healthcare 4.0 environment: opportunities and challenges. Comput. Electr. Eng. **72**, 1–13 (2018)
24. V. Jagadeeswari, V. Subramaniyaswamy, R. Logesh, V. Vijayakumar, A study on medical internet of things and big data in personalized healthcare system. Health Inf. Sci. Syst. **6**(1), 14 (2018)
25. P.K.D. Pramanik, B.K. Upadhyaya, S. Pal, Internet of things, smart sensors, and pervasive systems: enabling connected and pervasive healthcare, in *Healthcare Data Analytics and Management*, pp. 1–58, Elsevier, Amsterdam, Netherlands (2019)
26. S.M.R. Islam, D. Kwak, M.H. Kabir, M. Hossain, K.-S. Kwak, The internet of things for health care: a comprehensive survey. IEEE Access **3**, 678–708 (2015)
27. J. Maktoubian, K. Ansari, An IoT architecture for preventive maintenance of medical devices in healthcare organizations. Health Technol. **9**(3), 233–243 (2019)
28. G. Rathee, A. Sharma, H. Saini, R. Kumar, R. Iqbal, A hybrid framework for multimedia data processing in IoT-healthcare using blockchain technology. Multi. Tools Appl. **78**, 1–23 (2019)
29. V. Sindhura, P. Ramya, S. Yelisetti, An IOT based smart mobile health monitoring system, in *Proceedings of the 2018 Second International Conference on Inventive Communication and Computational Technologies (ICICCT)*, pp. 1186–1192, Coimbatore, India (2018)
30. H. Hamidi, An approach to develop the smart health using Internet of Things and authentication based on biometric technology. Future Gener. Comput. Syst **91**, 434–449 (2019)
31. S. Zemmoudj, N. Bermad, M. Omar, Context-aware pseudonymization and authorization model for IoT-based smart hospitals. J. Ambient Intell. Human. Comput. **10**(11), 4473–4490 (2019)
32. S. Zeadally, O. Bello, Harnessing the power of internet of things based connectivity to improve healthcare. Internet of Things, Article ID 100074 (2019)
33. Y. Yuehong, Y. Zeng, X. Chen, Y. Fan, The internet of things in healthcare: an overview. J. Indus. Inf. Integr. **1**, 3–13 (2016)
34. H. Elazhary, Internet of things (IoT), mobile cloud, cloudlet, mobile IoT, IoT cloud, fog, mobile edge, and edge emerging computing paradigms: disambiguation and research directions. J. Netw. Comput. Appl. **128**, 108–140 (2018)
35. A. Arfaoui, A. Kribeche, S.-M. Senouci, Context-aware anonymous authentication protocols in the internet of things dedicated to e-health applications. Comput. Netw. **159**, 23–36 (2019)
36. A.A. Abdellatif, M.G. Khafagy, A. Mohamed, C.-F. Chiasserini, EEG-based transceiver design with data decomposition for healthcare IoT applications. IEEE Internet Things J. **5**(5), 3569–3579 (2018)
37. P.P. Ray, D. Dash, D. De, Edge computing for internet of things: a survey, e-healthcare case study and future direction. J. Netw. Comput. Appl. **140**, 1–22 (2019)
38. R. Rashid, M.A. Shah, EK-Healthcare: effectiveness of IoT in the medical field with enhanced features, in *Proceedings of the 2018 24th International Conference on Automation and Computing (ICAC)*, pp. 1–6, Newcastle upon Tyne, UK (2018)
39. M.G.R. Alam, M.S. Munir, M.Z. Uddin, M.S. Alam, T.N. Dang, C.S. Hong, Edge-of-things computing framework for cost-effective provisioning of healthcare data. J. Parall. Distrib. Comput. **123**, 54–60 (2019)

Classification of Multiple Steganographic Algorithms Using Hierarchical CNNs and ResNets

Sanghoon Kang, Hanhoon Park, and Jong-Il Park

Abstract In general, image deformations caused by different steganographic algorithms are extremely small and of high similarity. Therefore, detecting and identifying multiple steganographic algorithms are not easy. Although recent steganalytic methods using deep learning showed highly improved detection accuracy, they were dedicated to binary classification, i.e., classifying between cover images and their stego images generated by a specific steganographic algorithm. In this paper, we aim at achieving quinary classification, i.e., detecting (=classifying between stego and cover images) and identifying four spatial steganographic algorithms (LSB, PVD, WOW, and S-UNIWARD), and propose to use a hierarchical structure of convolutional neural networks (CNN) and residual neural networks (ResNet). Experimental results show that the proposed method can improve the classification accuracy by 17.71% compared to the method that uses a single CNN.

Keywords Steganalysis · Image steganography · Convolutional neural network · Residual neural network · Hierarchical structure · Quinary classification

S. Kang · H. Park (✉)
Department of Electronic Engineering, Pukyong National University, 45, Yongso-ro, Nam-gu, Busan 48513, South Korea
e-mail: hanhoon.park@pknu.ac.kr

S. Kang
e-mail: totohoon01@pukyong.ac.kr

J.-I. Park
Department of Computer Science, Hanyang University, 222,Wangsimni-ro, Seongdong-gu, Seoul 04763, South Korea
e-mail: jipark@hanyang.ac.kr

P. K. Pattnaik et al. (eds.), *Proceedings of International Conference on Smart Computing and Cyber Security*, Lecture Notes in Networks and Systems 149,
https://doi.org/10.1007/978-981-15-7990-5_36

1 Introduction

Image steganography is a technique that covertly embeds secret information that is intended to be passed on to media, such as photos and videos. In the image steganography, the media before data embedding is called cover image, and the media after data embedding is called stego image. Basically, in order to hide the existence of secret information, steganographic algorithms minimize visual and statistical changes in images caused by the data embedding. However, since LSB [1] and PVD [2], the representatives of the early steganographic algorithms, sequentially embedded data to all the pixels without considering the relationship between adjacent pixels, statistical changes caused by data embedding were relatively large, making it easy to detect whether the data were embedded or not. Therefore, subsequent algorithms, such as HUGO [3], WOW [4], and S-UNIWARD [5], embedded data only into the pixel positions that minimize the image deformation caused by data embedding, significantly reducing the likelihood of detection of the data embedding.

Image steganalysis is a technique for detecting images with steganographic algorithms applied, usually by analyzing visual and numerical changes caused by the application of steganographic algorithms. Statistical methods such as RS-analysis [6] were able to effectively detect the early steganographic algorithms, such as LSB and PVD, which involve significant statistical changes. To detect the sophisticated steganographic algorithms, such as WOW and S-UNIWARD, which minimized statistical changes and distortions caused by data embedding, machine learning techniques, such as SPAM [7] and SRM [8], were proposed. Recently, inspired by the great success of AlexNet [9] in image classification, steganalytic methods using convolutional neural networks (CNN) have begun to be intensively studied. The deep learning methods enabled to automatically learn the optimal features for classification through convolution operations, rather than heuristically designing the features to be extracted as in the machine learning methods. This made deep learning methods achieve much higher accuracy.

Since the first steganalytic method using CNN, XuNet [10], has been introduced, a number of variants have been proposed [11–13], focusing on optimizing the network structure to improve the classification accuracy of XuNet. As an impressive one, the method using a deep residual neural network (ResNet) which has a path directly connecting the input to the output of the convolution block to prevent loss of gradient caused by convolution operations achieved a detection accuracy of 95% and 94% for WOW and S-UNIWARD, respectively [14]. In spite of high accuracy of steganalytic methods using CNNs or ResNets, they focused on a binary classification between cover and stego images. This indicates that we can know whether steganography has been applied to a given input image but do not know which algorithm was applied. However, in order to extract and restore data from images with steganography applied, it must first be possible to identify what steganographic algorithms have been applied to the images.

In this paper, we propose a steganalytic method that does not only detect the application of steganography but also identify what algorithm has been applied.

Unlike the conventional methods of performing binary classification, the proposed method was designed to enable multi-class classification. A similar previous method, aiming at ternary classification [15] of cover, WOW stego and S-UNIWARD stego images used a single CNN structure. However, it was found that as the number of classes increased, the accuracy of the single CNN structure drastically dropped. In this paper, we aim to classify the five classes, i.e., cover, LSB stego, PVD stego, WOW stego, and S-UNIWARD stego images. To this end, we propose to use a hierarchical structure of CNNs, splitting classes into hierarchically connected multiple CNNs to be classified step by step, rather than the multi-class classification on a single CNN. The method of using multiple networks can reduce the number of classes that need to be classified in each network, resulting in reducing the classification complexity. In addition, different networks that specialized to classify specific classes can be used at each stage. Therefore, we deployed deep ResNets to improve the classification accuracy of WOW and S-UNIWARD. Finally, the optimal hierarchical structure of multiple networks for multi-class classification and the optimal configuration for each network are intensively explored.

2 Classification of Multiple Steganographic Algorithms

In order to detect and classify multiple steganographic algorithms, it is not desirable to simply increase the output number of existing CNNs originally designed for binary classification. Indeed, the change in images due to the application of steganography is so tiny that CNN-based stegianalysis becomes more difficult as the number of classes increases. This section presents methods to consider for the identification of multiple steganographic algorithms. First of all, Sect. 2.1 describes a method that modifies the structure of a single CNN designed for binary classification and increases the number of its outputs. Next, Sect. 2.2 describes a method that deploys CNNs designed for binary classification in a hierarchical manner. Finally, Sect. 2.3 describes the proposed method that deploys CNNs and ResNets in a hierarchical manner.

2.1 Using a Single CNN

As mentioned, we would like to classify five classes of cover, LSB stego, PVD stego, WOW stego, and UNIWARD stego images. To this end, we change the structure of a CNN designed for binary classification and increase the number of its outputs to enable quinary classification. With 10 SRM filters (that are confirmed to better detect tiny variation on images [15]) as a pre-processing high-pass filter (HPF), we add one convolution layer in the XuNet structure as shown in Fig. 1. When an image with a resolution of 256 × 256 is entered as input, it becomes 10 pre-processed images via SRM filters. In the first convolution operation (with 5 × 5 filters), the number

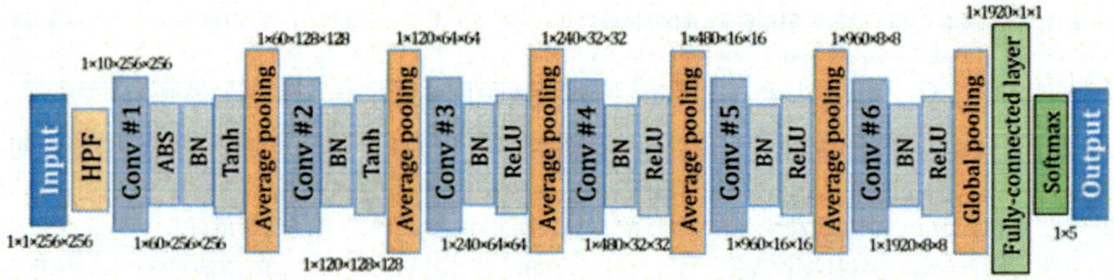

Fig. 1 XuNet with an additional convolution layer. The number of feature maps and the status of images in each layer are expressed in a four-dimensional matrix. Each number means the number of images, the number of feature maps, the height of the images, and the width of the images

of feature maps will increase to 60, and the average pooling will reduce the size of the images by a quarter. Later, the number of feature maps in the subsequent layers doubles, and the size of the images decreases by a quarter. The final number of feature maps is 1920, and the probability values for the five classes are derived through the fully connected layer.

2.2 *Using Hierarchical CNNs*

Given that image classification utilizes differences between images, images belonging to different classes should have meaningful differences, either arithmetic or statistically. However, the change in images due to the application of steganography is very small. The difference between different stego images as well as the difference between cover and stego images is very small. Therefore, it is not easy to identify multiple steganographic algorithms simply by increasing the number of outputs on CNN.

In order to more effectively identify multiple steganographic algorithms, we attempt to adopt hierarchical binary classification rather than single quinary classification. Hierarchical binary classification classifies the group of steganographic algorithms at the front end and classifies each individual algorithm at the back. As shown in Fig. 2, a total of four binary CNNs are used for the hierarchical binary classification of five classes (cover, LSB stego, PVD stego, WOW stego, and S-UNIWARD stego images). Net #1 classifies whether the input image is a cover image or a stego

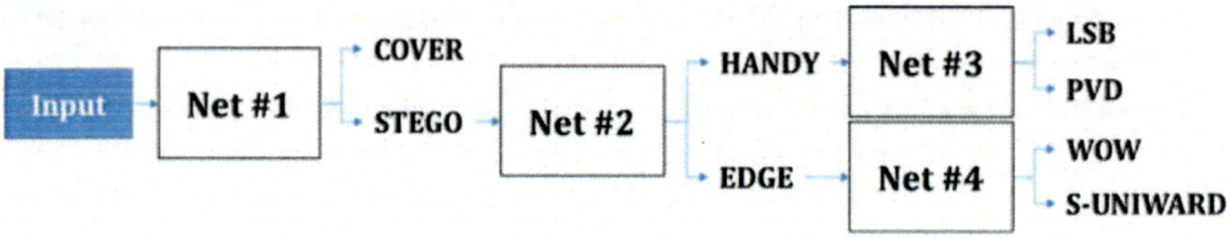

Fig. 2 Hierarchical binary CNNs (Net #*n*) for quinary classification

image. If the input image is classified as a cover image, it will no longer proceed to the later Nets. On the other hand, when classified as a stego image, Net #2 classifies steganographic algorithms into two groups, HANDY and EDGE. HANDY includes LSB and PVD, while EDGE includes WOW and S-UNIWARD. Depending on the group classified at the former Nets, the subsequent networks Net #3 and Net #4 classify LSB and PVD, WOW and S-UNIWARD, respectively.

2.3 *Using Hierarchical CNNs and ResNets: A Proposed Method*

In the hierarchical structure in Fig. 2, each Net #*n* can be different and specialized to better classify specific classes. Therefore, we employ a ResNet that is much deeper and can be more effective for the classification between very similar images, such as WOW stego and S-UNIWARD stego images. The ResNet has the same structure as the XuNet except for four non-bottleneck residual blocks (Fig. 3), each following the average pooling layers. It has five convolution layers, and the residual blocks contain two convolution operations, but retain the number of feature maps. Therefore, input images with a resolution of 256 $\times$ 256 generate 960 feature maps through the convolution layers, and 960 feature maps are fed into the fully connected layer, deriving the probability values corresponding to the five class.

In the previous study [14], more residual blocks with three convolution operations were used to achieve high accuracy. However, since the proposed method requires more classes to be classified than the previous study, there is a significant increase in computational complexity, which limits the use of more residual blocks or convolution operations.

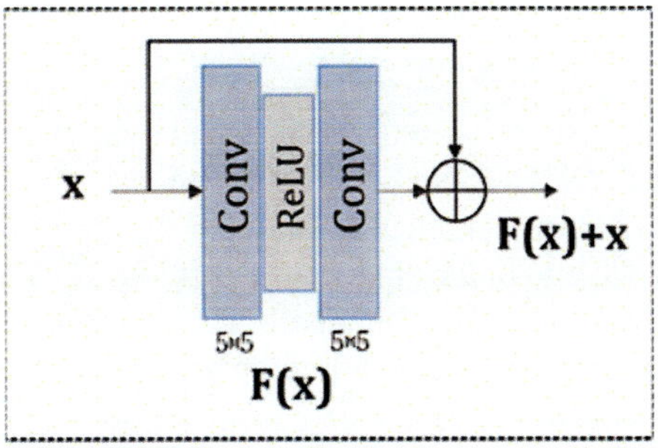

Fig. 3 Residual block added to the XuNet for improving the classification between WOW and S-UNIWARD stego images

Table 1 Accuracy in % of binary classification using a single CNN

Algorithms	Cover	Stego	Average
LSB	97.07	97.32	97.20
PVD	99.62	99.98	99.80
WOW	83.26	75.17	79.22
S-UNIWARD	76.47	77.44	76.96

Table 2 Accuracy in % of quinary classification using a single CNN

Cover	LSB	PVD	WOW	S-UNIWARD	Average
88.45	64.80	0.00	99.87	30.58	56.74

3 Experimental Results

For the experiment, 10,000 grayscale images of 512 × 512 resolution of BossBase1.01 [16] were used. Each image was partitioned into non-overlapping segments to produce a total of 40,000 256 × 256 images and then grouped into 30,000 for learning and 10,000 for verification. Using the images for learning and verification, we created stego images with each steganographic algorithm applied. For WOW and S-UNIWARD, we used the C++ and MATLAB code provided in [17] and set the payload to 0.4. The Pytorch library [18] was used to implement each network structure described in Chap. 2, and each network was trained 200,000 times using a momentum optimizer with a momentum value of 0.9 using cross-entropy as a cost function.

3.1 Using a Single CNN

Tables 1 and 2 show the results of binary classification and quinary classification using a single CNN. Despite the use of the same structure, as the number of classes increased, the quinary classification accuracy decreased by more than 30% compared to the average 88.30% of the binary classification. The PVD stego and S-UNIWARD stego images were not classified at all. This indicates that it is difficult to classify multiple steganographic algorithms simply by increasing the number of classes for a single CNN structure.

3.2 Using Hierarchical CNNs

Table 3 shows the accuracy of each CNN's binary classification in Fig. 2. Net #1 had

Table 3 Classification accuracy in % of each CNN in the hierarchical CNN structure

Net #*n*	Class 1	Class 2	Average
$n = 1$	78.14	86.46	82.29
$n = 2$	98.45	97.60	98.03
$n = 3$	99.75	99.95	99.85
$n = 4$	42.94	94.01	68.47

Table 4 Accuracy in % of quinary classification using hierarchical CNNs

Cover	LSB	PVD	WOW	S-UNIWARD	Average
78.14	88.00	98.34	32.15	64.81	72.29

an average classification accuracy of 82.29% for classifying cover and stego. Net #2 classified HANDY and EDGE with a high accuracy of 98.03%, showing that the difference between the two groups of steganographic algorithms is relatively large. Net #3 classified LSB and PVD, and Net #4 classified WOW and S-UNIWARD. Net #4 had a relatively low accuracy of 68.47% due to the high similarity between the two steganographic algorithms.

Table 4 shows the results of the quinary classification through hierarchical binary classification. Compared to the results of using a single CNN, the classification accuracy of cover and WOW decreased, but the classification accuracy of LSB, PVD, and S-UNIWARD increased significantly. As a result, the average classification accuracy increased by 15.55%, and the rest of the classes, except for WOW, could be classified with acceptable accuracy. The WOW's low classification accuracy will be due to the poor performance of Net #4 and because the hierarchical classification further reduces the classification accuracy at the back by accumulating misclassifications at the front end.

3.3 *Using Hierarchical CNNs and ResNets: A Proposed Method*

Table 5 compares the accuracy of the classification between WOW and S-UNIWARD

Table 5 Accuracy in % of the classification between WOW and S-UNIWARD when using the CNN and ResNet

	WOW	S-UNIWARD	Average
CNN	42.94	94.01	68.47
ResNet	82.02	72.70	77.36

Table 6 Accuracy in % of quinary classification using hierarchical CNNs and ResNets

Cover	LSB	PVD	WOW	S-UNIWARD	Average
78.14	88.00	98.34	54.09	53.69	74.45

when using the CNN and ResNet. The use of ResNet reduced the classification accuracy for S-UNIWARD, but greatly increased the classification accuracy for WOW. As a result, the average classification accuracy was about 9% better than the case with CNN.

Table 6 shows the results of the quinary classification when substituting the CNN of Net #4 with the ResNet. Due to the better performance of the ResNet in classifying between WOW and S-UNIWARD, the classification accuracy for WOW greatly increased. Although the classification accuracy for S-UNIWARD decreased, the average accuracy for both increased by 5.41%. As a result, the proposed method achieved a classification accuracy of 74.45% on average.

4 Conclusion and Future Work

In this paper, CNN-based steganalytic methods for classifying five classes (cover, LSB stego, PVD stego, WOW stego, and S-UNIWARD stego images) were explored. First, it was impossible to classify five classes in a single CNN structure. So, we considered deploying multiple CNNs in a hierarchical manner, but the high similarity between WOW and UNIWARD made it difficult to recognize WOW stego images properly. Finally, we introduced a ResNet with high classification performance for WOW and S-UNIWARD, enabling to classify five classes with 74.45% accuracy.

However, the classification accuracy for WOW stego and S-UNIWARD stego images was still low. Therefore, we plan to seek ways to resolve the problem of accumulating the misclassification by the former networks in the hierarchical structure.

Acknowledgements This work was supported by the research fund of the Signal Intelligence Research Center supervised by the Defense Acquisition Program Administration and Agency for the Defense Development of Korea.

References

1. C.K. Chan, L.M. Cheng, Hiding data in images by simple LSB substitution. Pattern Recogn. **37**, 469–474 (2004)
2. D. Wu, W. Tsai, A steganographic method for images by pixel-value differencing. Pattern Recogn. **24**, 1613–1626 (2003)
3. T. Pevný, T. Filler, P. Bas, Using high-dimensional image models to perform highly undetectable steganography. LNCS, Vol. 6387 (Springer, Berlin, Heidelberg, 2010)
4. V. Holub, J. Fridrich, Designing steganographic distortion using directional filters, in *Proceedings of IEEE International Workshop on Information Forensics and Security*, pp. 234–239 (2012)
5. V. Holub, J. Fridrich, T. Denemark, Universal distortion function for steganography in an arbitrary domain. EURASIP J. Info. Security **1** (2014)
6. S. Manoharan, An empirical analysis of RS steganalysis, in *Proceedings of Third International Conference on Internet Monitoring and Protection*, pp. 172–177 (2008)
7. Y. Hou, R. Ni, Y. Zhao, Steganalysis to Adaptive pixel pair matching using two-group subtraction pixel adjacency model of covers, in *Proceedings of International Conference on Signal Processing*, pp. 1864–1867 (2014)
8. J. Fridrich, J. Kodovsky, Rich models for steganalysis of digital images. IEEE Trans. Inf. Forensics Secur. **7**, 868–882 (2012)
9. A. Krizhevsky, I. Sutskever, G.E. Hinton, ImageNet classification with DEEP convolutional neural networks. Commun. ACM **60**, 84–90 (2017)
10. G. Xu, H.-Z. Wu, Y.Q. Shi, Structural design of convolutional neural networks for steganalysis. IEEE Signal Process. Lett. **23**, 708–712 (2016)
11. Y. Yuan, Z. Wei, B. Feng, J. Weng, Steganalysis with CNN using multi-channels filtered residuals, in *Proceedings of International Conference on Cloud Computing and Security*, pp. 110–120 (2017)
12. B. Li, W. Wei, A. Ferreira, S. Tan, ReST-Net: diverse activation modules and parallel subnets-based CNN for spatial image steganalysis. IEEE Signal Process. Lett. **25**, 650–654 (2018)
13. X. Deng, B. Chen, W. Luo, D. Luo, Fast and Effective global covariance pooling network for image steganalysis, in *Proceedings of the ACM Workshop on Information Hiding and Multimedia Security*, pp. 230–234 (2019)
14. W. Songtao, S. Zhong, Y. Liu, Deep residual learning for image steganalysis. Multim. Tools Appl. **77**, 10437–10453 (2018)
15. S. Kang, H. Park, J.-I. Park, CNN-based ternary classification for image steganalysis. Electronics **8**, 1225 (2019)
16. P.Bas, T. Filler, T. Pevný, "Break Our Steganographic System": the Ins and outs of organizing BOSS. LNCS, Vol. 6958, pp. 59–70 (Springer, Berlin, Heidelberg, 2011)
17. https://dde.binghamton.edu/download/stego_algorithms/
18. https://pytorch.org

Author Index

P. K. Pattnaik et al. (eds.), *Proceedings of International Conference on Smart Computing and Cyber Security*, Lecture Notes in Networks and Systems 149,
https://doi.org/10.1007/978-981-15-7990-5